History, Philosophy and Theory of the Life Sciences

Volume 14

More information about this series at http://www.springer.com/series/8916

Peter Distelzweig • Benjamin Goldberg
Evan R. Ragland
Editors

Early Modern Medicine and Natural Philosophy

Springer

Editors
Peter Distelzweig
Department of Philosophy
University of St. Thomas
St. Paul, MN, USA

Benjamin Goldberg
Department of Humanities
and Cultural Studies
University of South Florida
Tampa, FL, USA

Evan R. Ragland
Department of History
University of Notre Dame
Notre Dame, IN, USA

ISSN 2211-1948 ISSN 2211-1956 (electronic)
History, Philosophy and Theory of the Life Sciences
ISBN 978-94-017-7352-2 ISBN 978-94-017-7353-9 (eBook)
DOI 10.1007/978-94-017-7353-9

Library of Congress Control Number: 2015955134

Springer Dordrecht Heidelberg New York London

Printed on acid-free paper

Springer Science+Business Media B.V. Dordrecht is part of Springer Science+Business Media (www.springer.com)

Acknowledgments

This volume grew out of the medicine, philosophy, and the 'scientific revolution' initiative funded in part by a Faculty Collaborative Research in the Humanities grant from the Humanities Center at the University of Pittsburgh. The project aimed to cultivate collaborative work on the interactions between medicine and natural philosophy during the 'scientific revolution' across departments and institutions and included a number of workshops and reading groups at the University of Pittsburgh. This volume grows most immediately out of the initiative's culminating conference on the theme hosted by the University of Pittsburgh Center for Philosophy of Science in November 2012. We are grateful to the Center; to Professor John Norton, its director; and to Joyce McDonald and all the wonderful Center staff for their support and help with the conference. We also thank the other conference cosponsors: the University of Pittsburgh's Humanities Center, World History Center, Department of History and Philosophy of Science, and Medieval and Renaissance Studies Program, along with the Department of History and Philosophy of Science at Indiana University, Bloomington. We thank all the conference presenters and discussants for making it such a fruitful occasion to confirm, explore, and articulate the importance of the medical context for early modern natural philosophy. It is our hope that this volume will be the same for its readers. We thank the series editors and staff at Springer for their support and patience in bringing it to completion.

Contents

Contributors

Tawrin Baker Department of History and Philosophy of Science, Indiana University, Bloomington, IN, USA

Antonio Clericuzio Dipartimento Studi Umanistici, Roma Tre University, Rome, Italy

Karen Detlefsen Department of Philosophy, University of Pennsylvania, Philadelphia, PA, USA

Peter Distelzweig Department of Philosophy, University of St. Thomas, St. Paul, MN, USA

Patricia Easton Department of Philosophy, Claremont Graduate University, Claremont, CA, USA

Melissa Gholamnejad Department of Philosophy, Claremont Graduate University, Claremont, CA, USA

Benjamin Goldberg Department of Humanities and Cultural Studies, University of South Florida, Tampa, FL, USA

Hiro Hirai Center for the History of Philosophy and Science, Radboud University, Nijmegen, Netherlands

Ashley J. Inglehart Department of History and Philosophy of Science, Indiana University, Bloomington, IN, USA

Cynthia Klestinec Department of English, Miami University of Ohio, Oxford, OH, USA

Craig Martin History Department, Oakland University, Rochester, MI, USA

Domenico Bertoloni Meli Department of History and Philosophy of Science and Medicine, Indiana University, Bloomington, IN, USA

Evan R. Ragland Department of History, University of Notre Dame, Notre Dame, IN, USA

Jole Shackelford Program in the History of Medicine, University of Minnesota, Minneapolis, MN, USA

Justin E.H. Smith Département Histoire et Philosophie des Sciences, Université Paris Diderot - Paris 7, Paris, France

Charles T. Wolfe Ghent University, Ghent, Belgium

Chapter 1
Introduction

Benjamin Goldberg, Evan R. Ragland, and Peter Distelzweig

There is no more fruitful occupation than to try to know oneself. And the benefit that one expects from this knowledge does not just extend to morals, as many may initially suppose, but also to medicine in particular. – René Descartes, Description of the Human Body.

(Descartes 1998, 170)

Many, perhaps even most, members of the early Royal Society of London were physicians (though not all of these were *practicing* physicians).[1] The Society could never have prospered without the support of wealthy physicians, the rolls of the organization were filled with doctors and surgeons, and medical and biological observations and analyses crowd the pages of its journal. Indeed, some members of the College of Physicians complained about the Royal Society's forays into medicine. Any simple skimming of the *Philosophical Transactions* will immediately reveal a network of men (and they were only men) persistently occupied with medical and biological problems—hardly an issue was printed without mention of various medicinal cures, surgical or medical procedures, or observations of strange and mysterious animals or plants. To take one small example, in the very first issue of the *Transactions* there is a brief article entitled, "An Account of a very odd Monstrous calf" (*Philosophical Transactions* 1665, 10), which describes a calf with various deformities, including having no joints and a triple ('Cerebus-like') tongue.

[1] Cook 1990.

B. Goldberg (✉)
Department of Humanities and Cultural Studies, University of South Florida, Tampa, FL, USA
e-mail: metabenny@gmail.com

E.R. Ragland
Department of History, University of Notre Dame, Notre Dame, IN, USA
e-mail: eragland@nd.edu

P. Distelzweig
Department of Philosophy, University of St. Thomas, St. Paul, MN, USA
e-mail: peter.distelzweig@stthomas.edu

P. Distelzweig et al. (eds.), *Early Modern Medicine and Natural Philosophy*, History, Philosophy and Theory of the Life Sciences 14,
DOI 10.1007/978-94-017-7353-9_1

Observations like this one—and much more detailed empirical and theoretical analyses—can be found throughout the early issues of the *Transactions*, and they were of vital importance to those working on various outstanding problems, in this case the problems of animal generation and of the origin of monstrosity. This news item was communicated by none other than the Honorable Robert Boyle, whose interests go well beyond the physical and chemical sciences for which we usually remember him.

This observation illustrates the core concern of this volume: to bring to the fore the medical context of natural philosophy—not only in England in the second half of the seventeenth century, but throughout Europe in the early modern period. While the papers in this volume range in approach and topic, they share a core background assumption, namely, that medicine and natural philosophy shaped and drove each other on multiple levels. This mutual influence took many forms and acted at numerous interfaces, including the institutional and (inter)personal. Of course, the universities constitute one major institutional interface, but others existed and developed in the period—as exemplified by the constitution and preoccupations of the early Royal Society. The mutual influence was driven both by overlapping traditions of learning and by a common imperative to understand, restore, and maintain human well-being. This influence ranged over shared theoretical concerns (for example, the nature of matter, the faculties of the soul, and the classification and operations of plants and minerals) as well as methodological debates on the appropriate way to gain, certify, and communicated knowledge of natural things. As Harold Cook has emphasized, medicine was often called 'physick,' and this term signals that the medical context was an important site where early moderns negotiated an understanding of *physis* (nature).[2] Medicine in the early modern period encompassed a much wider sphere of ideas and activities then it does today, and the relationship between natural philosophy and medicine was complex and substantial.

Attending to and articulating this relationship invites the reexamination of canonical actors of the Scientific Revolution, from Harvey, Boyle, and Locke to Descartes and Leibniz. But it also reveals connections with a wide variety of less canonical but historically important natural philosophers and physicians, such as Hieronymus Fabricius ab Aquapendente, Daniel Sennert, Pierre Gassendi, Louis de la Forge, and Petrus Severinus—all of whom shall be discussed in this volume. Our goal here is to expand the scope of who counts as a philosopher or physician important enough to study, as well as our conception of what debates and issues are important for a deeper understanding of early modern thought. We want to recapture something of the heterogeneity and interpenetration of early modern philosophy, medicine, and science—a complexity that can be obscured by our own disciplinary boundaries (e.g. between history of medicine and history of philosophy). A great deal of work needs to be done in unpacking the concepts and terminology of early modern actors in such a way as to ensure that our conceptual schemata do not distort them. This need is seen clearly in terms such as 'experiment' and 'mechanism,' and the medical perspective we aim to explore is central for their clarification and

---

[2] Cook 1990.

contextualization. In the early modern period, experiments on living things were widespread and complex, and likely influenced linguistic tactics in other disciplines—yet these developments have not been studied in comprehensive detail. The term 'mechanism,' meanwhile, was often contrasted with the spontaneous, responsive phenomena of life. Even for those wishing to extend mechanistic explanations into the living world, older accounts of what properly characterized living things—the presence and activity of a soul or the operation of faculties—remained a contrast class for their own replacement programs and vocabularies.

We hope to emphasize in this volume the myriad ways in which the intellectual training and disciplinary structure of medicine were congenial to the development of early modern science. For instance, medicine included both *theoria* and *practica*—training in medicine was based around both deep study of philosophical and medical texts (especially Galen and Aristotle) and practical anatomical and therapeutic experience and instruction which aimed at curing human bodies.[3] Medicine also surveyed a wide scope of divisions, often five in number.[4] The 1620 textbook *Institutiones* of the Aristotelian chymist Daniel Sennert of Wittenberg can give us one snapshot of the scope of medicine from near the middle of the chronology surveyed in this volume. First, he attenuates the distinction between *theoria* and *practica* by asserting that medicine is a unitary art, with even *theoria* aimed at the common end of health. This is consonant with his studied rejection of medicine as proper *scientia*, since the physician *qua* physician does not reach to first principles.[5] Of course, Sennert and other physicians were also philosophers, and used their findings from chymistry, anatomy, and natural history to build and critique philosophical claims.[6] Sennert, following Alexandrian tradition, then divides medicine into *physiology*, which treats the constitution, actions, and uses of the parts; *pathology*, which deals with the nature, differentia, and causes of diseases and symptoms; *semiotics*, which handles the signs by which the hidden causes of disease can be known; *hygiene*, which teaches rules for conserving health and, as much as possible, forewarns of disease; and *therapeutics*, which shows how the physician can restore lost health and eliminate diseases, their causes, and symptoms.[7] That Sennert was a committed and careful Aristotelian who combined humoral theory with sophisticated experiments pointing to ensouled chymical corpuscles illustrates the dynamism and diversity of the period.

[3] The relative status of medical *theoria* and *practica* courses changed over time. Taddeo Alderotti in the thirteenth century sought to elevate the status of medicine by associating its *theoria* with contemplative natural philosophy. Yet many physicians across the sixteenth century concentrated on the importance of medical *practica*, even to limiting *theoria* to mere introductory instruction. Siraisi 2001, 215; Maclean 2002, 68–9.

[4] Though there was no strict orthodox division. Maclean 2002, 69.

[5] Here he follows the strong subalternation of medicine to natural philosophy proposed by Avicenna, *Canon* 1.1.1.2. Siraisi 2001, 86.

[6] Newman 2006; French 1994; Findlen 1994.

[7] Sennert 1620, 3–7.

The fact that physicians were trained to bring their manipulations of the natural world into connection with a body of learned theory is a promising starting point for understanding how scientific experimentation and theorizing developed and changed over the early modern period.[8] In this regard, tantalizing leads remain to be explored and rendered with more exact content and more satisfying context. The works of Roger French and Robert Frank, for example, have stressed the importance of Harvey's approach to the circulation and the ensuing controversies and research traditions to the emergence of experimental practices and experimental philosophy.[9]

Recent scholarship has opened up new views of the multivalent nature of early modern medicine. We know much more about the widespread sharing of concepts and practices in natural history, medical case histories, and humanist erudition.[10] We have a much better sense of the permeability, overlaps, and dynamism of the communities of physicians and their places in early modern learned culture. Nonetheless, we still lack a complete picture of the relations between philosophy and medicine in the early modern period. This lacuna is problematic for reasons important both to historical actors and historians themselves; given that so many early moderns justified their systems on the basis of their ability to lead to medical knowledge, the lack of sophisticated and detailed historiography on the importance of medicine in early modern science and philosophy likely reflects more the biases and interests of modern historians than the people and events under discussion. Descartes' intense and prolonged interest in medicine and the significance of medical thought for his general philosophy is now coming into focus.[11] Other figures, from Locke to Leibniz, have also begun to be reevaluated from this biological and medical perspective.[12] This volume aims to benefit from and continue this effort. In addition, it seeks to place recent historiographical breakthroughs in richer and broader contexts of early modern philosophy and medicine. The renaissance of the study of the history of alchemy or 'chymistry' of the past two decades[13] should be brought into a mutually enlightening conversation with the history of medicine. Chymical endeavors from pharmaceutical remedies to the search for universal solvents and the elixir developed within medical traditions and vied for patients in the medical marketplace. The recent proliferation of studies concerning non-traditional actors and objects in this period—notably, *medical* actors and objects—demonstrates that there is still much to understand.

This volume attends to these historiographical concerns especially, but not exclusively, by providing detailed studies of key figures, keeping the intellectual content and context of their work in focus. While interest in the history of medicine from historians of philosophy may be increasing, leading historians of science have

[8] See Wolfe and Gal 2010.

[9] French 1994; Frank 1980.

[10] See, for instance, Ogilvie 2006; Pomata and Siraisi 2005.

[11] See, for instance, Aucante 2006; Manning 2008. For earlier treatments see Lindeboom 1979; Bitbol-Hespériès 1990.

[12] For example, in Anstey 2011 and Smith 2011.

[13] As represented in, e.g., Newman and Principe 2002.

begun to downplay the role of philosophy in early modernity. There has been a distinct move away from discussion of the content of philosophical and scientific theories as explanatory of—and in some cases, even *important to the understanding of*—scientific change. In its place, there is an increasing tendency towards explanations involving only economic and social factors. For instance, the prominent historian of science and medicine Harold Cook has recently been quite explicit about looking *away* from early modern philosophy for the motivating values and even basic ideas of the Scientific Revolution, such as the concept of 'objectivity.'[14] The work presented in this volume keeps intellectual content and context at the center. This is not, in any way, to deny that social and economic factors are important facets of our understanding of the Scientific Revolution, but simply to point out that we have much still to learn about this period through careful attention to intellectual content and context. We hope that this volume can begin to demonstrate that philosophy and medicine were in deep *theoretical* and *methodological* dialogue, as well as establish the fundamental importance of this dialogue for understanding the history of early modern philosophy, medicine, and science.

Chronological and thematic considerations have shaped the organization of this volume into four parts. The three papers in Part I (*Philosophy, Medicine and Method in the Renaissance*) each address ways in which disciplinary boundaries between medicine and philosophy were negotiated and renegotiated, and how such negotiations affected the goals, methods, sources, resources, and of course, the content, of the resulting work. Taken together, these chapters suggest that we must pay close attention to this process of negotiation between philosopher and physician in order to understand the changing methodological, epistemological, and social statuses of both philosophers and physicians.

In his contribution, Craig Martin argues that Renaissance physicians, marked by humanist attitudes and approaches to knowledge and the recovery, evaluation, and assimilation of ancient texts, attempted to assimilate Hippocratic and Galenic works with the Aristotelian *Problemata*. He attends especially to the ways Lodovico Settala, among others, attempted to integrate the Aristotelian *Problemata* and the Hippocratic *Airs, Water, Places*. This story vividly illustrates interaction between natural philosophy and medicine arising out of the negotiation of overlapping traditions of learning. Thus philosophers could draw on Hippocrates for understanding the soul, especially in response to Galen's apparent agnosticism about its immortality; physicians, changing their scholarly hats, could find in Aristotle more resources for discussing the nature of the heart, plagues, semen, and bodily spirits.

Cynthia Klestinec considers a rather different kind of dynamic at work among medical practitioners, concentrating on changing attitudes to the relationship between anatomy and surgery among learned surgeons in Renaissance Italy. Focusing on the works of Giovanni Andrea della Croce and Leonardo Fioravanti, Klestinec suggests that debates between learned surgeons and empirics problematized

[14] See Cook 2007, which won the Pfizer Award from the History of Science Society in 2009. Cook does discuss philosophy and philosophers, but his focus is clearly on other aspects of early modernity.

"the authority of anatomy and the legitimacy it offered to learned surgery." She illustrates how, by the second half of the sixteenth century, anatomy, which in learned settings was connected closely to natural philosophy, had become a conflicted resource in the medical marketplace. There, anatomical expertise was in danger of being connected not to the reduction of clinical errors, but to a practitioner's violent approach to the living body of the patient. Klestinec shows how, in this context, Croce is concerned to distinguish learned surgery from anatomy and to assimilate the practice of the learned surgeon to the visual arts.

In the final chapter of this section, Tawrin Baker shifts our focus from anatomy and surgery to anatomy and natural philosophy. He provides a detailed, careful account of their close interaction in the works on vision of logician and natural philosopher Jacopo Zabarella and physician and anatomist Hieronymus Fabricius ab Aquapendente. Both thinkers integrate philosophical accounts of light, color, and vision with anatomical accounts of the structure, action, and usefulness of the parts of the eye. Baker's meticulous study of their works demonstrates important interaction between the two thinkers, between broadly medical and natural philosophical approaches, and between anatomical and experimental research and ancient authority. Finally, Baker emphasizes the influence of these efforts at integrating anatomical and philosophical accounts of vision on Kepler's work. Baker's study nicely captures mutual influence between late Renaissance medicine and natural philosophy at personal, theoretical and methodological levels.

The papers in Part II (*Life and Mechanism*) focus firmly on the seventeenth century and on the place of mechanism in that period's investigation of living things. These chapters explore a range of conceptual, explanatory, and methodological issues surrounding the application of mechanical or mechanistic perspectives to understand the complex causal and ontological systems of living things. Here the interaction between anatomy (particularly post-Harveian anatomy) and Cartesian philosophy looms large.[15] However, as these chapters make clear, seventeenth-century mechanism was a multifaceted phenomenon and cannot be identified with the Cartesian program.

This important point is reflected in the first chapter of the section. Domenico Bertoloni Meli provides a wide-ranging and stimulating exploration of mechanistic anatomy broadly conceived—of efforts to provide machine-like explanations of bodily operations. Bertoloni Meli begins the important process of *asking questions* about the interaction between mechanisms and mechanical constructions and the study of life in early modernity, ultimately forcing us to think about what the mechanization of living bodies really meant. What did 'mechanical' mean, and when? The relevant conceptual and manual resources for thinking and working with bodies changed. There were many new machines in the seventeenth century, and new uses of machine-behavior and analogies to understanding living structures and actions. Thus Robert Hooke used a new microscope to observe drop-like structures hanging under the wings of some insects. He conjectured they might be used in these minia-

[15] Matter theory and mechanism, while making a supporting appearance in these chapters, takes a more central role in Part IV of the volume.

ture flying entities to regulate flying motions, just as pendulums act in human-made machinery. Bertoloni Meli here explores these complex processes and ends by investigating the role of dead bodies and body parts as tools of investigation and experimentation, a sort of boundary object between the world of living organism and that of artificial machines. Thus anatomists could work on the blood and vessels of the body—compared to sluice gates, mills, and pumps from Harvey to arch-mechanists—as on other objects of study. Johannes Walaeus produced support for Harvey's circulation by pressing blood from an artery into an emptied vein that ought to have been continuous with the artery by the invisible connecting vessels. Like pendulums and sluice gates, dead bodies have no life in them. Yet their structures are much the same as living bodies.

Peter Distelzweig picks up on the role of machine analogies in William Harvey, in particular, and places them within a broader analysis of the nature and role of mechanism and mechanics in Harvey's thought. He distinguishes six meanings of 'mechanical' relevant to understanding Harvey's work and argues that Harvey has a consistent, stable understanding of the place of mechanism within his broadly Galeno-Aristotelian anatomical project—a project much influenced by the works of Hieronymus Fabricius ab Aquapendente. To this end, Distelzweig examines Harvey's published work, as well as methodological and programmatic remarks found in his lecture notes and his unpublished working notes on the organs of local motion.

Karen Detlefsen's paper brings us to Descartes' mechanistic project. Detlefsen is concerned with understanding whether and how Descartes could articulate a coherent theoretical conception of living things to delineate them as an object of study, given his austere mechanistic ontology and rejection of final causal explanations in natural philosophy (because of our ignorance of God's ends). She develops an account of Descartes' theoretical conception of life, and, in doing so, demonstrates that Descartes does not eliminate the class of living bodies from his natural philosophy. He is a reductionist with respect to explanation but not an eliminativist with respect to life. However, Detlefsen argues further that the best theoretical account of living beings available on Cartesian terms needed to make reference to God's ends, and she explores the possibility that, while he in fact rejects such a move, there is room within Descartes' system for employing such teleological explanations as merely hypothetical.

Evan Ragland explores the ways in which philosophy, anatomy, and chymistry were inextricably bound together in lively, late seventeenth-century Dutch debates over the action of the heart. The chapter explores the shifting nature and use of mechanical explanation in the realm of living things in the wake of Harvey's anatomical demonstration of the circulation of the blood and Descartes' provocative but error-prone anatomical speculations. Ragland shows how Dutch physicians adopted varied positions on the sources and status of anatomical knowledge, focusing on Franciscus Sylvius' central place in this history. Sylvius and his colleagues were generally comfortable with mechanical explanations, which they had already met in Galen's depictions of the mechanical anatomy of Erasistratus, but only as far as they squared with sensory experience. Even prominent mechanistic anatomists such as

Sylvius' student Nicolaus Steno would accept ideals and methods of mechanistic explanation, while rejecting particular proposed mechanisms for their sensory and experimental inadequacy. Our own sense of early moderns' errors may be of little use to historical understanding, but tracing *their* perceptions of error, especially in the autoptic anatomical tradition, is essential.

The discussion of post-Cartesian, mechanistic philosophy and medicine continues in the chapter by Patricia Easton and Melissa Gholamnejad examining the work of the French physician Louis de la Forge. They trace how La Forge, in his *Remarks* in the French edition of Descartes physiological works, advanced Descartes' account of the generation and the working of the animal spirits in the human body-machine. They examine similar themes in La Forge's *Treatise on the Human Mind*, in which he explained the functions of the soul while defending dualism and the mechanism of the body machine. Their discussion of his reception, development and revision of Descartes's physiology shows that Descartes' mechanical model of the body provided La Forge a scientific framework for reasoning about and testing the operations of the body. It also corrects for a tendency in the history of philosophy to attend only to La Forge's work on causation.

The papers in Part III (*Matter and Life, Corpuscles and Chymistry*[16]) explore sixteenth- and seventeenth-century thinkers writing on these subjects. The authors here demonstrate how vital it is for our histories of matter theory, corpuscularianism, and philosophical medicine to include chymical traditions. The rise of corpuscular thinking, so characteristic of seventeenth-century natural philosophy, is inexplicable without looking to traditions and figures such as those analyzed here. These papers also enrich the discussion of mechanism begun in the previous section by approaching it from a different perspective—one centered on questions of matter theory and generation and developed in relation to traditions of learning distinct from the anatomical context discussed there. Once more our attention is turned to varied and changing definitions of 'mechanical'—from Severinus' *semina* generating material beings according to divine plans to Boyle's material explanations of generation with plastic powers—and to the relevance of the medical context for understanding these variations. All the authors point out the importance of early modern chymistry to the rise of new matter theories and key problems such as generation and fermentation.

One major source for corpuscular thinking appears in Jole Shackelford's discussion of Petrus Severinus' *semina*, semi-material locations for development with ineliminable vital properties of development. Shackelford's chapter provides a systematic treatment of Severinus' doctrine of transplantation. Liminal between material and immaterial entities, semina connect impressions or ideal influences and material generations. At a general level, they draw on Neoplatonic ideas of the generation of material being and Aristotelian natural teleology. Distinct from transmutation or transformation, transplantation depended on the transference of semi-material seeds from place to place. In contrast to the later mechanical philosophies, semina had intrinsically temporal properties. Transplantation explained timed

[16] On the use of the term 'chymistry,' see Newman and Principe 1998.

development, development informed by the original seeds but altered by external influences from stars and elements. Severinus' vital semina—always as nondimensional locations which then put on material bodies—provided a corpuscular alternative to Lucretian atomism.

Complementing Shackelford's discussion of Severinus' complex doctrines of semina and transplantation, Hiro Hirai's chapter analyzes the role of seeds or living corpuscles in the accounts of generation and animal life in the work of Daniel Sennert, Pierre Gassendi, and Athanasius Kircher. For the physician and philosopher Sennert, living beings reproduce through an internal principle hidden in matter, a "seminal principle" or "soul." The soul informs the body, vivifies the body, but can also exist in a third mode, that of a latent soul residing in a body as if in a container. Hirai identifies some of the seeds of Sennert's own views in a little-known treatise on spontaneous generation, written by the Paduan professor of philosophy Fortunio Liceti. Liceti provided Sennert with the ideas that a soul can reside in a single atom and that the souls of many atoms can gather together under a ruling form or soul. Many of Boyle's deep debts to Sennert are fairly well-known from William Newman's recent work,[17] but Hirai adds another dimension, connecting Liceti, Sennert, and Boyle's interest in seminal principles. Gassendi, too, argued for the propagation of souls, though animal souls for him were closer to those of Democritus. These corporeal "little flames" composed of tiny, mobile corpuscles were endowed by God with *scientia* to form regular structures and species. As Hirai shows, Gassendi borrowed much of this notion of working seeds from Severinus, but casts it into a more materialist, atomist model. Hirai turns finally to Athanasius Kircher, focusing on his account of semina in spontaneous generation. Loosely following Thomas Aquinas, Kircher held that the substantial forms of living beings were drawn from the potentiality of the matter. But, as Hirai nicely traces, Kircher develops a view of seminal corpuscles and material spirit to account for spontaneous generation that draws on a diverse range of corpuscularian and chymical resources. And here, too, the influence of Liceti's account of spontaneous generation can be seen.

Antionio Clericuzio's study of fermentation, especially the context and content of Robert Boyle's account of fermentation, draws on and develops a number of themes in Shackelford's and Hirai's chapters. Attending to Boyle's medical interests and focusing on fermentation, Clericuzio can trace in Boyle's explanation of vital phenomena the interplay between chymistry, corpuscularianism, and experiment. Just as yeast worked real changes in bread and beer, so active ferments wrought alchemical transmutation, according to influential writers from the Middle Ages on. Paracelsus, especially, embraced ferments as agents of change throughout the body and in metallic transmutation. His heirs, especially Van Helmont, elaborated and spread the notion of active, spiritual ferments. Like his colleagues Thomas Willis and Ralph Bathurst, Boyle initially (if cautiously) allowed for the action of ferments for causing changes in bodies. Later, following chymical experiments into the nature of fermentation and the blood, Boyle, Willis, and others dropped talk of ferments. For Willis, John Mayow's nitre theory was more attractive, though Boyle

[17] Newman 2006.

remained hesitant to commit himself to a new chymical doctrine. Throughout, though, Boyle and his colleagues remained committed to material explanations for fermentation and processes supposedly caused by ferments.

Few problems in the interrelation of philosophy and medicine were as vexing as generation.[18] The last chapter in this section, by Ashley Inglehart, analyzes the work on generation of Robert Boyle and its reception by Marcello Malpighi. Boyle's appeal to 'plastic powers' to organize the process of generation has smacked of the vestiges of Galenism or Aristotelian thinking to some scholars. However, Inglehart argues that, while he sometimes used similar terminology, Boyle never advocated the existence of something like Galenic faculties. Boyle's explanations of animal generation, in contrast to Harvey's search for the organizing activity of the soul, remained mechanical since he adhered to material explanations of *how* generation unfolded, rather than *why*. For Boyle and Malpighi, even granting the existence of directing souls, such souls would still be bound to work mechanically, by arranging matter in motion. This matter and its motions were the proper subject of inquiry, not the activity of the soul. Specifically, Boyle applied his researches into the mechanical explanation of the formation of stones and gems to the phenomena of animal generation. Malpighi closely followed Boyle's language and explanations and added experiments and mechanisms of his own. In the end, Boyle appropriated and re-shaped traditions of chymical investigation in terms of ensouled or *scientia-bearing* corpuscles to push material explanations as far as possible.

The final section of the volume (*Medicalizing Philosophy?*) takes a broader view on the relations between natural philosophy and medicine. This section contains two wide-ranging papers that explore different ways in which the interactions between medicine and philosophy affected the goals and larger social image of physician and philosopher. The first finds Justin E. H. Smith forcing us to rethink what being a philosopher in early modernity meant, arguing that we must take seriously the medical or therapeutic goal of philosophy. Smith explores how Gottfried Leibniz's medical, dietetical, and pharmacological concerns and endeavors were intimately linked with his philosophical ideas concerning the metaphysics of corporeal substance. Smith argues then that the proper maintenance of the human corporeal substance constitutes a sort of corporeal flip-side of morality and was thus a central concern to a philosopher. According to Smith, the primary concern of the physician, health, is also of deep metaphysical importance to the natural philosopher and, for Leibniz, could offer the possibility of harmonizing rationalism and empiricism.

The final chapter of the book has Charles T. Wolfe considering the social and epistemological implications of medicine and the resulting cultural conception of the physician in early modern Europe. Wolfe focuses on the image of the physician as an atheist and explores the origin of this image in a certain sort of medical philosophy. Wolfe calls this *radical medicine* – a medical precursor of the Radical Enlightenment, symbolized by the slogan, *tres medici, duo athei*: medicine as a basis for atheism. This theme runs through various medical and medico-theological

[18] Smith 2006. See also Roger 1963.

works, such as Thomas Browne's 1643 *De religio medici*, which begins with Browne regretting rumors of doctors being atheists as the "general scandal of my Profession." But these are examples of the *fear* of a radical medicine – a medicine that denies the existence of an immortal soul, or even defends materialism and atheism. Are there positive statements of this doctrine? Indeed, as Wolfe demonstrates, attacks on it were much more common than statements identifying with it.

The chapters in this volume examine figures from the sixteenth century to the mid-eighteenth, and across this breadth there are a number of trends and themes we want to emphasize. First, there is a strongly suggestive trend across the sixteenth and seventeenth centuries of greater interrelation of medical and philosophical concerns, perhaps even a cross-disciplinary unification of methods and modes of explanation. This may be part of the larger expansion and reorganization of natural philosophy across these centuries, as evidenced by classifications from the end of the seventeenth century that include medicine, natural history, mathematical disciplines, and mechanical arts in the category of 'natural philosophy.'[19]

We can observe illuminating moments in this gradual, though not universal, trend of integration. Our second chapter shows how learned physicians across the sixteenth century such as Girolamo Cardano understood the Hippocratic text *Airs, Waters, Places* to share topics and even methods with natural philosophy, especially in reasoning from effects to causes. The fourth chapter shows the philosopher Zabarella and the physician Fabricius experimenting around 1600—almost certainly, together—to understand the *usus* of the vitreous humor of the eye.

In Harvey's work, and the ensuing controversy with Descartes over the action of the heart, we find anatomical observations and arguments used as key components in comprehensive philosophical systems and debates. For Descartes, accounting for living bodies was difficult, given his own austere ontology and his rejection of claims to knowing God's ends. Yet it also presented him with a problem he could have domesticated by accepting bodies as having simple natures with ends as a working hypothesis. La Forge, discussed in Chap. 9, following Descartes, attempted to extend and refine the application of Descartes's simple but comprehensive principles, keeping his accounts of animal spirits, the pineal gland, generation, and memory squarely within Cartesian philosophy. We should note counterexamples, though, and the chapters by Klestinec and Ragland remind us that the integration of medicine and natural philosophy was neither complete nor uncontested.

Second, as is well-known, mechanism in all its meanings engaged medical topics in productive and complicated ways.[20] Fully half of our chapters grapple with mechanism and living things. Taken together, they showcase some of the diversity of meanings embraced by the term 'mechanical.' They also outline some key

[19] Daston and Park 2006, 3. Gregor Reisch's important 1503 *Margarita philosophica*, in contrast, placed the operative part of medicine under the headings of practical and factive philosophy, and the *theoria* of medicine under divisions of theoretical, real, and physical or natural philosophy. Cf. Bylebyl 1990; Mikkeli 1999.

[20] For recent work, see Bertoloni Meli 2011; Smith 2006 and 2011; Wolfe and Gal 2010; Manning 2008; Cook 1990.

problems or areas of investigation. The chapters by Bertoloni Meli, Distelzweig, Shackelford, and Inglehart explicitly treat different meanings of the term 'mechanical.' Distelzweig and Bertoloni Meli, in particular, survey a range of meanings from the machine-like composition and interaction of parts to the rejection of souls and Galenic faculties as explanatory principles.

These chapters dealing with mechanical approaches to medical themes and phenomena largely agree with the view that 'the mechanical philosophy' dealt in restricted ontologies and means of explanation. But in most of the chapters, mechanical philosophers—perhaps even Descartes, as Detlefsen argues in Chap. 7—needed to adopt hypotheses that reached beyond utterly inert extended matter to account for the details of disease and generation, the seemingly obvious view that living bodies have natures, and the regularity of living forms and kinds. But if Boyle, Malpighi, and other illustrious proponents of 'the mechanical philosophy' adopted such tools as active chymical powers, it seems that much mechanical philosophizing in the seventeenth century slipped more neatly into the outlines of an eclectic materialism.

Third, in terms of chymistry and life, integration proceeded in at least two directions: philosophical explanations of living things informed chymical theory while chymical practice and ontology informed philosophical doctrines. Hirai's chapter shows Liceti's novel account of spontaneous generation bearing fruit in Sennert's chymistry of ensouled corpuscles. He also shows chymical theory and practice shaping metaphysical doctrines of souls. The Paracelsian Severinus, struck by the temporal emergence and development of diseases, plants, and animals, made chronological development according to divinely-implanted knowledge or *scientia* a central feature of his doctrine of *semina.* Drawing on Severinus and the philosopher Fortunio Liceti, Sennert, Gassendi, and Kircher combined observations of the organization of living entities and chymistry to frame new philosophical accounts of souls and matter. Phenomena of fermentation—from brewing to blood—became resources for the articulation of an array of chymical accounts of digestion, disease, and metallic transmutation. Some thinkers, such as Van Helmont, opposed materialist principles and cast ferments as spiritual agents shaping corporeal matter. Others, such as Boyle and colleagues, generally sketched material accounts of fermentation.

Inevitably, it seems, we return to mechanism. The chapters here should help to clarify our understanding of its meanings for the historical actors. Looking to philosophical categories, it seems that Boyle's nescience on the *nature* of the seminal principles or plastic powers should leave his ontology open to something like the souls in Sennert's chymical corpuscles. After all, Sennert's explanations of qualitative chymical change in terms of the association and dissociation of corpuscles deeply informed Boyle's chymical program, and Boyle's explanations of generation seem to reach for some sort of organizing principle. Yet, as Inglehart stresses, Boyle strove to shift the mode of explanation from one of understanding causes in terms of natures and ends to one of explaining natural events in terms of *how* material constituents interact. Whatever the organizing cause was, the materials of generating gemstones and chicks moved about in ways the mechanical philosopher could investigate. In this respect, Boyle appears closer to a methodological materialism than Gassendi, who drew on Severinus's doctrine of *scientia*-bearing corpuscles to describe his own seminal *moleculae,* which acted according to God's plans to dispose the corpuscular elements and principles in the distinctive order and regular succession of living species. For Gassendi, the souls and semina of plants and animals remained corporeal, yet he explicitly adopted impressed divine *scientia* which exceeded the limits of strictly inert mechanism.

Most of the time, our chapters expand and refine our understanding of early modern accounts of the constitution, action, and ends of living bodies. In terms of Sennert's division of medicine into physiology, pathology, semiotics, hygiene, and therapeutics, we can note that physiology and philosophy receive the lion's share of our attention. However, hygiene and therapeutics are not absent. They make leading appearances in Smith's stimulating chapter. For Leibniz, in particular, learning *how* to care for and cure the body was not only the corporeal counterpart of ethics, but furnished notions of *appetitus* important for his later thought about perceptive monads. The question of *why* humans, uniquely among creatures, had to learn how to preserve and restore health was also a pressing philosophical problem, with consequences for notions of human-animal distinctions, ethics, and epistemology. Clericuzio's contribution points out that Boyle hoped to concoct beneficial foods and medicines through the study of ferments inside and outside the body. And mechanistic physicians could reap new rewards in pathology and therapeutics, as Bertoloni Meli points out. The influential seventeenth-century anatomist and physician Marcello Malpighi defended the medical utility of mechanical approaches to the body and health by citing the origins of gout in excess acidity. Mixing "mechanically" spirit of vitriol or another strong acid with other fluids produces similar effects *in vitro*.

Objects, especially new ones, were important things to think about and think with in early modern philosophy and medicine. More than others, Bertoloni Meli's chapter illuminates the productive interworking of new machines and experiments with philosophical questions about the soul and medical goals of healing. He stresses the swiftly-changing flow of resources investigators had on hand with which to think about and work with bodies. Microscopes, pendulum devices, and barometers were new to the seventeenth century, as were new ways of thinking about simple machines, such as Hooke's law of the spring. In Baker's chapter, philosophers and physicians think in strikingly similar ways about eyes, lenses, diagrams, and *camerae obscurae*. And Inglehart demonstrates similar mechanisms of ontology and explanation Boyle and Malpighi applied to gemstones and embryos.

Finally, in terms of institutions, we find the sort of variations in human interaction one might expect. Institutional or geographic proximity could foster productive collaboration, as Baker's study shows in the teamwork of Zabarella and Fabricius at Padua, and as we find in Ragland's examples of teaching experimentation at Leiden. But institutional sharing can also become crowded and even antagonistic. Thus Klestinec points to the association of natural philosophy and anatomy in the universities, an integration that contrasted sharply with learned surgeons' distancing of surgery from university anatomy. And in Leiden, Ragland argues, anatomist-physicians repeatedly objected to the perceived anatomical errors of Cartesian philosophers by partitioning disciplinary identities and trumpeting their own reliance on their senses. Even in a period in which philosophers and physicians enlarged the borders of natural philosophy, different social groups could survey the intellectual and institutional landscape along different lines and stake claim to their own territories.

Our historiographic stances are most squarely historicist; each author aims primarily to articulate concepts and explicate texts with fidelity to the arguments and contexts of the historical actors. Thus Smith urges us to reconsider the aims of philosophers in terms of body-soul *eudaimonia* and Inglehart explains how seminal principles could remain properly mechanical. Sometimes, though, translation and understanding calls for present-day terms and speculations. For Shackelford, calling the divine *scientia* in Severinus' semina 'programming' helps us to understand the regulated, temporal developments so important to his thought. And Detlefsen offers scholars a novel suggestion not only for what Descartes *could* have argued in order to secure the seemingly robust natures of living bodies in health and disease, but also what he *should* have argued, given his resources and commitments. Attempting to think along with our subjects, we hope, can help us to craft historical interpretations of their texts and thought that they might have recognized and perhaps even found akin to their own.

This volume provides strong evidence of the indispensability of medical concerns and contexts to the history of early modern philosophy. It also provides ample evidence that philosophy was integral to early modern learned medicine.

References

Anstey, Peter. 2011. *John Locke and natural philosophy*. Oxford: Oxford University Press.

Aucante, Vincent. 2006. *La philosophie médicale de Descartes*. Paris: Presses Universitaires de France.

Bertoloni Meli, Domenico. 2011. *Mechanism, experiment, disease: Marcello Malpighi and seventeenth-century anatomy*. Baltimore: The Johns Hopkins University Press.

Bitbol-Hespériès, Annie. 1990. *Le principe de vie chez Descartes*. Paris: Vrin.

Bylebyl, Jerome J. 1990. The medical meaning of *Physica*. *Osiris* 6: 16–41.

Cook, Harold J. 1990. The new philosophy and medicine in seventeenth-century England. In *Reappraisals of the scientific revolution*, ed. Roberts S. Westman and David C. Lindberg, 397–436. Cambridge: Cambridge University Press.

Cook, Harold J. 2007. *Matters of exchange: Commerce, medicine, and science in the Dutch Golden Age*. New Haven: Yale University Press.

Daston, Lorraine, and Katharine Park. 2006. Introduction: The age of the new. In *The Cambridge history of science, vol. 3: Early modern science*, ed. Katharine Park and Lorraine Daston, 1–17. Cambridge: Cambridge University Press.

Descartes, René. 1998. *The world and other writings*. Trans. and ed. Stephen Gaukroger. Cambridge: Cambridge University Press.

Findlen, Paula. 1994. *Possessing nature: Museums, collecting, and scientific culture in early modern Italy*. Berkeley/Los Angeles: University of California Press.

Frank, Robert G. 1980. *Harvey and the Oxford physiologists: A study of scientific ideas and social interaction*. Berkeley/Los Angeles: University of California Press.

French, Roger. 1994. *William Harvey's natural philosophy*. Cambridge: Cambridge University Press.

Lindeboom, G.A. 1979. *Descartes and medicine*. Amsterdam: Rodopi.

Maclean, Ian. 2002. *Logic, signs and nature in the renaissance: The case of learned medicine*. Cambridge: Cambridge University Press.

Manning, Gideon. 2008. Naturalism and un-naturalism among the Cartesian physicians. *Inquiry* 51(5): 441–463.

Mikkeli, Heiko. 1999. The status of the mechanical arts in the Aristotelian classifications of knowledge in the early sixteenth century. In *Sapientiam amemus: Humanismus und Aristotelismus in der Renaissance: Festschrift für Eckhard Kessler zum 60*, ed. Paul Richard Blum, 109–126. Munich: Wilhelm Fink.

Newman, William R. 2006. *Atoms and alchemy: Chymistry and the experimental origins of the scientific revolution*. Chicago: University of Chicago Press.

Newman, William R., and Lawrence M. Principe. 1998. Alchemy vs. chemistry: The etymological origins of a historiographic mistake. *Early Science and Medicine* 3(1): 32–65.

Newman, William R., and Lawrence M. Principe. 2002. *Alchemy tried in the fire: Starkey, Boyle, and the fate of Helmontian chymistry*. Chicago: University of Chicago Press.

Ogilvie, Brian. 2006. *The science of describing: Natural history in renaissance Europe*. Chicago: University of Chicago Press.

Pomata, Gianna, and Nancy Siraisi (eds.). 2005. *Historia: Empiricism and erudition in early modern Europe*. Cambridge, MA: MIT Press.

Reisch, Gregor. 1503. *Margarita philosophica*. Freiburg: Schott.

Roger, Jacques. 1963. *Les Sciences de la vie dans le pensée françaese du XVIIIe siècle*. Paris: Armand Colin.

Sennert, Daniel. 1620. *Institutionum medicinae libri V*. Wittenberg: Zacharias Schurerus.

Siraisi, Nancy G. 2001. *Medicine and the Italian Universities, 1250–1600*. Leiden: Brill.

Smith, Justin E.H. (ed.). 2006. *The problem of animal generation in early modern philosophy*. Cambridge: Cambridge University Press.

Smith, Justin E.H. 2011. *Divine machines: Leibniz and the sciences of life*. Princeton: Princeton University Press.

Wolfe, Charles, and Ofer Gal (eds.). 2010. *The body as object and instrument of knowledge: Embodied empiricism in early modern science*. Dordrecht: Springer.

Part I
Philosophy, Medicine and Method in the Renaissance

Chapter 2
Lodovico Settala's Aristotelian *Problemata* Commentary and Late-Renaissance Hippocratic Medicine

Craig Martin

Abstract Renaissance physicians, influenced by humanism and spurred by their increased knowledge of Hippocratic and Galenic writings, attempted to assimilate these medical works with Aristotelian thought. The similarities between the Aristotelian *Problemata* and the Hippocratic *Airs, Waters, Places* allowed Girolamo Cardano and Lodovico Settala, among others, to blur the distinctions between natural philosophical and medical authorities. Philological and historical considerations of these texts as well as judgments about authenticity were colored by the belief that these works were useful for humoral physiology and offered insights into the unity of ancient and modern knowledge.

Keywords Aristotelian *Problemata* • Hippocratic *Airs, Waters, Places* • Renaissance humanism • Lodovico Settala • Girolamo Cardano

2.1 Introduction

Late-Renaissance Italian intellectual debate often involved attempts to change or defend the status of particular disciplines. The hierarchy of subjects was frequently a matter for dispute, and leading intellectual figures attempted to raise the status of their particular fields. Just as this was true for mixed mathematics, it was also true for medicine. A number of physicians attempted to promote the status of medicine by defining it as part of natural philosophy, even though some philosophers and humanists insisted that medicine was an art not a *scientia*.[1] To the contrary, well known professors of philosophy at Bologna and Padua, including Alessandro Achillini, Pietro Pomponazzi, Lodovico Boccadiferro, Giacomo Zabarella, and Cesare Cremonini, maintained that medicine was subaltern and thus inferior to

[1] For the view that medicine was an art see Averroes 1564, 4r; Achillini 1548, 148v; Salutati 1947, 2224; Mikkeli 1992.

C. Martin (✉)
History Department, Oakland University, Rochester, MI 48309, USA
e-mail: martin@oakland.edu

P. Distelzweig et al. (eds.), *Early Modern Medicine and Natural Philosophy*, History, Philosophy and Theory of the Life Sciences 14,
DOI 10.1007/978-94-017-7353-9_2

philosophy.[2] During the sixteenth century, philosophy and medicine became separated to a greater degree institutionally at Padua and Bologna, where professors in the faculty of arts and medicine were increasingly specialized in either philosophy or medicine.[3] This institutional division of philosophy and medicine likely engendered a competitive atmosphere in which professors sought to defend or raise the status of their fields.

The attempt to raise medicine's status is well known for the field of anatomy, where its practitioners, drawing from ancient sources, increasingly presented themselves as creating a proper philosophical *scientia*, not merely a craft, during the second half of the sixteenth century. For example, Andreas Vesalius advocated anatomy as natural philosophy, perhaps inspired by Galen's methodological treatise, *De anatomicis administrandis*, which staked a similar claim.[4] Later in the century, Girolamo Fabrici used public anatomies in Padua to investigate topics of natural philosophy.[5]

Links between medicine and natural philosophy extended beyond anatomy, as physicians and philosophers alike investigated dietetics and temperaments. Despite disparaging his physician predecessors, Pomponazzi examined in detail the subject of digestion in his commentary on *Meteorology* IV, blurring the lines between philosophical and medical knowledge.[6] Francisco Vallés wrote a comprehensive tome that aimed to reconcile disagreements between philosophers and physicians on numerous physiological topics in his *Controversiae medicarum et philosopharum*.[7] While Vallés's work undermined distinctions between medical and philosophical knowledge, Girolamo Cardano went so far as to claim that medical knowledge was more certain than natural philosophy, which he maintained derives causes from effects, while medicine often infers effects from causes.[8]

As medical treatises and philosophical treatises, such as Vallés's and Cardano's, made a greater attempt to improve natural philosophy through medical knowledge, Aristotle, still extremely dominant in natural philosophy, grew in importance for the field of medicine during the sixteenth century. A number of Aristotle's writings, such as his zoological works and *Meteorology* IV, were potentially relevant to medicine. The sixteenth century also witnessed the rise in the number and influence of commentaries on the Aristotelian *Problemata*. Interpretations of the *Problemata* became a touchstone for those who wanted to blur the boundaries between Aristotelian philosophy and erudite medicine. For example, Cardano argued that it was possible to use medical principles to investigate issues of natural philosophy

---

[2] Martin 2002, 10–14; Mikkeli 1992, 159–177; Schmitt 1985; Agrimi and Crisciani 1988, 21–47; Bylebyl 1990.

[3] Lines 2001; Bylebyl 1979, 338.

[4] Carlino 1999, 125–128.

[5] Klestinec 2007.

[6] Pomponazzi 1563, 27r-30r.

[7] Vallés 1591.

[8] Cardano 1663, 8:585. "Et ob hoc intelligimus, Medicinam esse certiorem naturali philosophia, cum naturalis philosophia semper procedat ab effectibus ad causas, Medicina vero persaepe a causis supra effectus."

that were not directed toward medical purposes, and cited the third book of the *Problemata* that concerns drunkenness as an example of such an investigation.[9] Gabriele Falloppio (1523–1562), a professor of surgery at Padua best known for his anatomical research and the eponymous tubes, integrated material about teeth from the *Problemata* in a commentary on the Galenic *De ossibus*.[10]

The emergence or reemergence of the *Problemata* as a source for medical and philosophical commentary in the late sixteenth century stemmed from the values of medical humanism that prized ancient sources and philological investigations. Learned physicians integrated their interest in the *Problemata* with reconsiderations of Hippocratic writings and a broader knowledge of the Galenic corpus. The best example of this integration is found in Lodovico Settala's 1200-page commentary on the *Problemata* that was printed in the first decades of the seventeenth century.[11] Philological and historical investigations form a significant part of Settala's considerations of the *Problemata*. They were part of his goal of applying Aristotle's writing to issues of medicine and philosophy, including importantly the relation between temperament and the human soul. Settala described his work as flowing "across the banks into the open field of philosophy and philology."[12]

Rising interest in the *Problemata* occurred simultaneously with the development of an Aristotelian medicine that was at times at odds with long-standing Galenic views that were often transmitted in Avicenna's *Canon*, still the most important book for university instruction of medicine.[13] The medical reading of Aristotle also coincided with the growth of Hippocratism and humanist medicine in general, which grew slowly from the new editions and translations first printed by the Aldine press in the 1520s.[14] Ancient sources grew in value, while medieval sources were discounted. The *Problemata* was particularly valuable because of its links to the Hippocratic text *Airs, Waters, Places* (*AWP*), a work that, despite being available in Latin from the fifth or sixth centuries, had no commentary tradition until the 1570s.[15] *AWP*, which examines the effects of climate and diet on temperament and health, became one of the more influential Hippocratic texts during the seventeenth century.[16] Correspondences between portions of the *Problemata* and *AWP* made the two texts useful for forging considerations of temperaments and the effects of climate on health into knowledge that could be seen as appropriately authoritative for both philosophy and medicine. Moreover, the correspondences between the texts suggested that the blurred boundaries between philosophy and medicine had its roots in the writings of the most ancient authoritative authors of those respective fields, Aristotle and Hippocrates.

---

[9] Siraisi 1997, 52–57.

[10] Falloppio 1570, 40v.

[11] Settala 1632.

[12] Settala 1632, 1:4r.

[13] Siraisi 1987.

[14] Nutton 1989.

[15] Kibre 1975, 123–126.

[16] Wear 2008.

2.2 The Aristotelian *Problemata*

It is difficult, if not impossible, to summarize the contents of the *Problemata*. It contains a series of questions without manifest solutions to these queries. The proposed answers can be interpreted as definitive or tentative. The work was written in the format of: "Why does …?" followed by "Is it because …? or is it because …?," a format common to its genre as whole. Works such as the twelfth-century Salernitan medical questions as well as a host of other problem literature that was produced or diffused during the Middle Ages and Renaissance followed this format, comprising a body of literature that, according to Ann Blair, multiplied during the Renaissance as the result of a growing desire for encyclopedic reference material in both high and low print cultures.[17] Grouped into 38 books or *particulae*, each of which is further divided into questions or problems, the *Problemata* is hardly comprehensive despite the wide number of subjects it tackles. While medical topics are frequently discussed, the work also addresses some assuredly non-medical themes such as mathematics (15), music (19), and justice (29), and others that are only tangentially related to medicine or to humoral physiology such as the nature and characteristics of winds (26), the root of courage (27), and self-control (18). Others subjects are either explicitly medical (1, 10, 14, 22) or require little imagination to connect them to medicine, such as the nature of shrubs and herbs (20), the powers of the hot and the cold (8), and the characteristics and effects of odors (12, 13). In general, the books dedicated to medicine regard health as being determined by climate (14) and diet (22). The arrangement of the books, as well as the material within them, is haphazard. Problems are repeated nearly word-for-word. There are no thematic transitions between either *particulae* or problems; and, books that share similar themes are not always close to each other.

Most of the problems address natural phenomena that are recalcitrant and defy obvious explanation. The solutions are almost always found in material and efficient causation: in the actions and powers of the four elements, the four qualities, and in human physiology. Many of the dilemmas posed are what the modern mind might consider trivial or even dubious. They are often concerned with exceptions rather than general rules, such as "Why are humans the only animal that stutters? (10.40)"; "Why do eunuchs have no or few varicose veins? (10.37)"; "Why do fewer things smell in the winter? (12.6)"; "Why are those who shed their eyebrows given to sexual excesses? (4.18)"; or "Why do some men enjoy the passive sexual role? (4.26)."[18] The phenomena are treated as natural, not as miraculous, marvelous, or preternatural. They are, however, by and large, purposeless. The formal and final causation that looms so large in Aristotelian natural philosophy seldom appears, although the coherency of the natural world is maintained. Although a number of these problems have had little influence, the problem (30.1) that asked: "Why are

[17] Lawn 1963; Blair 1999b.

[18] Cadden 1997.

all men, who are distinguished in philosophy, poetry, politics, or other arts, melancholic?" served as an authoritative discussion of melancholy in the Middle Ages and Renaissance. Pietro d'Abano's comments on this passage gave a theoretical basis to connections between excessive black bile and creative inspiration.[19]

Even though the *Problemata* often jumps from one subject to another without giving exhaustive explanations, it could be thought of as providing insights into the oddities and particulars that were not explicitly explained in Aristotle's more theoretical works, such as the *Physics* and the *De anima*, which formed the basis of medieval and Renaissance university instruction in philosophy. Pietro d'Abano, admiring the wide scope of the work, maintained, perhaps implausibly, that it treated nearly all philosophy and therefore it could be considered as an encyclopedic guide to the seemingly intractable issues found in diverse subjects, such as humoral physiology and ethics.[20] Francis Bacon praised the *Problemata,* along with the zoological works, as being the best parts of the Aristotelian corpus because of their reliance on experience, unlike the *Physics*, which was, in Bacon's view, a compilation of vain dialectical exercises.[21] Yet his Aristotelian contemporaries were not prone to consider this work a Baconian *historia.* Settala, for example, disagreed with Pietro d'Abano that it treated all of natural philosophy, yet saw this work as concerned with causal knowledge for a range of subjects including natural and moral philosophy.[22]

Unlike medieval and Renaissance thinkers, few, if any, twentieth-century scholars considered Aristotle to be the true author of the *Problemata*, although it is widely accepted to be a product of the Peripatos of the third century B.C.E. Indeed statements in the *Problemata* appear to contradict well-established Aristotelian positions, in its apparent advocacy of light as a material substance (11.33.903a12 15)[23] and the entire body as the source of sperm (4.6.877a17-18).[24] In recent times it has been attributed to direct followers of Aristotle, such as Theophrastus, and to unknown authors in late antiquity. Unlike most of Aristotle's extant works and like many late-Peripatetic works, contemporary historians of philosophy rarely consider the *Problemata*. It has contributed little to modern philosophical debate or treatments of ancient Aristotelian thought, and indeed many of its subjects are no longer considered to be under the rubric of philosophy.[25] One twentieth-century reader, J. L. Stocks, after suggesting that the *Problemata* are among the "weakest, least philosophical treatises found in the Aristotelian corpus," concluded that, "Even if

[19] Klibansky et al. 1964, 68, 72, 119.

[20] Pietro d'Abano 1482, prologue, sig. a2r. "In hoc libro inveniuntur fere totius phylosophie per modum cuisdam alligationis sermonis compilati." Klemm reasonably substitutes "colligationis" for "alligationis." Klemm 2006, 307.

[21] Bacon 2004, 11:98–99.

[22] Settala 1602, vii.

[23] Cf. Aristotle, DA 2.7,.418b13-16; 2.12.424a17-b20.

[24] Cf. Aristotle, GA 1.18.723b23-724a1.

[25] An exception is Lennox 1994.

the *Problems* were in bulk Aristotelian, which they certainly are not, they could do no more than illustrate by occasional sidelights Aristotle's point of view."[26]

In contrast to the modern negative assessments, during the Renaissance determining the authenticity of the *Problemata* required not only philological examination but depended, at least partly, on finding its value for medicine. Its authenticity was questioned widely during the Renaissance and possibly during the Middle Ages, but the stakes differed from those of the past century.[27] Leading Renaissance scholars questioned its provenance. The philologist Juan Luis Vives maintained that the work was a collection of discussions among those who listened to Aristotle's lectures. The result, in his eyes, was a work unworthy of the weight of Aristotle's genius since it provides only doubts without definitive solutions.[28] In the 1550s, Francesco Vimercati, a translator, commentator on Aristotle, and professor at the Collège royal, contended that Theophrastus wrote the *Problemata* because the section on winds was more similar to the Theophrastean *De ventis* than to the second book of the *Meteorology* where Aristotle tackled the same subject.[29] The Platonist Francesco Patrizi, a tireless interrogator of Aristotelian texts, also doubted its authenticity in his *Discussiones peripateticae,* 1571, because it does not conform to Diogenes Laertius's list of Aristotle's works.[30]

Others found evidence for the *Problemata's* authenticity. In the preface to his 1608 commentary on the first ten books of the *Problemata,* Giulio Guastavini marshaled an impressive list of Aristotle's citations of the *Problemata* in other works as well as citations from ancient authors, including Aulus Gellius, Plutarch, Athenaeus, Diogenes Laertius, and Macrobius.[31] Guastavini's position, while based on philological evidence, is inevitably related to his perception of the utility of the work. Because the *Problemata* was seen as helpful in determining truths about medicine and the natural world, Guastavini wrote a commentary on this work, aimed at a medical and philosophical, not purely antiquarian, audience. In the circle of learned physicians, ancient writings gave evidence not just about the past but nature as well. Therefore, its purported genuine provenance gave authority to its arguments. In a book dedicated to clarifying obscure doctrines found in the Aristotelian corpus (1590), Felice Accoramboni maintained that citations of the *Problemata* in *De generatione animalium* and the fact that the "style and method of finding causes for these questions smell of Aristotle's style and doctrine" make it difficult to doubt that Aristotle is the author. Nevertheless, Accoramboni admitted that there are many problems that have been added that are "foreign to the science of Aristotle."[32]

For Patrizi, who mustered up all possible arguments to denigrate Aristotle, lack of authenticity suggested worthlessness. It is unclear, however, to what extent the

---

[26] Stocks 1930, 21.

[27] Williams 1995, 45.

[28] Vives 1538, 5r-5v.

[29] Vimercati 1556, 220.

[30] Patrizi 1571, 25.

[31] Guastavini 1608, 3.

[32] Accoramboni 1590, 742.

supposed spuriousness of the work guided the opinion of those more faithful to a given author, if the work was determined to be ancient and derivative of the author. The famed physician and medical author, Girolamo Mercuriale (1530–1606), for example, devised a hierarchy for Hippocratic works based on the likelihood that Hippocrates was the author, in order to evaluate the merits of each work and their proximity to the "mind" of Hippocrates, but not to further the goal of outright dismissal of those treatises that were penned by an acolyte rather than the supposed father of medicine.[33] Similarly, Settala, although noting the uncertainty of the authorship of the *Problemata* in his commentary on *Airs, Waters, Places*, continued to cite it as authoritative. In any case, by the time he wrote the commentary on the *Problemata* such worries had apparently diminished and the text held authority nearly equal to the rest of the Aristotelian corpus, even though at times he questioned whether Aristotle was the true author,[34] and at other times specifically states that certain problems (e.g., 7.8 and 7.9) are Aristotelian but not by Aristotle himself.[35] Settala evaluated the authenticity of other writings as well. For example, he dismissed the *Problemata* attributed to Alexander of Aphrodisias as inauthentic.[36] While Settala was concerned with philological issues these investigations informed and were informed by his understanding of Hippocrates' and Aristotle's authority. Late-Renaissance Aristotelianism and medical humanism conditioned his judgment on the genuineness of the *Problemata.* His medical humanism and his conception of the *Problemata* built on the techniques yet diverged from the interpretations of the preceding generations.

2.3 Renaissance Aristotelianism and Medical Humanism

The Renaissance Aristotelian tradition with its numerous strands and camps included professors of medicine and natural philosophy and humanists interested in the *Ethics* and *Politics,* ancient languages, and issues of translation.[37] During the fifteenth and sixteenth centuries, scholars, enchanted by newly available ancient works and having taken up the task of learning ancient Greek, made new translations of Aristotelian works, criticized the medieval intellectual tradition, and polished their Ciceronian Latin prose in invectives against rivals.[38]

Humanism, especially its uncovering of new sources and its privileging of ancient authors as models and authorities, had a noticeable impact on interpretations of Aristotle. Jacques Lefèvre d'Etaples and Ermolao Barbaro made paraphrases that

[33] Mercuriale 1588, 1:46; Siraisi 2003.

[34] Settala 1590, col. 407. "Aristoteles etiam (si modo libri illi sunt Aristoteli tribuendi, quod non facile affirmarem) in Problem. sect. 4. problem. 16."

[35] Settala 1632, 1:383.

[36] Settala 1632, 3:348.

[37] Schmitt 1983.

[38] Kraye 1996.

imitated Themistius's,[39] and Agostino Nifo took Alexander of Aphrodisias, whose authority was bolstered by his being the earliest commentator on Aristotle, to be his guide in some of his commentaries.[40] Despite the viciousness of some humanists' attacks on the Middle Ages, the medieval tradition in several ways carried on. Even as late as the turn of the seventeenth century, commentaries on Aristotle used translations made in the thirteenth century, preferring interpreters of Aristotle included Albertus Magnus, Averroes, and Thomas Aquinas.[41] Nevertheless, humanists scrutinized Aristotelian works with the tools of philology, just as they did the entire available corpus of ancient writings, trying to free them from what they saw as linguistic errors.

Renaissance commentaries on the *Problemata* built on and reacted to humanist evaluations and transformations of this work. The scrutiny that the *Problemata* endured in the fifteenth century was in several ways exceptional. Translations of this work provoked more controversy and contention than did those of many Aristotelian works. Bartholomew of Messina's translation, which was the only Latin version of this work until the 1450s, suffers from what cannot be considered anything else but numerous mistakes, probably far more than in most medieval translations of Aristotelian works.

The causes of the mistranslations were both intrinsic and extrinsic to the text. Unlike most Aristotelian works, there was only one thirteenth-century translation of this text. It did not, like much of the corpus, first make the transition from Arabic to Latin, accompanied by Averroes' commentary, before it was translated a second time a few decades later from Greek to Latin. Rather, Bartholomew made the first translation from the Greek, without the aid of any commentary, paraphrase, or other self-standing interpretative guide.[42] The intrinsic cause is found in the nature of the structure and content of the *Problemata* that hardly promotes ready comprehension. The long-lived jest that Aristotle was a cuttlefish who obscured himself with his own ink was perhaps nowhere more evident than in the *Problemata*, for those who thought it was genuine.[43] Rare vocabulary frequently describes accidental and oftentimes strange subjects, whose existence at times is a matter of conjecture rather than universal assent. The unsystematic nature of the text and its lack of organization limited the ability of potential interpreters to predict accurately the likely meaning of unclear passages, thereby forcing educated guesses. Thus understandably Bartholomew's translation and Pietro d'Abano's commentary that used his translation contain interpretations that are so distant from those based on modern editions of the text that if they are not considered mistakes they must be considered perversely bizarre.

As a result of the difficulties of interpreting this work, fifteenth-century investigations into the *Problemata* focused on translation and philology. Renaissance

---

[39] Rice 1970.

[40] Nifo 1552, sig. ***ii [5]; Nifo 1551, 1r.

[41] Mahoney 1980; Cranz 1978; Burnett 1999.

[42] For the Latin translations of the *Problemata* see: Ventura 2008.

[43] Schmitt 1965.

humanists were rarely if ever forgiving over perceived linguistic mistakes, especially those found in the works of university professors and the translations they used. In the first years of the 1450s, two Greek emigrants to Italy, George of Trapezuntius and Theodore of Gaza, made the first translations of the *Problemata* into Latin since Bartholomew's. Gaza's work is noteworthy for its anticipation of modern methods of philology. He used the technique of *emendatio* and compared multiple manuscripts in an attempt to establish a more accurate version of the original text. Gaza had little sympathy for the scholastic tradition and his version altered the earlier translation to an astonishing extent. He changed the vocabulary, eliminated graecisms, replacing them with words found in classical Latin sources, and styled his Latin with Ciceronian flourishes, demanding elegance for his Latin rather than word-for-word fidelity. More significantly, in an attempt to improve the organization of the *Problemata*, he changed the structure of the text, deleting repetitive problems and reordering it.[44]

Gaza's editorial liberties, his word choice, and his prose style met opposition almost immediately. Humanist rhetoricians were as unkind to their own ilk as they were to their scholastic predecessors. In either 1453 or 1454 George Trapezuntius, in an invective against Gaza, criticized his Latin vocabulary, his interpretation of Aristotle, and his alleged "inept garrulousness."[45] Trapezuntius, defending Albertus Magnus, Giles of Rome, Walter Burley, and especially Thomas Aquinas as accurate and theologically correct interpreters of the Stagirite,[46] took issue with Gaza's attempts of eloquence and translations that strayed far from Aristotle's text. At that time, Trapezuntius was working on his own translation that surfaced in 1454. A year later he added scholia, primarily concerned with language and the choice of vocabulary. Unlike Gaza's translation, which became the standard of incunables and early sixteenth-century Latin printings, Trapezuntius's translation was never printed and circulated in a relatively small number of manuscripts, none of which later Renaissance scholars, such as Settala, appeared to consult. Trapezuntius was not alone in attacking Gaza's translation. Angelo Poliziano, perhaps best known for his role in developing modern methods of classical editing,[47] without adopting the excessively polemical style of Trapezuntius, praised Gaza as learned but criticized his translation of what Bartholomew's usage of *melancholica* instead of the transliterated *biliosa atra*, a criticism that Trapezuntius also leveled in his invectives.[48]

While humanist scholars debated the nature of translation and the interpretation of the *Problemata*, medical authors consulted the text and corrected medieval interpretation. Humanists' inquiries into ancient writing changed learned medicine in the first decades of the sixteenth century, as new texts were discovered, edited, translated, and diffused. The first Greek edition of Galen's *Opera omnia* was printed

[44] Monfasani 1999; Perfetti 1995.

[45] Trapezuntius 1967, 3:280; Monfasani 2006.

[46] Trapezuntius 1967, 3:341.

[47] Grafton 1977.

[48] Poliziano 1498, cap. 90, sig. I iiiir-I iiiiv; Trapezuntius 1967, 3:285–286; Olivieri 1988, 147–153.

in 1524. Two years later, an edition and Latin translation of the Hippocratic corpus followed. These works informed the Renaissance appropriation of the *Problemata* because physicians, influenced by humanism, interested in philology, and absorbed in integrating newly available ancient works into their thought, were among the most frequent readers of the *Problemata*. For example, Antonio Musa Brasavola (1500–1555), a professor of medicine at Ferrara, a center of early medical humanism, added the entire twentieth book of the *Problemata*, which treated plants and shrubs, to his seemingly exhaustive description of what he maintained were all simple medicines.[49]

In general, Brasavola followed the Ferrarese tradition of medical humanism, first promoted there by Nicolò Leoniceno (1428–1525), which contended that the Arabico-Latin tradition should be entirely replaced by Greek authorities. Leoniceno collected manuscripts and made translations of Galen. Giovanni Manardi (1462–1536) continued this tradition, advocating the use of Greek among physicians to avoid terminological confusion. Similarly, Brasavola embraced Galen as an authority, making an index of the Galenic corpus and promoting Galen's commentaries on Hippocratic works such as *Regimen in Acute Diseases*, *Epidemics*, and the *Aphorisms*.[50] He integrated his interest in textual studies with empirical research. He directly observed living plants, comparing their structures and characteristics to what was described in ancient botanical works by Dioscorides and Theophrastus. Thus for him the *Problemata* was one more Greek source that could aid in the identification of the species of flora with healing properties.[51]

The Ferrarese school did much to promote the availability of accurate versions of Galenic and Hippocratic sources that became extremely influential. While slow to spread, Hippocrates gradually matched and, for some, overcame Galen as an authority in medicine. The oracular and aphoristic style of many Hippocratic writings lent the works *gravitas* in the eyes of Renaissance physicians.[52] Moreover, the interpretation of the Hippocratic writings demanded little rigidity, because of their obscurity and frequent vagueness, so that they could accommodate a wider range of positions and more new discoveries than Galen's prolix, detailed, and polemical prose could.[53] Accordingly the newly translated Hippocratic works seeped into the prevailing Aristotelian and Galenic foundations of medicine, throughout Europe. In this light, the humanist scholar J. J. Scaliger promoted the practical treatise *De vulneribus capitis*.[54] Others, such as Gemma Frisius combined Hippocrates with Plato and the *prisca theologia*.[55]

While some, such as Scaliger, continued to promote Leoniceno's strict stance of using only ancient sources, a number of sixteenth-century medical authors, just as

[49] Brasavola 1544, 518–530.

[50] Nutton 1997.

[51] Reeds 1991, 536–537.

[52] Mercuriale 1588, 1:56.

[53] Nutton 1989.

[54] Nutton 1985; Hippocrates 1578.

[55] Hirai 2011, 104–122.

Trapezuntius a century before, did not wish to eliminate the entire medieval tradition but hoped to integrate the new Greek sources with earlier medieval works. Cardano, who commented on Hippocratic works such as *AWP* and *De alimento*, maintained that those, such as Manardi and Leonhart Fuchs, who rejected all Arabic authors and their experiences, should stick to grammar and leave medicine to physicians. While he reacted against late-medieval scholastic physicians, such as Jacopo Forlì, Ugo Benzi, Gentile da Foligno, he nevertheless maintained the necessity of reading Averroes, al-Razi, Avicenna, and Pietro d'Abano, even if he harshly criticized Pietro d'Abano at times.[56] Thus among some Renaissance medical authors who did not wish to reject the entire medieval tradition, Pietro d'Abano was an acceptable guide to medicine. In the sixteenth century, Pietro d'Abano's *Conciliator* was a standard reference for those interested in medical topics and was printed at least 19 times in between 1472 and 1595.[57] Similarly Pietro d'Abano's commentary on the *Problemata* was frequently consulted, being the only printed line-by-line commentary on the work until Settala's. It was printed eight times from 1475 to 1582.[58]

The usefulness of his commentary on the *Problemata* was tempered by its dependence on an unreliable translation. In order to remedy the unreliability of Pietro d'Abano's *Problemata* commentary, Antonio Luiz (d. 1565), a Portuguese physician, wrote a short treatise that listed what he saw to be Pietro d'Abano's mistakes, due to "the poor quality of the old translation,"[59] and then gave corrections. Luiz, while pointing out the limitations of Bartholomew's efforts, also found faults with Gaza's, although in this work he was primarily interested in improving the interpretation of the *Problemata* found in Pietro d'Abano's comments. For example, he noted that in 12.8, the question asks: "Why do roses on a sharp stem (umbelicus asper) have a greater perfume?" whereas Pietro d'Abano thought the question read: "Why do men with sharp navels (*umbelicus asper*) smell roses better?" He then attempted to explain why this is in fact the case. Luiz explained that Pietro d'Abano's reading of the text did not fit with the rest of the question and then reasonably contended that any explanation of this supposed phenomenon would be just as absurd as presuming it exists.[60] In this vein he clarified a number of passages that can only be considered confusing if not downright confused. Luiz was far from hostile toward the *Problemata* tradition as a whole and wrote five books of his own problems.[61] His work suggests that he considered Pietro d'Abano's commentary useful to medical knowledge if one could avoid its pitfalls.

The inclusion of Pietro d'Abano among the trusted medieval authorities during the late Renaissance shows the importance of Aristotle for early modern physicians as well as high regard for Pietro d'Abano's goal of reconciling medicine and natural

56 Siraisi 1997, 48, 60; Giglioni 2008.

57 Norpoth 1930, 301.

58 Lohr 1972, 331.

59 Luiz 1540, 109r. "antiquae tralationis vitio."

60 Luiz 1540, 109v–110r.

61 Lawn 1963, 132.

philosophy. As physicians, such as Vallés and Cardano, attempted to advance natural philosophy through medical knowledge, Aristotle, still dominant in natural philosophy, grew in importance for the field of medicine. For example, Giambattista da Monte (1489–1551), a prominent professor of medicine at Padua, claimed to expound on the first fen of the first book of Avicenna's *Canon* by giving the views of Aristotle, his good commentators (most likely meaning Greek commentators), Averroes, and Galen, thereby relying on the "nature of things, not on the interweaving of obscurities."[62]

Late-Renaissance reception of the *Problemata* differed from the humanist inquiries in that, while still interested in philology, its interpretations more explicitly sought to use Aristotle's thought to resolve medical issues. The rise of Aristotelian medicine coincided with the climbing importance of Hippocrates as well as a growing knowledge of the entire Galenic corpus. Not surprisingly medical thought integrated and reconciled these three corpora. Because Galen explicitly claimed to be combining the concepts of Aristotle and Hippocrates and maintained that Aristotle appropriated Hippocratic material, no grand imaginative leap was necessary for sixteenth-century medical authors to link these authors.[63] The same scholars worked on both Hippocrates and Aristotle. Vallés translated and commented upon both Aristotle and Hippocrates; and, Andrea Cesalpino addressed Hippocrates' views on the role of the divine in natural philosophy in a work whose title described it as Peripatetic.[64]

2.4 *Problemata* in the Renaissance

It is in the context of rising Hippocraticism and Aristotelian medicine of the late Renaissance that Italian scholars and physicians gave attention to the Aristotelian *Problemata*. The *fortuna* of the *Problemata* stands apart from a large portion of Aristotelian works. Its commentary tradition, in both the Middle Ages and the Renaissance, is negligible compared to treatises, such as the *De anima, De caelo,* and *Meteorology* that were typically part of university instruction. Settala complained that if the *Problemata* were "read publicly, they would be understood better."[65] Because they were not part of university curricula, complete or near complete commentaries on this work number three, from the period between 1300 and 1632, even if the paucity of commentaries does not signify an absence of readers.

The Renaissance commentary tradition on the *Problemata* was a product of Italian erudite culture closely tied to universities and its vibrant Aristotelianism and

[62] Da Monte 1557, 2. "Tractabo autem; sicut docuerunt, & Aristot. & sui boni expositores, & Aver. & Galen. solvendo scilicet difficultates per naturam rerum, & non per ambagum implicationem." See also Siraisi 1987, 248–250.

[63] Galen 1996, 487, 559; Smith 1979, 61–176.

[64] Cesalpino 1580; Martin 2002.

[65] Settala 1632, preface, 4r.

medical education. This tradition culminated in the work of Settala, a physician who lived primarily in Milan, although he also taught medicine at Pavia. As Ann Blair has pointed out, his commentary at times has a modern feel because he discussed the issue of authenticity by comparing parallels in this text with other Aristotelian works and he attempted to give an accurate reading of the text's meaning, which would correspond to the real opinion of Aristotle.[66] Indeed, Settala engaged in these practices, and modern editors of ancient works have praised him for his skilled deciphering of the original Greek. Although he corrected Pietro d'Abano's translation errors just as Luiz had done, to see his goals in exclusively this light would be mistaken. While in a sense modern, Settala was also a product of his time and the motives for commenting on this text were not exclusively philological. Understanding the real meaning of Aristotle's texts had practical purposes. Skilled philological interpretations were not always the final goal, but rather a tool to find insights that were applicable to salient issues of the day. For Settala many of these issues related to contemporary debates in medicine.

The *Problemata's* value largely derived from both Aristotle's authority and from its correspondence to Hippocratic writings. Like other late-Renaissance physicians, Settala thought that Aristotle lifted doctrines from the Hippocratic corpus and thus made the case that Aristotle was a source for some of the oldest, thus most authoritative, views regarding human health and physiology. As a result of views such as Settala's, throughout the late-sixteenth and early-seventeenth centuries erudite physicians, such as Cardano, Domenico Montesauri, Baccio Baldini, Giovanni Battista Selvatico, and Eustachio Rudio, linked the *Problemata* to both *AWP* and Galen's treatise *Quod animi mores sequuntur temperamenta corporis* (*QAM*).[67] All three works address the relation between body and soul by considering the role of humoral physiology in the formation of differences in customs. All three works were classified as medical works that investigated principles and doctrines of natural philosophy.

AWP was most likely written in the fourth century B.C.E., and so probably predates the *Problemata*. Its first sections describe how locales, their climates, and the qualities of drinking water affect health and contribute to the varying characteristics of different peoples. The author, then, addressed why Asians differ from Europeans, concluding that the extremes and sudden changes in weather make Europeans varied in temperament and as a result susceptible to violent behavior. To the contrary Asians are mild, calm, and feeble as the result of the temperate climate and their political situation. Living under kings, Asians are convinced that they will not reap the rewards from war and thus are reluctant to engage in it. Similar arguments explain the customs and characteristics of Egyptians, Libyans, and Scythians. Within these discussions, the author contended that artifice could change the physical nature of ethnic groups.

---

[66] Blair 1999a, 194.

[67] Cardano 1663, 8:147; Montesauri 1546, 248v; Baldini 1586, 203; Selvatico 1601, 117; Rudio 1611, 47–82.

The author of *AWP* recounted the origins of a group called the Macrocephali, or "Big Headed People," who at one point in their history bound infants' heads so that they would grow in length. The Macrocephali supposedly prized long heads, equating them with nobility. Eventually, according to the author of the treatise, the characteristic was inherited by subsequent generations and while the practice became obsolete, the group's offspring were born with long heads naturally. This inheritance was possible, the author contended, because human seed comes from all parts of the body. Therefore, the seed, being influenced by the shape of the father's head, caused the offspring to resemble their parents in this respect. The author offered more familiar examples as evidence: bald fathers often produce bald children, and children often have the same-colored eyes as their parents. In sum, the treatise argues that environment affects the temperaments of people, which in turn explain not only their propensity to suffer various diseases but also the customs of different races. These changes in temperament, even if artificially induced, are passed on to later generations and thereby explain why and how ethnic groups differ from each other.

Many of the ideas of *AWP* are also found in the *Problemata*. For example, *Problemata* 1.3 discusses how the seasons and winds are factors in etiology; in 14.1, the author asks why those who live in conditions of excessive cold or heat suffer disturbances in both mind and body; and the entire *particula* 14 of the *Problemata* is dedicated to exploring the role that regions play in forming temperament and differences among races; *Problemata* 4.21 contends that semen comes from all parts of the body. Moreover, *particula* 30.1 of the *Problemata* explains that excellence in philosophy, politics, poetry, and art is related to possessing an atrabilious, that is melancholic, temperament, arguing that temperaments are responsible for intellectual as well as emotional dispositions.

Galen noted the similarities between the Aristotelian and Hippocratic texts, citing both the *Problemata* and *AWP*, in his small treatise *QAM*, or, *That the customs of the soul follow the temperaments of the body*. Here Galen argued that a balanced temperament is crucial not only to health but also to moral and intellectual excellence, arguing that this temperament can be altered through changes in regimen. This position exalts potentially the status of physicians, who accordingly have the ability to improve not just patients' health but also their capacity to think and act morally.[68] That the soul and body are interdependent was widely accepted by Renaissance and medieval physicians.[69] Controversially for Christian thinkers, Galen took an agnostic position toward the mortality of the intellect, claiming there was no firm evidence that the soul is capable of living after the death of the body. Rather all evidence suggests that the soul is dependent on the body and its temperaments for its intellective capacities.

There are broad similarities in not just the content of the *Problemata*, *AWP*, and *QAM* but also in the medieval and Renaissance reception of these treatises. While available, they were either infrequent or never the subject of commentaries in the

[68] Lloyd 1988.

[69] Galen 1997, 282–283; Park 1988.

Middle Ages, and as a result physicians only rarely addressed the interconnections between the works until the sixteenth century. As the Galenic and Hippocratic corpora spread throughout learned circles during the sixteenth century, these interconnections were thought to elucidate the historical relation between Aristotle and Hippocrates in addition to providing, for some, a basis for reconciling the views of three of the most trusted ancient sources for medicine and natural philosophy. Even while some found the positions regarding psychology problematic either on philosophical grounds, such as Cesare Cremonini, or theological grounds, such as Eustachio Rudio, other physicians and philosophers found in these texts a plausible way to diffuse debates over whether Aristotelian natural philosophy undermined or contradicted Galenic medicine, showing at least the resemblance of conciliatory positions.[70]

Connections between these three works were apparent to Domenico Montesauri, a physician based in Milan, who wrote a commentary on the *Problemata* in 1546. In his comments on 4.21, the passage that contends that male seed comes from the entire body, he wrote that, "The Philosopher follows Hippocrates in this question, who in his treatise on *AWP*, the fourth book of *De morbis*, and in his treatise *On the seed*, teaches that the seed comes from all parts of the body."[71] Later in his comments on 14.1, he noted that Galen's belief that, "Abundances of heat, arising from the presence of cold air, alter not only the temperament of the body but also that of the soul" was also true according to Hippocrates, Plato, and Aristotle.[72]

Cardano, in his commentary on *AWP*, however, was not so ready to accept Galen's contentions that his view of the soul is supported by *AWP*. Citing *Problemata* 1.3, in his discussion of the *Macrocephali,* he agreed with Hippocrates' and Aristotle's purported view that changes in an individual's natural temperament could be passed on to future generations. In Cardano's view, Galen grossly underestimated the difficulty in changing natural temperament. Only sustained disease, which could be provoked by changes in weather or seasons, could truly alter a natural temperament; and Galen's attribution to Hippocrates the position that dietetics or other alterations in regimen could change temperament was the result of a hallucination rather than an accurate reading of *AWP*. Cardano's familiarity with these texts, while used to promote his interpretation of *AWP*, also promoted his position regarding human temperaments while confirming his opposition to Galen.[73]

Although philologically astute, his method is not merely historical. Cardano understood *AWP* as integral to his attempt to make portions of medicine have the same status as natural philosophy. He asked rhetorically, "[If] we wish to philosophize,

---

[70] Cremonini 1598, 178r-195r; Rudio 1611, 72–82.

[71] Montesauri 1546, 138v. "Philosophus Hipp. Sequitur in hac quaestione qui in com. de aere aquis et locis et in com. de semine semen ab omnibus corporis membris procedere docuit."

[72] Montesauri 1546, 248v. "Haec enim temperies non solum corpori, sed animae protinus, exceptus autem caloris, ex frigoris aeris non solum corprois, sed et animi temperamentum pervertunt. Hanc sententiam ex mente Hipp. et Platonis, ac Aristotelis in commento supra citato Galenus diffuse declaravit."

[73] Cardano 1663, 8:147; Hirai 2011, 110–111.

who, I ask, is a better philosopher than Hippocrates?"[74] Dividing medicine into three categories, *scientia*, which pertains to natural bodies, *cognitio*, which concerns what is *contra naturam*, and *operatio*, which is knowledge of actions taken by physicians to restore health, Cardano concluded that *AWP* presents a contemplative science because it does not concern action. Rather, in this work Hippocrates applied both the resolutive and compositive methods of demonstration.[75] The resolutive method finds causes from effects, while the compositive method uses those causes to further understanding of the subject being investigated. Therefore, the book is useful not just for conserving or restoring health, but also for philosophy, geography, and astrology. Moreover, since this book's ability to explain how temperament is the cause of the "goodness of the soul," its contents are especially valuable not just because it potentially suggests cures but also because "knowing causes is praiseworthy."[76] Cardano's view corresponded to that of Adrien L'Alemant, a Parisian physician and commentator on *AWP*. He agreed with Cardano that Hippocrates used "doctrina resolutoria" in *AWP* because Hippocrates advocated physicians to first examine the various effects of the season and the differing qualities of winds and waters before making general conclusions.[77] As a result of his consideration of the nature of things, Hippocrates was the leader of "rational" medicine.[78]

Cardano in fact put forth causal explanations for natural phenomena, namely on the causes of winds, in his commentary on *AWP*. The discussion of winds in *particula* 26 of the *Problemata* provides another example where this work shares more similarities to Hippocratic writings than Aristotle's other texts. In *Problemata* 26.2 and 26.34 (940b58; 944a26-27) as well as in the Hippocratic *De flatibus* (3,2) wind is characterized as moving air, despite Aristotle's assertion in *Meteorology* 2.4 (360a2833) that the hot and dry exhalation, not simply moving air, is the matter of winds. Cardano accepted that wind was moving air and, using the resolutive method, mustered signs (*indicia*)—such as the supposed differences in the velocity of comets depending on their direction, the flowing of tides, and the supposed fact that wind always blows through fissures—that suggest that the wind constantly circles the earth generally moving from east to west.[79] In this manner Cardano used a number of effects to arrive at a general theory of the nature of the wind.

Other commentaries on *AWP* were also interested in its relation to natural philosophy.[80] Settala believed *AWP* discussed natural philosophy, in addition to medicine, cosmography, and astrology, pointing in particular to the section on winds as a prime example of Hippocrates' consideration of the causes of natural effects.[81]

[74] Cardano 1663, 8:12.

[75] Cardano 1663, 8:3.

[76] Cardano 1663, 8:2.

[77] L'Alemant 1557, 7r; Hippocrates 1894, 1.

[78] L'Alemant 1557, 5r.

[79] Cardano 1663, 8:67.

[80] Siraisi 2007, 93102.

[81] Settala 1590, col. 4, col. 10.

Baccio Baldini, a professor of philosophy and of medicine at Pisa, who wrote a commentary on *AWP* that was published in 1586, believed Hippocrates used the compositive method, whereby he began with knowledge of the causes of effects such as temperament and humors and through them explained the composite person that they form, thus beginning with more simple parts leading toward the whole substance. Baldini's view of Hippocrates' alleged method bolstered his general position toward medicine being a kind of natural philosophy. The method of applying basic principles, moving from simples to wholes, according to Baldini, is the one Aristotle used in his natural philosophy, where he started with matter, form, and privation. Consequently, Hippocrates and Aristotle shared the same philosophical method.[82]

Using this method Baldini showed how it is possible to understand the soul in terms of the simpler temperament, which causally underpins it. He endorsed the view he attributed to both Hippocrates and Galen, that changes in the air affect the mind of all men, and that because the *mores* of the soul follow the temperament of the body, "the soul, whether it should be mortal or immortal, is dependent on the health of the body, therefore should the body change, the soul also must necessarily change."[83] Baldini understood *mores* to come from the concupiscent potency of the soul, capable of being corrupted either through the practice of vice or through disease, and capable of being restored either by the nature of the temperament or through the practice of philosophy. Thus the soul depended on the body, yet choice and free will continued to play a role in the development of virtue, just as it had for Aristotle.[84]

While Baldini's endorsement of Galen's position might have helped physicians make medicine a part of natural philosophy, Galen's psychology was not without controversy, both theologically and philosophically. Attempts to treat medicine as natural philosophy provoked polemical reactions among some philosophers and physicians, who objected to materialistically deterministic aspects of Galenic psychology. Cremonini, a famed professor of philosophy at Padua, in a short treatise, *Quaestio de animi moribus et facultatibus*, written in 1598, attacked Galen's position. Cremonini opposed Galenism and its incursions into natural philosophy, writing treatises that defended the Aristotelian view on the centrality of the heart in human physiology, and on the nature of innate heat.[85] He went so far as to write comic poetry that accused Galen of numerous errors.[86] In the case of *QAM*, he reduced Galen's position to: the soul is a temperament, therefore the soul follows the faculties of the temperament. Objecting to the direction of causation, he contended that Aristotle held that form has a greater explanative power than matter. Form endows diversity to matter, rather than matter causing diversity in form.

---

82 Baldini 1586, 45.

83 Baldini 1586, 204. "videlicet animam sive mortalis sive immortalis sit sanitati corporis ancillari, cum ergo corpus mutatur animum quoque mutari necesse est."

84 Baldini 1586, 237.

85 Ongaro 2000.

86 Nacattel 1645, 7273.

Therefore, it is the soul, which he explicitly claimed is immortal, that explains temperament, rather than vice versa.[87] While Cremonini attacked Galen because he thought his views were philosophically incoherent, others found *QAM* potentially dangerous because of its materialistic view of the soul. Nicolas de Nancel contended that Galen's opinion of the soul was "false, impious, full of error and pernicious danger."[88] Two decades later, Eustachio Rudio, a professor of medicine at Padua and, according to John Aubrey, one of William Harvey's teachers, attacked the psychological views found in *AWP*, *QAM*, and the *Problemata*.[89]

Others took a more pragmatic position, hoping to reject sufficiently Galen's agnostic view towards the immortality of the rational soul, yet maintaining that his work could be useful to medicine. For example, Giovanni Battista Persona, a professor of medicine at Bergamo and the author of the sole commentary on *QAM* in the Renaissance, printed in 1602, tried to diffuse the controversy surrounding this book by contending that Galen's view towards the immortality of the soul was impious and contrary to the Christian faith. Nevertheless the doctrines contained in *QAM*, were, according to Persona, essential to understanding natural temperament, which in turn was key to preserving health.[90]

Increased awareness of *AWP* and these controversies over Galen's view of the soul informed interpretations of the *Problemata*. Settala, in his *Problemata* commentary, relied on Hippocrates' and Aristotle's views about the relation of the human soul to temperament. In his comments to 14.1, which asks "Why those who live in excessive heat and cold are wild in appearance and customs?," he addressed the relation between climate and human intelligence. The author of this question tentatively answered, "moderation confers intelligence, while excesses harm the body and the temperament of the mind."[91] Settala linked this question to *AWP*, altering the terms of the argument and maintaining that the mild climate of Europe has conferred not just intelligence on its inhabitants but liberty as well, in contrast to Asia. The causal relation between weather, bodily temperament and intellect outlined in this question and *AWP* correspond to Galen's teaching in *QAM*. Here and in *Problemata* 14.8, Aristotle confirmed not only that *mores animi* follow the body but also that "the universal cause of these passions of different souls goes back to the active qualities of the hot and dry."[92] While this position might suggest determinism or a materialist interpretation of the human soul, Settala outlined that intelligence or *mores* should not be taken as equivalent to reason. Recognizing that free will is the doctrine of the Church as well as of philosophy, Settala concluded that humans, unlike animals, "act beyond custom and nature because of their reason."[93] Thus Galen's teachings about customs of the soul and Aristotle's views of the origins of

[87] Cremonini 1598, 186v.

[88] Nancel 1587, 1v.

[89] Rudio 1611,41–48; 72–82; Woolfson 1998, 89.

[90] Persona 1602, 4, 9.

[91] Settala 1632, 2:271.

[92] Settala 1632, 2:273.

[93] Settala 1632, 2:274–275.

human intelligence are not meant to include the capacity for reason. Yet, the relation between temperament and soul necessarily places the mind dependent on the body.

Settala further explained his views on the soul in his comments on *Problemata* 30.1, the famed question on melancholy. While Marsilio Ficino in 1.5 of the *De vita* reconciled Plato's *Timaeus* with Aristotle, and Democritus, Settala, perhaps doubtful of Neoplatonism, dismissed such a syncretic approached and held that only Hippocrates and Aristotle "reached the truth in this matter."[94] Unlike Ficino and later physicians François Valleriola and Giovanni Battista Selvatico, Settala held that Aristotle's understanding of melancholy did not correspond to Plato's.[95] He dismissed Plato's understanding of form and soul, rejecting the belief that knowledge is the recollection of preformed ideas.[96] Rather, he wrote that the intellective faculty of the soul is posterior to the soul's other faculties, those of growth, sensation, and locomotion. As a result the intellective capacity is dependent on sensation, which has its seat in the heart. Therefore, Settala concluded, "the place of the mind will be the heart itself."[97] Deviating from Galen, who believed that the brain had primacy, Settala used the *Problemata* to endorse the Aristotelian view that saw the heart as the central governing organ of the body.

Locating the soul within the heart allowed Settala to make sense of question 4.21 of the *Problemata* (which Settala numbered as 4.22), the question in which the author endorses the view that male seed comes from the entire body, a view that corresponds to *AWP* yet is in potential disagreement with *De generatione animalium*.[98] The problem asks "Why do those who have sexual intercourse generally feel tired and weaker? Perhaps, is it because the seed is a secretion that comes from all the parts of the body?" Settala, in apparent agreement with this solution, argued that soul, with its base in the heart, "operates throughout the entire body, not directly but by intermediary spirits."[99] This spirit, directed by the soul in the heart, extends throughout the body, "so that matter transmitted to the testicles, just as what is expelled in sleep, is filled with spirit and innate heat, which is drawn in through the friction during the act of sex, transformed by the spirit from the heart." Male seed, therefore, does not act through heat, but rather through the "spirit, which is in the semen, contained in the foamy body, and the nature, which is in the spirit, that corresponds in respect to proportion to the element of the stars."[100]

Settala's belief that Aristotle's *Problemata* borrowed from Hippocrates underpinned not just his interpretations of psychology and human generation, issues of natural philosophy rather than medicine, but also his views on problems specifically about health and disease.[101] Perhaps most notable is his discussion of the contagion

---

[94] Settala 1632, 3:355; Ficino 2012, 58; Gowland 2011, 58.

[95] Valleriola 1554, 428–430; Selvatico 1601, 4.

[96] Settala 1632, 2:197.

[97] Settala 1632, 3:360.

[98] Aristotle *GA* 1.18.723b-724a.

[99] Settala 1632, 280–281.

[100] Settala 1632, 281.

[101] Settala 1632, 1:3; 1:40.

of plague and other diseases. In between the time he published the first two volumes and the third and final volume of the *Problemata* commentary, Settala also wrote a plague treatise (1622) and served as *protofisico* of Milan during the disastrous plague of 1630.[102] Manzoni rendered an unsympathetic portrait of Settala, acting in this capacity, in his *I Promessi sposi*. In *De peste,* Settala reaffirmed his contention that plagues spread through corrupted vapors, defining contagion as "the transit or communication by likeness of a particular corruption of mixture according to substance from one body into another."[103] This was the same definition that he used in the *Problemata* commentary, where he specified that the communication occurred through the putrefaction of vapors caused by active qualities, in particular heat. Seeing that disease was transmitted through the vapors and exhalations, he saw no need for Girolamo Fracastoro's view that contagion happens through seeds or corpuscles. There is no difference between corpuscles and vapors, which themselves are bodies that do not have a specific mixture.[104]

2.5 Conclusion

Philological considerations informed those interested in the *Problemata* during the early seventeenth century, even if they did not relive the polemics over language witnessed in the fifteenth century. Leading commentators, such as Settala, were experts in the Greek language and knowledgeable about a wide range of ancient literature. Yet their considerations were by no means purely historical. Their philology was tempered by external considerations of a different sort than those influencing modern commentators. Research into the past was not merely an abstract consideration of antiquity but a source for knowledge of nature and medicine. After Settala, the practical medical considerations derived from the *Problemata* continued to recommend it to his successors, such as Giovanni Manelfi, a professor of medicine and *protomedico* at Rome, who in his 1646 annotations to the Hippocratic *Aphorisms,* made frequent references to the *Problemata,* especially to the portions of the third book that deal with the relations between weather and health. His knowledge of the *Problemata* resulted from his work on a commentary on the first book of this work, in which he addressed the question of contagion and epidemic disease.[105]

Determining the authenticity of the treatise related to the perceived quality or genius of the content. Investigations into the relation of ancient texts influenced their reception. The conviction that Aristotle borrowed material from Hippocrates for the *Problemata* increased the authority of that work as well as that of *AWP*. Both works were evidence of agreement among the most important authors of their

[102] Ripamonti 1841, 41–44.

[103] Settala 1622, 88.

[104] Settala 1632, 1:12–14.

[105] Manelfi 1646, 30, 34–35, 90.

respective fields. Thus Hippocrates could become an authority for natural philosophy, helping raise the status of medicine to that of *scientia* for some, and Aristotle became a greater authority for medicine.

While the correspondences between the *Problemata* and Hippocratic writings are real, perhaps the correspondences between Renaissance writings on the *Problemata* and *AWP* are even more evident. The goal of late-Renaissance reconciliation of ancient authors was more precise and textually astute than grand fifteenth-century attempts of philosophical reconciliation, such as that of Giovanni Pico della Mirandola, yet the association of Hippocrates with Aristotle illuminates the extent to which ancient texts continued to drive intellectual endeavors. By enlarging the circle of texts that were subject to commentary to include the *Problemata* and *AWP*, physicians and philosophers found new ways of interpreting Aristotle and Hippocrates. The already great degree of flexibility that their writings allowed became even greater, and Hippocrates became an authority on the human soul and Aristotle an expert on plagues, the nature of the heart, and a proponent of the idea that the male seed derives from a spirit that circulates throughout the body.[106]

References

Accoramboni, Felice. 1590. *Interpretatio obscuriorum locorum & sententiarum omnium operum Aristotelis, & praecipuorum dubiorum*. Rome: Santi.

Achillini, Alessandro. 1548. *De subjecto medicinae*. Venice: Scoto.

Agrimi, Jole, and Chiara Crisiciani. 1988. *Edocere medicos: medicina scolastica nei secoli XII-XV*. Naples: Guerini.

Averroes. 1564. *Colliget libri vii*. Venice: Giunta.

Bacon, Francis. 2004. In *The Instauratio magna: Part II Novum organum. Vol. 11 of The Oxford Francis Bacon*, ed. G. Rees and M. Wakely. Oxford: Oxford University Press.

Baldini, Baccio. 1586. *In librum Hyppocratis de aquis, aere, et locis commentaria*. Florence: Sermartelli.

Blair, Ann. 1999a. Authorship in the popular "Problemata Aristotelis.". *Early Science and Medicine* 4: 139.

Blair, Ann. 1999b. The Problemata as a natural philosophical genre. In *Natural particulars: Nature and the disciplines in early modern Europe*, ed. Anthony Grafton and Nancy Siraisi, 171–204. Cambridge, MA: MIT Press.

Brasavola, Antonio Musa. 1544. *Examen omnium simplicium medicamentorum, quorum in officinis usus est. Addita sunt Aristotelis Problemata quae ad stirpium genus, & oleracea pertinent*. Lyon: Pullon.

Burnett, Charles. 1999. The second revelation of Arabic philosophy and science. In *Islam and the Italian renaissance*, ed. Charles Burnett and Anna Contadini, 185–198. London: Dorset.

Bylebyl, Jerome J. 1979. The school of Padua: Humanistic medicine in the sixteenth century. In *Health, medicine and mortality in the sixteenth century*, ed. Charles Webster, 335–370. Cambridge: Cambridge University Press.

[106] I thank the editors and the anonymous readers for their comments, suggestions, and corrections. Research for this chapter was supported by a FWO grant administered by Katholieke Universiteit Leuven, De Wulf-Mansion Centre, Institute of Philosophy.

Bylebyl, Jerome J. 1990. The meaning of *physica*. *Osiris* 6: 16–41.

Cadden, Joan. 1997. Sciences/silences: The natures and languages of "sodomy" in Peter of Abano's *Problemata* commentary. In *Constructing medieval sexuality*, ed. Karma Lochrie, Peggy McCracken, and James A. Schulz, 40–57. Minneapolis: University of Minnesota Press.

Cardano, Girolamo. 1663. *Opera omnia*, ed. C. Spon. 10 vols. Lyon: Huguetan and Rivaud.

Carlino, Andrea. 1999. *Books of the body: Anatomical ritual and renaissance learning*. Trans. J. Tedeschi, and A.C. Tedeschi. Chicago: University of Chicago Press.

Cesalpino, Andrea. 1580. *Daemonum. Investigatio peripatetica. In qua explicatur locus Hippocratis in Progn. si quid divinum in morbis habetur*. Florence: Giunta.

Cranz, F.E. 1978. The publishing history of the Aristotle commentaries of Thomas Aquinas. *Traditio* 34: 157–192.

Cremonini, Cesare. 1598. *Quaestio de animi moribus et facultatibus*. Biblioteca Universitaria Padova MS 2075.

Da Monte, Giambattista. 1557. *In primi libris canonis Avicennae primam fen, profundissima commentaria*. Venice: Valgrisi.

Falloppio, Gabriele. 1570. *Expositio in librum Galeni de ossibus*. Venice: Galignani.

Ficino, Marsilio. 2012. *De vita libri tres*, ed. M. Boenke. Munich: Wilhelm Fink.

Galen. 1996. *On the elements according to Hippocrates*, ed. P. De Lacy. Berlin: Academie Verlag.

Galen. 1997. *Selected works*. Trans. P. Singer. Oxford: Oxford University Press.

Guastavini, Giulio. 1608. *Commentarii in priores decem Aristotelis Problematum sectionis*. Lyon: Cardon.

Giglioni, Guido. 2008. Nature and demons: Girolamo Cardano interpreter of Pietro d'Abano. In *Continuities and disruptions between the Middle Ages and the Renaissance*, ed. Charles Burnett, José Meirinhos, and Jacqueline Hamesse, 89–112. Louvain-la Neuve: FIDEM.

Grafton, Anthony. 1977. On the scholarship of Politian and its context. *Journal of the Warburg and Courtauld Institutes* 40: 150–188.

Gowland, Angus. 2011. Melancholy, imagination, and dreaming in renaissance learning. In *Diseases of the imagination and imaginary diseases: Disease in the early modern period*, vol. 2, ed. Yasim Haskell, 53–102. Turnhout: Brepols.

Hippocrates. 1578. *De capitis vulneribus liber,... Eiusdem Hippocratis textus graecus a Iosepho Scaligero Iul Cae. F. castigatus, cum ipsius Scaligeri castigationum suarum explicatione*. Paris: Stephanus.

Hippocrates. 1894. *Opera quae feruntur omnia*, vol. 1, ed. H. Kuehlewein. Leipzig: Teubner.

Hirai, Hiro. 2011. *Medical humanism and natural philosophy: Renaissance debates on matter, life and the soul*. Leiden: Brill.

Kibre, Pearl. 1975. Hippocrates Latinus: Repertorium of Hippocratic writings in the Latin Middle Ages [II]. *Traditio* 31: 99–126.

Klemm, Matthew. 2006. Medicine and moral virtue in the *Expositio Problematum Aristotelis* of Peter of Abano. *Early Science and Medicine* 11: 302–335.

Klestinec, Cynthia. 2007. Civility, comportment, and the anatomy theater: Girolamo Fabrici and his medical students. *Renaissance Quarterly* 60: 434463.

Klibansky, Raymond, Erwin Panofsky, and Fritz Saxl. 1964. *Saturn and melancholy: Studies in the history of natural philosophy, religion and art*. New York: Basic.

Kraye, Jill. 1996. Philologists and philosophers. In *Cambridge companion to renaissance humanism*, ed. Jill Kraye, 142–160. Cambridge: Cambridge University Press.

L'Alemant, Adrien. 1557. *De aere, aquis, & locis commentarius*. Paris: Gorbinus.

Lawn, Brian. 1963. *The Salernitan questions: An introduction to the history of medieval and renaissance problem literature*. Oxford: Clarendon.

Lennox, James G. 1994. Aristotelian problems. *Ancient Philosophy* 14: 53–77.

Lines, David. 2001. Natural philosophy in Renaissance Italy: The University of Bologna and the beginnings of specialization. *Early Science and Medicine* 6: 267–320.

Lloyd, G.E.R. 1988. Scholarship, authority and argument in Galen's *Quod animi mores*. In *Le opere psicologiche di Galeno*, ed. Paola Manuli and Mario Vegetti, 11–42. Naples: Bibliopolis.

Lohr, Charles H. 1972. Medieval Latin Aristotle commentaries (Authors Narcissus-Richardus). *Traditio* 28: 281–396.

Luiz, Antonio. 1540. *Liber de erroribus Petri Apponensis in Problematibus Aristotelis exponendis*. Lisbon: Rodriguez.

Mahoney, Edward P. 1980. Albert the Great and the *Studio patavino* in the late fifteenth and early sixteenth centuries. In *Albertus Magnus and the sciences*, ed. James A. Weisheipl, 537–563. Toronto: Pontifical Institute of Mediaeval Studies.

Manelfi, Giovanni. 1646. *Hippocratis Aphorismi, una cum annotationibus quibusdam, & circa textum praecipue*. Venice: Turrini.

Martin, Craig. 2002. Francisco Vallés and the Renaissance reinterpretation of Aristotle's *Meteorologica* IV as a medical text. *Early Science and Medicine* 7: 130.

Mercuriale, Girolamo. 1588. *Censura operum Hippocratis*. Venice: Giunta.

Mikkeli, Heikki. 1992. *An Aristotelian response to Renaissance humanism: Jacopo Zabarella on the nature of arts and sciences*. Helsinki: SHS.

Monfasani, John. 1999. The *Problemata* and Aristotle's *De animalibus*. In *Natural particulars: Nature and the disciplines in Renaissance Europe*, ed. Anthony Grafton and Nancy G. Siraisi, 205–247. Cambridge, MA: MIT Press.

Monfasani, John. 2006. George of Trebizond's critique of Theodore Gaza's translation of the Aristotelian *Problemata*. In *Aristotle's* Problemata *in different times and tongues*, ed. Pieter De Leemans, and Michèle Goyens. Leuven: Leuven University Press.

Montesauri, Domenico. 1546. *In sectionem primam problematum Aristotelis commentarium. Eorum ad medicinam pertinent cuius quaestiones liii.* Biblioteca Ambrosiana Milan MS A 113 inf.

Nacattel, Lootri [Troilo Lancetta]. 1645. *Raccolta medica, et astrologica*. Venice: Guerigli.

Nancel, Nicholas de. 1587. *De immortalitate animae, velitatio adversus Galenum*. Paris: Richerius.

Nifo, Agostino. 1551. *In libris Aristotelis meteorologicis commentaria*. Venice: Scoto.

Nifo, Agostino. 1552. *Expositio super octo libros de physico auditu*. Venice: Giunta.

Norpoth, Leo. 1930. Zur Bio-, Bibliograpahie und Wissenschaftslehre des Pietro d'Abano, Mediziners, Philosophen und Astronomen in Padua. *Kyklos: Jahrbuch für Geschichte und Philosophie der Medizin* 3: 292–353.

Nutton, Vivian. 1985. Humanist surgery. In *The medical renaissance of the sixteenth century*, ed. Andrew Wear, Roger K. French, and Iain M. Lonie, 75–99. Cambridge: Cambridge University Press.

Nutton, Vivian. 1989. Hippocrates in the renaissance. In *Die hippokratischen Epidemien: Theorie, Praxis, Tradition*, ed. Gerhard Baader and Rolf Winau, 420–439. Stuttgart: Steiner.

Nutton, Vivian. 1997. The rise of medical humanism: Ferrara, 1464–1555. *Renaissance Studies* 11: 2–19.

Olivieri, Luigi. 1988. *Pietro d'Abano e il pensiero neolatino*. Padua: Antenore.

Ongaro, Giuseppe. 2000. La controversia tra Pompeo Caimo e Cesare Cremonini sul calore innato. In *Cesare Cremonini: Aspetti del pensiero e scritti*, ed. Ezio Riondato, and Antonino Poppi, 1:87–110. 2 vols. Padua: Accademia Galileiana.

Park, Katharine. 1988. The organic soul. In *The Cambridge history of renaissance philosophy*, ed. Charles B. Schmitt et al., 464–484. Cambridge: Cambridge University Press.

Patrizi, Francesco. 1571. *Discussiones peripateticae*. Venice: De Franceschi.

Perfetti, Stefano. 1995. "Cultius atque integrius" Teodoro Gaza, traduttore umanistico del *De partibus animalium. Rinascimento* 35, series 2, supplement: 253–286.

Persona, Giovanni Battista. 1602. *In Galeni librum cui titulus est Quod animi mores corporis temperiem sequuntur commentarius singularis*. Bergamo: Ventura.

Pietro d'Abano. 1482. *Expositio problematum Aristotelis*. Venice.

Poliziano, Angelo. 1498. *Centuria miscellaneorum*. Venice: Manuzio.

Pomponazzi, Pietro. 1563. *Dubitationes in quartum Meteorologicorum*. Venice: De Franceschi.

Reeds, Karen. 1991. *Botany in medieval and renaissance universities*. New York: Garland.

Rice Jr., Eugen F. 1970. Humanist Aristotelianism in France: Jacques Lefèvre and his circle. In *Humanism in France*, ed. A.H.T. Levi, 132–149. Manchester: Manchester University Press.
Ripamonti, Giuseppe. 1841. *La peste di Milano del 1630. Libri cinque cavati dagli annali della città*. Milan: Pirotta.
Rudio, Eustachio. 1611. *Liber de anima*. Padua: Bertellum.
Salutati, Coluccio. 1947. *De nobilitate legum et medicina*, ed. E. Garin. Florence: Vallecchi.
Schmitt, Charles B. 1965. Aristotle as cuttlefish: The origin and development of a renaissance image. *Studies in the Renaissance* 12: 60–72.
Schmitt, Charles B. 1983. *Aristotle and the renaissance*. Cambridge, MA: Harvard University Press.
Schmitt, Charles B. 1985. Aristotle among the physicians. In *The medical renaissance of the sixteenth century*, ed. Andrew Wear et al., 1–15. Cambridge: Cambridge University Press.
Selvatico, Giovanni Battista. 1601. *Controversiae medicae numero centum*. Milan: Bordone.
Settala, Lodovico. 1590. *In librum Hippocratis Coi de aeribus, aquis, locis, commentarii*. Cologne: Ciotti.
Settala, Lodovico. 1622. *De peste, & pestiferis affectibus libri quinque*. Milan: Bidelli.
Settala, Lodovico. 1632. *In Aristotelis Problemata commentaria*. 3 vols. Lyon: Landry.
Siraisi, Nancy G. 1987. *Avicenna in the renaissance: The Canon and medical teaching in Italy after 1500*. Princeton: Princeton University Press.
Siraisi, Nancy G. 1997. *The clock and the mirror*. Princeton: Princeton University Press.
Siraisi, Nancy G. 2003. History, antiquarianism, and medicine: The case of Girolamo Mercuriale. *Journal of the History of Ideas* 64: 231–251.
Siraisi, Nancy G. 2007. *History, medicine, and the traditions of renaissance learning*. Ann Arbor: University of Michigan Press.
Smith, Wesley D. 1979. *The Hippocratic tradition*. Ithaca: Cornell University Press.
Stocks, J.L. 1930. Review of *The Oxford Aristotle. Classical Review* 44: 20–21.
Trapezuntius, George. 1967. *Adversus Theodorum Gazam in perversionem Problematum Aristotelis*. In *Kardinal Bessarion als Theologe, Humanist und Staatsmann: Funde und Forschungen*, ed. Ludwig Mohler. Paderborn: Schöningh. Reprint Aalen: Scientia-Verlag.
Valleriola, François. 1554. *Ennarationum medicinalium libri sex*. Lyon: Gryphius.
Vallés, Francisco. 1591. *Controversiae medicarum et philosophicarum*. Padua: Meietti.
Ventura, Iolanda. 2008. Translating, commenting, re-translating: Some considerations on the Latin translations of the Pseudo-Aristotelian *Problemata* and their readers. In *Science translated: Latin and vernacular translations of scientific treatises in medieval Europe*, ed. Michèle Goyens, Pieter De Leemans, and An. Smets, 123–154. Leuven: Leuven University Press.
Vimercati, Francesco. 1556. *In quatuor libros Aristotelis Meteorologicorum commentarii*. Paris: Vascosan.
Vives, Juan Luis. 1538. *De libris Aristotelicis censura*. Basel: Oporinus.
Wear, Andrew. 2008. Place, health, and disease: The *Airs, waters, places* tradition in early modern England and North America. *Journal of Medieval and Early Modern Studies* 38: 443–465.
Williams, Steven J. 1995. Defining the corpus Aristotelicum: Scholastic awareness of Aristotelian spuria in the High Middle Ages. *Journal of the Warburg and Courtauld Institutes* 58: 29–51.
Woolfson, Jonathan. 1998. *Padua and the Tudors: English students in Italy, 1485–1603*. Cambridge: Clarke.

Chapter 3
Renaissance Surgeons: Anatomy, Manual Skill and the Visual Arts

Cynthia Klestinec

Abstract By the second half of the sixteenth century, anatomy had become a conflicted resource for surgeons. Emphasized in a clinical context, anatomical experience was connected not only to less error, but to a practitioner's violent approach to the living body of the patient. Taking the case study of two practitioners in late sixteenth-century Venice, this essay explores the problem of anatomy and the emergence of a more robust language of manual skill, with terms drawn from the visual arts.

Keywords Anatomy • Surgery • Manual skill • Arts • Post-vesalian

3.1 Introduction

In his large volume, *The Universal and Perfect Surgery (of all the parts necessary for the optimal surgeon)* (1574), the Venetian surgeon Giovanni Andrea della Croce appended the traditional definition of surgery in order to emphasize the differences between anatomy and surgery, differences we would see as obvious.[1] Croce was a successful learned surgeon, for many years the Prior of Venice's college of surgery, and the author of several treatises on surgery topics in Latin and Italian.[2] In his

[1] Croce, early in his career, spent time as a surgeon for the naval fleet of the Republic of Venice, when the fleet was providing protection to its own vessels and to the Mediterranean trade routes against attacks by the Ottomans. Subsequently in 1532, Croce received his license and was accepted as a member of the medical college, and in the late 1540s and throughout the 1550s, he served as Prior of the college (1548, 1550, 1551, and 1558). In this capacity, he would not only have provided patient care and often collaborated with physicians but also overseen examinations, apprenticeships, and appointments to government boards and other lucrative posts. See Bernardi 1826.

[2] Croce first published two treatises on wounds in Giovanni de Vigo 1560, *La prattica universale in cirugia*. He then published a Latin surgery, 1573, and an Italian surgery in 1574 and again in 1583. My translations come from the 1583 edition.

C. Klestinec (✉)
Department of English, Miami University of Ohio, Oxford, OH 45056, USA
e-mail: klestic@miamioh.edu

P. Distelzweig et al. (eds.), *Early Modern Medicine and Natural Philosophy*, History, Philosophy and Theory of the Life Sciences 14,
DOI 10.1007/978-94-017-7353-9_3

book, he felt compelled to distinguish surgery according to its goals—to restore unity and unite parts that are broken, cut, destroyed, or otherwise divided:

> I say in the human body (to show the differences between the art of surgery of *medici* and that art of *marescalchi*, who work on bodies that are inhuman and animal); I say living to make it understood that surgery is very different from the anatomical activities, which are done solely on dead bodies... the anatomical art is different from surgery, since surgery works on the living human body and anatomy on the dead body; and because of the goal, since surgery works to unite the parts that are separate or divided in the human body, while anatomy seeks to separate and divide the parts that are continuous and united.[3]

Croce's remarks attempt to correct a somewhat ominous double vision, one that blurs the distinction between the anatomist and the surgeon with respect to his knives, cutting, and objectified bodies.

Earlier surgeons did not elaborate these distinctions. Giovanni de Vigo, whose volume on surgery was the standard for the entire sixteenth century and served as the template for Croce's own volume, began with a first chapter on anatomy, indicating its fundamental importance to the discipline of surgery. Vigo cited Galen and the importance of knowing anatomy in order to understand the particular conditions of the body affected by disease and to understand the disposition of the body (in health). He then explained:

> Not only does it go well for the surgeon but also for the physician to know anatomy. For he who does not know anatomy, as Albucasis has demonstrated, does not finish the work in human bodies, cutting, giving fire, sewing, and doing those activities required of the office [of a surgeon], and if making an error, you will kill someone.[4]

Vigo connected the fatal errors on the part of surgeons to a lack of anatomical knowledge. He meant not only that a surgeon must know the location and function of the parts, but also that he must have an understanding of anatomical parts in both a normal or healthy state and a compromised or diseased state. Anatomy, for Vigo, was the rational foundation for a surgeon's ability to diagnose and treat, which he inherited from the rational surgeries of the middle ages. Anatomy was also necessary for knowing where and how to cut the body, sew it up, and cauterize its parts. For Croce, however, the security of that foundation had been compromised. He limited anatomy to an introductory or initial area of study, which allowed him to retain its significance and to clarify the distinctions set out above between anatomy, surgery, and animal care.

---

[3] Croce 1574, preface 1...5. "dico, nel corpo humano (per dimostrar la differenza fra l'arte Chirurga de'Medici, e quella de'Marescalchi, che operano ne' corpi inhumani, e bruti), dico vivente per far conoscer la Cirugia esser molto diversa dall'attioni anatomiche, che operano solamente nei corpi morti... Perche è diversa l'arte anatomica della Chirurga, per ragion del soggetto, operando la Chirugia nel corpo humano vivente, e la anatomica nel corpo morto, per ragione del fine, perche si affatica la Chirurga a unire le parti, che sono separate, ò divise, e la anatomica a separar, e divider le parti continue e unite."

[4] Vigo 1560, A3. "La onde non sola stà bene a Cirugici, ma ancora à Fisici saper l'anotomia. Però che chi non sà anotomia, come dimostra Albuc[asis] non finisce di operare ne'corpi humani, tagliando, dando il fuoco, cucendo, e cosi fatti uffici facendo, si che per errore ne ammazzi qualchuno."

By the second half of the sixteenth century, anatomy had become a conflicted resource for surgeons. It remained the means for demonstrating one's knowledge of the physical body. However, its clinical advantage was more frequently rendered, in the texts that described it, as a problem: anatomical experience was now connected not to less error, as Vigo had surmised, but to a practitioner's violent approach to the living body of the patient. One precondition for this development was a flourishing anatomical culture, where anatomy lessons routinely took place, as was the case in Venice and Padua, and where books, broadsides, and pamphlets on anatomical topics circulated widely.[5] The double vision or conflict of perspectives surely grew out of the diffusion and circulation of anatomical information to readers and viewers with different backgrounds, learning, and conceptual frameworks. Rather than pursue such an extended set of phenomena, this essay will follow the conflict down a narrower path, one that takes us to Venice, to the stimulating debates among its medical practitioners, and to the second half of the sixteenth century. In this setting, the authority of anatomy and the legitimacy it offered to learned surgery and learned practitioners did not go unchallenged. Rather, stark distinctions were made between an anatomist's cut and a surgeon's cut; and against the critiques of empiric-surgeons such as Leonardo Fioravanti, learned surgeons such as Croce inserted new terms for their handiwork, terms that derived from Hippocratic and Galenic sources and were associated with the visual arts. In order to render these distinctions and promotions visible, this essay offers a fine-grained analysis of the terms of a surgeon's handiwork and seeks to position debates about anatomy, surgery, and health care in the context of the medical marketplace, where such debates reflected the opposition and exchange between learned and vernacular healers.

Recent historiography has focused attention on empirics, once thought to be outliers of Renaissance medical culture.[6] Although historians have come to accept the presence of empirics in the main lines of medical care, questions remain about how learned and vernacular practitioners were related—if not always or solely antagonistically. While we know that learned practitioners, sitting on health boards, sought to curtail the activities of empirics, we understand much less about how learned medical values shifted in response to empirics. If empirics criticized surgeons for their reliance on anatomy, they did so by degrading the anatomist's work, the physical toil of dissection. Surgeons, such as Croce, recast their own laborious efforts, generating alternative terms for their physical labor. Though the learned surgeon was an educated professional, the development of terms and associations for his labor participates in a more widespread appreciation of artisans in the period. As the studies of Pamela H. Smith and Pamela O. Long have indicated for both northern and southern European locales, artisans and artisanal values were increasingly visible and valuable for the production of goods, natural knowledge, and political power in the early modern period.[7] Moreover, in Venice, between the late 1570s and

[5] The bibliography on this topic is extensive and growing. See Park 1994, 1–33; 1995, 111–132; and 2006; Siraisi 1990, 153–86; Carlino 1994, and especially 1999; and most recently, Kusukawa 2012.

[6] Gentilcore 2006; Eamon 1994 and 1993, 29–44.

[7] Smith 2004 and Long 2011.

the early 1590s, the importance of skilled labor increased: population decreases because of plague, the sluggishness of immigration from country to city, and the ineffectiveness of state regulation made skilled workers (masters rather than apprentices or journeymen) more valuable; their wages, for example, kept time with rising grain prices until the early 1590s.[8] This context was one in which skilled workers were being revalued (and paid more), and it invites us to consider how learned practitioners advertised or represented their technical skill in more elaborate and positive expressions. Technical skill had long been a part of learned medicine, but its place in the development of a more sustained relationship between anatomy and natural philosophy at the university and in the medical curriculum remains unclear as does the role it acquiredin a thriving medical marketplace, with its competitive and allying forces.

Responding to concerns about anatomy, especially the colorful critique of anatomy given by his contemporary, the empiric-surgeon, Fioravanti, Croce did not retreat into the philosophical corners of learned medicine. Instead, as this essay will attempt to show, Croce emphasized manual, technical skill with terms that were associated with the visual arts and an artisan's manipulation of matter. Although the association implied aesthetic qualities and superficial operations, Croce focused it less on the patient's body or experience and more on the surgeon's work, the surgeon's subjective experience of doing operations.[9] With Croce's text at the center, this case study offers an example of more subtle shifts taking place in the emphasis and representation of medical labor and surgical work in this period.

3.2 Anatomy and *Ars*

The practical intellect provides a framework for medical ideas related not only to ethical considerations but also to technical skill. According to Aristotle, the practical intellect referred to *prudentia* and *ars*.[10] If prudence was everywhere (and interesting for its wide diffusion), *ars* had a more selective pattern of development.[11] One place where *ars* appeared in learned medical writings was in discussions of competence and expertise; here, somewhat standard, *ars* would often be reduced to manual skill. It might serve in a critique of physicians (following Petrarch); or it might

---

[8] Pullan 1964, 407–28. This article adjusts the view of general deterioration offered by Braudel 1949.

[9] Masciandaro 2007, introduction, suggests that the terms of labor represent the experience of work in its subjective, effortful dimensions or in its objective dimensions: in medieval English, his area of expertise, *werk* and *craft* emphasize the product while *travail* and *labour* emphasize the effort and livelihood that went into making the product.

[10] For artisans, see Smith, 3–30, 59–94; and Long, 30–37; for the technical arts, Whitney, 1990; and Roberts et al. 2007; and in the context of medicine, von Staden 2007, 21–49; and Lawrence 1999, 156–201.

[11] Prudence was a fixture of the textual tradition of learned medicine—present in Cardano's autobiography, in the genre of "rules of caution" for physicians, and in the many medical topics that were influenced by *Nicomachean Ethics* (book 6).

boost the efforts of medical authorities to police the boundaries between medical occupations.[12] Vesalius, to use a well known example, placed the study of anatomy on the branch of natural philosophy, but his description of the decline of medicine hinged on the failed transmission of manual skills: fashionable doctors "despising the work of the hand, began to delegate to slaves the manual attentions which they judged needful for their patients, and themselves merely to stand over them like master builders."[13] Though the concoction of drugs had been handed over to apothecaries and druggist shops "were filled with barbarous terms and false remedies," the real loser was anatomy: "But this perverse distribution of the instruments of healing among a variety of craftsmen inflicted a much more odious shipwreck and a far more cruel blow upon the chief branch of natural philosophy…[that is] it began to perish miserably when the doctors themselves, by resigning manual operations to others, ruined Anatomy."[14] Although Vesalius' remarks may reflect his experiences in Paris where opposition between physicians and surgeons was more pronounced (and, as one reviewer of this essay pointed out, his own memory of Galen's own remarks in *De anatomicis administrationibus*), his remarks reveal one of the routine ways in which manual operations entered the discourse of learned medicine—through the category of *ars*, the failed transmission of manual skill, and the subsequent decline and ruination of anatomy.

Writing in 1542 from Padua, Vesalius was perhaps a little disingenuous when he painted such a bleak picture of anatomical training in the Veneto. In Venice, anatomy was a routine area of study for physicians and learned surgeons as well as the students and apprentices they trained. In Venice, the medical colleges of physicians and surgeons organized annual anatomies, which were held at various locations in Venice, including the church of S. Paternità, which is no longer extant, the church of San Stefano, and the church of San Giovanni and Paolo.[15] By 1603, the medical colleges petitioned the Great Counsel of the Republic for funding that would provide for the materials for the dissection and the burial. Although the petition covered "physicians, surgeons, and barbers," it was directed primarily at barbers: "There are many barbers that treat in Venice and let blood, many of them have never cut nor seen the cutting in an anatomy of the human body, and so they do not know how the veins and the other parts of our body [go], because of this, many errors are committed."[16] Anatomy, as a set of procedures (in addition to a conceptual training), was thus available, perhaps more than ever, and as a feature of a practitioner's preparation, it would work "to the benefit of everyone." It was associated with learned and non-learned practitioners, with manual operations of cutting, with observation, with procedures of bloodletting, and with the correction of error.

---

[12] Carlino 2005, 559–82; Gentilcore 1997, 75–110.

[13] Farrington 1932, 39–48.

[14] Ibid, 41–42.

[15] BMV, It.VII 2370 (9668), Cap XXXVI, 81, *De anotomia*. Between 1550 and 1605, the records indicate anatomies took place in 1574, 1585, 1594, 1602, 1603. See BMV, It. VII.2327-2335 (9721–9729) Collegio medico chirurgico di Venezia: Atti (1476–1805), Libro D: 1549–1628 (cc.171). In addition, Palmer 1979, 451–460.

[16] BMV, xxxvi, ibid.

This medical context presented surgeons, including learned surgeons, with a valuable resource. Like their medieval predecessors, Renaissance surgeons located themselves within the tradition of rational surgery, where anatomical knowledge was fundamental, and aligned themselves with physicians in theory and in practice.[17] Anatomy had been essential to the assimilation of surgery as a university subject; given that most learned surgeons did not complete their university training and receive a diploma, anatomy was an important symbol of what made the learned surgeon *learned* (in addition to the ability to read Latin, cite learned sources, and make reasoned, prudent judgments). Anatomy grounded the discipline and the surgeon's claims to manual expertise. It is somewhat surprising, then, to find learned surgeons, such as Croce, distancing their work from anatomy. One reason they needed to do this was a pervasive concern about the ways that anatomy conditioned the surgeon to see and engage the patient. This concern lay at the heart of Fioravanti's critique of anatomy.

3.3 Against Anatomy

In the second half of the sixteenth century, the medical marketplace in Venice was thriving, as it was elsewhere. Barbers, barber-surgeons, empirics, empiric-surgeons, the general handyman and women were treating patients both on the streets and in domestic settings. According to the records of the medical colleges of physicians and surgeons, in 1574, surgeons were upset that some barbers were not content with traditional activities such as bloodletting and beard cutting; instead "unskilled and inexpert barbers [Barbieri imperiti et inesperti]...give oral medicine and... they perform every other operation that would be for the excellent and expert *fisico in cirugico*."[18] In 1601, these surgeons sought to pass new provisions against those practitioners—barbers, women, handymen—who "go about destroying human bodies."[19] These "ignorant" healers did not remain in the public spaces of the city; instead they moved indoors, to apothecary shops as well as to the rooms of patients. These records indicate the dynamic environment of healthcare in Venice in this period—in which learned physicians and learned surgeons encountered not only barbers and empirics, but also rustic healers and women. In this context, learned practitioners began to debate with their "ignorant" counterparts the nature and significance of manual skill and to entertain new or revised claims to medical expertise.

---

[17] See McVaugh 2006; Siraisi 1990; and Nutton 1985, 75–99.

[18] BMV, Libro D: 1549–1628 (cc.171). "li barbieri ossino medicare burschi, sgraffadure, macedure, feri de, et altri simili casi...senza alcuna altra licentia...Ma si fanno lecito etsi Barbieri imperiti et inesperti...terminatione, medicare o operar non solamente nelli casi ditti in detta termination; la danno medicare per bocca, se fanno ogn'altro operation che ad eccelense e esperto fisico in cirugico converebbe."

[19] Ibid, 113, July 3, 1601. "si potra dire con verità che...sono molti barbieri, donne, fachini et altra simil gente che vanno distrugendo li corpi humani et con la lor morte occidendo ancor le misere et infelici famiglie che restano oppresse per la morte de Padri et loro beneffattori."

Coming to Venice in the winter of 1558, the empiric-surgeon, Leonardo Fioravanti discovered a city only partially ravaged by the plagues and typhus of 1555 and 1556 and in the midst of a population boom. There were anxious cries for land reclamation projects that would diminish the famines that struck in the early and mid 1550s and would strike again in 1558–1559. In 1564, with the publisher Valgrisio, Fioravanti saw to press *Dello specchio di scientia universale*, which covered the arts and professions, reordering them, commenting on their moral value and social utility, and criticizing corruption.[20] For medical professions, Fioravanti was especially motivated to promote the utility of his own treatments and to criticize the corruption and general ineffectiveness of learned medical practitioners.[21] For Fioravanti, anatomy played a pivotal role. He used anatomy rhetorically as the occasion to celebrate God's handiwork and to deny the authority of learned medicine and especially learned surgeons: "But I myself always have seen that these surgeons, who are such good anatomists, when they treat patients, they wish always to make their anatomy with knives, cutting the poor bodies [of their patients] as if they were chops of a pig, they wish to scrape the bones for the fire."[22] This passage is usually read to confirm the assumption that all kinds of surgery were barbaric; yet the object of Fioravanti's critique was only the learned surgeon, whose pretense was to be a participant in the academic traditions of anatomy and dissection. Fioravanti added: "these surgeons were the inventors of this anatomy, alleging the continuation [of the tradition] at the university, they cut dead bodies, making anatomies in order to teach the students the composition of human bodies, that is, so that then they will know how to treat patients, when they will practice their Surgery."[23] Although Croce explicitly refuted it, Fioravanti encouraged the double vision, connecting the learned surgeon and the anatomist to a set of deforming activities, involving skinning, cutting, and scraping and to the dead rather than living body.[24] Fioravanti subsequently linked the work on dead bodies to the disruption of burial: "wolves never give such discomfort to other dead wolves, dogs, cats, birds, living fish, never torment the dead bodies of their own."[25]

[20] Fioravanti 1564. This text was republished several times and by several printers in the subsequent decades. I quote below from the Sessa edition of 1572.

[21] Eamon 2010, calls this Fioravanti's anti-establishment rhetoric.

[22] Fioravanti 1572, 49r-50v. "Ma io per me ho sempre veduto, che i cirurgi, che sono buoni anatomisti, quando medicano piaghe, sempre vogliono fare la loro anatomia coi ferri tagliando le povere carni humane, come se fossero brasuole di porco, vogliono raschiare gli ossi, dare fuoco."

[23] Fioravanti 1572, 49v. "...i Cirugici, i quali volgiono sostenare, che loro sono stati gli inventori di questa anatomia, allegando che di continuovo ne'studii publici tagliano huomini morti, facendo notomia di essi, per insegnare alli Scolari, come sta la compositione dei corpi humani, accioche poi sappino medicare, quando eglino pratticaranno la Cirugia."

[24] Carlino 1994, 118–119, 219, traces the historical evidence for the general 'disgust' for dissection in the anatomical literature of the early period. Anatomists reflect on the superstitions about the dead body as a contaminating object.

[25] Fioravanti 1570, 215. "E questa è la prova, che la notomia è contra l'ordine di natura, che ella sia contra conscienza lo dirò io, ma poi lo lasciarò giudicare ad altri, truovo io per le cose che hò viste, e provate, che i lupi non danno mai fastidio alli corpi di lupi morti, i cani, i gatti, gli uccelli, pesci vivi, mai molestano i corpi morti della loro generatione."

If this criticism of anatomy and surgery came into view against a more familiar set of ideas about professions and adjustments to the hierarchy organizing them—as reflected in Fioravanti's *Dello specchio di scientia universale* (1564) as well as Tommaso Garzoni's *Piazza universale* (1585) and Fabio Glissenti's *Discorsi morali* (1596)—Fioravanti provided a more sensitive and potentially damaging assessment of the anatomical training of surgeons in *La cirugia* (1570).[26] Drawing attention to the work of the hand, Fioravanti acknowledges the practice of widening wounds in order to ascertain the injured part, citing its origin in the Hippocratic text, *De vulnibus capitis* (included in the Aldine Greek edition which appeared in 1526 and translated into Latin and commented upon by the Florentine surgeon, Guido Guidi, in 1544.). He then coordinates the anatomical cut (that uncovers parts), the theme of going against nature, and the failure of *medici* (a general category including surgeons) to produce effective treatments:

> When someone is given a wound in the head, they [surgeons] immediately give another transverse one and uncover the bone, and if the bone is cut, they'll uncover the *dura mater*, that which nature has used so much artifice to cover, and the surgeon who is the minister of nature, prescribes that everything be done contrary to nature, and where the sword has made three bad cuts, the *medico* wants to make ten; I don't know how this can be, I marvel at these *medici*, who do this, and I do not know with what reason they can support it, nor with what experience they can prove it; but even more I wonder at those who are wounded, that they let themselves be so tortured without any probable reason, and when the surgeons have made the dilating cuts and uncovered bone and uncovered *dura mater*, they hardly know what to do with it, even though they have medicines with which they can conserve the injured parts, liquefy the blood, dress the wound, and clean it without danger to the wounded person.[27]

In this extended passage, Fioravanti expresses his doubts about the anatomically informed practices of surgeons, who cut and enlarge as they seek to treat wounds in the head. Fioravanti acknowledges the ancient practice, but the procedures are depicted as harmful and tortuous. He indicates that the anatomical training distracts the surgeon from applying real treatments, from applying dressings rather than painful cuts. In book 2 of *The Surgery*, when Fioravanti returned to the issue, he mentioned that "it would have been better to learn agriculture and [from it] to make

[26] This work, published several times in the sixteenth century, was an extension of Fioravanti's tracts on surgery, first published by Pietro and Lodovico Rostini 1561.

[27] Fioravanti 1572, 20. "Percioche tutti quelli, che hanno scritto de vulnibus capitis consegliano i cirugici, che quando sarà data una ferita in testa ad alcuno, che subito gli ne dieno un'altra in traverse, e scoprino l'osso, e se l'osso è tagliato scoprino la dura madre; cose che la natura hà usato tanto artificio in coprirle, e il cirugico che è ministro della natura, ordina che si faccia contra l'ordine della natura, e dove la spade hà fatto trè caratti di male, il medico ne vuol far dieci; cosa che non sò come possi stare, mi maraviglio assai de'medici, che lo fanno, e non sò con qual ragione lo possino sostentare, nè con quale esperientia lo possino dimostrare; ma molto più mi maraviglio, di quei, che son feriti, che si lasciano cosi tormentare senza alcuna ragione che sia probabile, e poi quando i cirugici hanno tagliato dilatator, e scoperto l'osso, e scoperta la dura madre, non sanno quasi ciò che fare, però che hanno medicamenti, coi quali possano conservare le parti offese, prohibire l'alterationi, liquefare il sangue, assotgliar la Marcia, incarnar la ferita, e sanarla senza pericolo del ferito…."

remedies for the treatment of wounds and other sores, with more facility and less torment for the wounded or diseased" than remedies attempted with anatomy.[28]

Fioravanti might have indicated that the surgeon's judgment rather than his manual technique was at fault. Note, for example, that widening wounds was a standard procedure, and often the place for considering the surgeon's judgment. In *Nova selva di cirugia* (Venice 1596), Camillo Ferrara (alias Gabriele Ferrara) recommends talking to the patient with the head wound and then judging the seriousness of the wound, how it might be enlarged and "regarded diligently" to see if it has a fracture and where the end of the fracture might be so that treatment could be provided.[29] Unlike Ferrara, Fioravanti did not include reflective terms associated with judgment, such as "looking diligently," "using every sort of *accuratezza*," or prudence and its cognates. Surgeons were used to relying on anatomy as the source and training for manual skill—assuming its necessity and effectiveness—but Fioravanti's attack identified problems with that model. Whether Fioravanti was the originator of this critique or merely one of its more colorful promoters, Croce addressed the concern explicitly (in the passage quoted at the beginning of this essay) by emphasizing the surgeon's skill at unifying the separated parts and not the practice of widening wounds.

Shifting from the problems posed by anatomy, Fioravanti turns to the overly subtle knowledge generated by it. He uses the anatomist's reliance on animal anatomy—all anatomists dissected animals (though their reasons for doing so differed)—as the occasion to highlight the basic knowledge that dissection yields rather than the subtleties that from his outsider perspective, would seem to inform the philosophical accounts of anatomy as a subject or area of academic research. He then casts further doubt on the usefulness of anatomical knowledge. Having acknowledged his desire to glorify God, Fioravanti explains:

> We are better composed than other animals, because a castrated bull has all that we have, he has blood, flesh, nerves, veins, muscles and bone, and inside, he has a liver, lungs, a heart, a spleen, and all that we have, and he generates, is born, grows, lives, and dies without the use of anatomy, unless the butcher galls him, and the cook cuts him up to cook him, our conclusion then will be that anatomy is a very subtle art, of great ingenuity but of little necessity to the world, as I have demonstrated in diverse places in my writings.[30]

---

[28] Ibid, 216. "che quando medicano un ferito di testa, che hà una ferita per il longo, gli ne danno un'altra per traverse, e se l'osso è coperto, lo scoprono, e se hà un picciol taglio, li fanno una gran rassatura, e di questo ne hò parlato a sofficienza nel capitol delle ferrite di testa, e quando uno hà una picciola stoccata, la vogliono aprire, e dilatare, e cosi sempre in tutti i casi cirugicali, vanno essercitando la anatomia, che hanno imparata, ma quando impariamo la anatomia, saria molto meglio d'imparar l'agricoltura, e di fare rimedii da sanare le ferite, e altre piaghe, con più facilità, e manco tormento del ferito…."

[29] Ferrara 1596, n.p. "Per grave o leggier anche sia la ferita consideri il Cirugico ben la sua qualità, e giudicando che si debbia di'atar, e allargar la cutica sia presto a far quanto si deve, e guardi diligentemente se vi sarà alcuna frattura per usar ogni sorte di accuratezza per trovar il fine d'essa frattura con li suoi Roini taglienti, e accommodate a tal effetto."

[30] Fioravanti 1572, 227. "che noi siamo meglio composti de gli altri animali, perche un castrato hà tutto quello c'habbiamo noi, hà sangue, carne, nervi, vene, muscoli, e ossa, e interiormente hà fegato, polmone, cuore, milza, e tutto quello ch'habbiamo noi, e si generano, nascono, crescono,

Much more elaborate than his earlier criticism, where the anatomical knowledge of a surgeon predisposes the surgeon to make additional cuts on the body rather than close existing ones, Fioravanti here emphasizes the uselessness of anatomy. Anatomy becomes useless once anatomical knowledge extends beyond the general or basic knowledge of animal and human bodies known to everyone, even butchers and cooks. Moreover, the comparison between the anatomist and the butcher/cook refuses to acknowledge the anatomist as a skillful cutter, one whose work can be linked to health. At least the butcher and the cook can produce food, nourishment, a theme echoed in Fioravanti's earlier recommendation that the surgeon would be more useful if he studied agriculture.

For Fioravanti, anatomy was "un'arte molto sottile" and of "grande ingegno" but unnecessary. From the Latin *ingenium*, *ingegno* was used in a variety of senses to indicate the cognitive power of human beings to make connections between different areas of knowledge (a lower level of cognition than the faculties enabling logical thinking). For example, for rhetoric, poetry, and the visual arts, *ingegno* was linked to invention in rhetoric, the ability to generate metaphors, the talent of artists to imitate (or surpass) nature; and for the mechanical arts and architecture, it indicated the inventive ability to conceive of new architectural layouts and mechanical designs.[31] A contemporary of Fioravanti, Tommaso Garzoni (*Piazza universale*, 1565) uses *ingegno* in all of these ways: rhetoricians *ingeniously* amplify their material; geometers display *ingegno* with their instruments; philosophers embody "the universality of *ingegno*."[32] Garzoni, however, reserves an equivocal use of *ingegno* for lawyers, who can stupefy the world with their "sublime *ingegno*." Highlighting the uselessness of anatomy, Fioravanti draws on this association, hinting that the anatomist-surgeon, like the lawyer, is too interested in subtleties and prone to use them for deceptive purposes. Why else, he wonders, would the wounded "let themselves be tortured" by learned practitioners?

3.4 Refining *Ars*

Surgeons extended the terms and ideas around manual and technical skill beyond those used conventionally for anatomy (and its decline). Whether in descriptions of instruments and their use or in descriptions of procedures, these terms were understood within the traditional framework of the practical intellect and in the context of *ars*. In his surgery texts (1573, 1574), Croce included the traditional definition of surgery. The first part comes from the etymology of *chiros* and the end or goal of

vivono, e muorono senza usare la notomia, se non quando il beccaro il scortica, e il cuoco li smembra per cucinarli, la conslusione nostra adunque sarà, che la notomia sia un'arte molto sottile, e di grande ingegno, ma poco necessaria al mondo, come in diversi luoghi de'miei scritti hò dimostrato."

[31] Lewis 2014, 117–124.

[32] Ibid. See also, Cherchi 1980.

surgery (to recuperate the lost unity of a particular part).[33] This part emphasizes the "habit" of the practical intellect and the idea that surgery is governed by rules and practical experience. The second part focuses on the manual operations of a surgeon, and it contains a variation. In his Latin edition of 1573 and his Italian edition of 1574, Croce describes surgery as an "artful operation done by the hand of the *medico*" and different from the other operations done with the intellect, such as "seeing, composing, resolving, defining, [and] demonstrating." The two editions differ, however, in their list of descriptive terms for these manual operations. The Latin edition indicates that the work is done "with order, art and prudence" while the Italian edition specifies that the work is done "with order, art, prudence, and not without gracefulness [*leggiadria*]."[34] Surgery texts, including Vigo's text, typically describe the work of the surgeon with terms, clearly related to the practical intellect, such as order, art and prudence. Unusual, however, is Croce's decision to add "gracefulness" to the list—and to add it only in the vernacular edition.

While the other terms—order, art, prudence—have a place in treatments of Aristotelian virtues (commentaries on *Nic. Ethics* and in late Scholastic logic), 'gracefulness' is tied more closely to vernacular traditions, as its inclusion in the Italian edition would suggest. These are traditions related to both word and image. The triptych of *ordine*, *arte*, *leggiadria* finds a place in Renaissance rhetorical theory from which it was developed in the literature on comportment, such as Castiglione's *Courtier* (1528), and in the debate on the arts, called the *paragone*.[35]

[33] Croce 1574, preface 1. "La cirugia à la più vecchia e la più certa parte di tutte la medicina, e è un'habito dell'intelletto prattico, acquistato con molte regole, e isperimenti, accioche con artificiosa operatione delle mani, e stromenti accommodati, uniendo, separando, e togliendo via molti affetti nel continuo delle parti del corpo humano: presto, sicuramente, e con poco dolore danar possi dico, che è artificiosa operatione fatta con le mani del Medico, a differenza di molte altre operationi fatte da lui con l'intelletto, dividendo, componendo, risolvendo, diffiniendo, dimostrando, o altramente operando con le parti dell'anima: Et è artificiosa operatione, cioè fatta con ordine, con arte, con prudenza, e non senza leggiadria, e è regolata dall'anatomia, e da una lunga prattica, laqual consiste in quelle cose, che con certe ragioni sono approbate, e confirmate con frequente esercitio, et operatione, et anche da natural ragione; imperoche essendo arte, e operatione, che cura alcun morbo, necessariamente piglia l'ingegno della cura dalla essenza di quello, laqual'è dimostrata con scienza, e natural ragione dal Theorico."

[34] See Croce 1573. "Opus nimirum artificiosum ipsarum manuum est, ab operationibus intellectus admodum di versum…quae fiunt dividendo, componendo, definiendo, demonstrando, sive quovis alio modo operando: *est tamen opus, quodmethodo, idest ordine, arte, et prudentia perficitur*; ac ab anatome, et longa praxi, quae in iis consistit, quae certa ratione excogitate sunt, et frequenti usu confirmata, dirigitur." See Croce 1574, 1583. "che è artificiosa operatione fatta con le mani del Medico, a differenza di molte altre operationi fatte da lui con l'intelletto, di videndo, componendo, risolvendo, diffiniendo, dimostrando, o altramente operando con le parti dell'anima: *Et è artificiosa operatione, cioè fatta con ordine, con arte, con prudenza, e non senza leggiadria*, e è regolata dall'anatomia, e da una lunga prattica, laqual consiste in quelle cose, che con certe ragioni sono approbate, e confirmate con frequente esercitio, et operatione, et anche da natural ragione." My emphasis.

[35] These terms originate to some extent in Petrarch, for whom *leggiadria* is connected to *honore*, *virtute*, *honestate*, and *bellezze in forme*. In the *Vocabolario della Accademia della Crusca* (Venice 1724), the entry for *leggiadria* cites a number of other texts including the *Galatea* and M. Francesco

As Fosca Mariani-Zini has explained, *leggiadria* expressed "an almost natural grace that was in no way divine but anchored in worldly reality, situated at the point of equilibrium in a tension between the natural and the artificial."[36] In the debate on the arts, writers took up the relative merits of the arts, whether and why painting might be superior to sculpture, or architecture, and they used these terms to describe and dignify the artisan's work, his ability to conceptualize and produce a work of art.[37] As Croce moved away from anatomy as the main symbol of manual expertise, he turned to the visual arts. He imported additional vocabulary for manual skill from vernacular debates on art, where the subjective dimensions of artisanal work were being captured by new, loftier terms for physical labor and manual techniques. These terms refer less to the patient's body as an objected re-crafted into health by the surgeon's labor, though that is certainly implied, and more to the subjective dimension of the surgeon's work, his experience of manual operation.

The *paragone* generated in particular a way of talking about the differences between Florentine and Venetian art, differences which art historians refer to as the *disegno/colorito* debate. Florentine art came to privilege design and conceptualization—something made possible by the medium of fresco. Discussions of Venetian art tend to emphasize the nature of color and its application—something encouraged by oil and its application in stages to canvas. Engaging this debate, Leon Battista Alberti used the term *leggiadria* for architecture (1452), emphasizing the "miraculously resplendent" light of a stone façade; and the contradictory nature of that shining stone was referred to as *leggiadria*."[38] Giovanni Paolo Lomazzo (*Trattato dell'arte de la pittura*, 1585) used *leggiadria* also in an architectural context: describing the Corinthian order, he aligned it with virginal gracefulness (*gracilità*) and with "members subtle and graceful, more minutely carved, that ties together flowers, fronds and leaves of every kind."[39] Like Alberti, Lomazzo used *leggiadria* to acknowledge the workmanship that made an artifact seem to over-

---

Alunno da Ferrara's "observations" on Petrarch, which provide additional terms in the gloss on Petrarch. See especially M. Francesco Alunno da Ferrara (Venice, 1550) and his entry on *leggiadria*, 240.

[36] Eds. Barbara Cassin, Emily Apter, Jacques Lezra, Michael Wood, 2014, 559-560. This text was first published in French as *Vocabulaire européen des philosophies: Dictionnaire des intraduisibles* (Paris: Editions de Seuil, 2004).

[37] On this debate, see Roskill 1968; Rosand 1988; and especially, Summers 1981 and 1987.

[38] Alberti 1550, 257. "per il quale tutta la faccia della bellezza risplende miracolasamente, ilche appresso di noi si chiamera leggiadria; la quale certamente noi diciamo che è la nutrice d'ogni gratia, e d'ogni bellezza, e è l'officio della leggiadria."

[39] Lomazzo 1584, 411. "L'ordine Corinthio richiede molto più che l'ordine Ionico le membra sottili e leggiadre, intagliate più minutamente di lavori, che tirano à legami, fiori, frondi, e foglie d'ogni maniera." Enhancing the link between *leggiadria* and *vaghezza*, Lomazzo devoted a chapter to *dei moti della vaghezza*, where he explains: "La leggiadria fà gl'atti vaghi…in tutte le cose sono desiderati, si come quelli che generano ammiratione, e sono il proprio ornamento delle cose, facendo comparire il leggiadro giovane, ò verginella nel più gratioso habito, e meglio concertato che si possa cosi per sua convenienza, come per diletto dell'occhio, che solo delle bellezze, e cose ben fatti si appaga. Però questi moti leggiadri difficilmente possono risplendere in un corpo brutto, è scomposto. La gentilezza fa gl'atti gratiosi, cortesi, nobili, e virtuosi" (146).

come its material constraints: stone that shined as if by a miracle, and flowers and fronds carved so carefully that they flowed together.

In the treatment of painting rather than sculpture or architecture, *leggiadria* is even more strongly associated with workmanship, and the hand of the artisan. In his *Lives of the Artists* (1550-), Giorgio Vasari uses *leggiadria* in his evaluation of the ability of painters: "they lost the gracefulness [*leggiadria*] to make svelte and gracious [*graziose*] all the figures."[40] Here, the term *leggiadria* is a feature of the artisan and his experience of making an object; the term is an index of the subjective experience of the artisan and thus distinguished from the object produced, that is, the gracefulness of the figures (for which the adjective, *graziose*, is used). Speaking about Venetian art, Carlo Ridolfi situated *leggiadria* in a discussion of color. In *Le maraviglie dell' Arte ovvero, Le vite degli Illustri Pittori Veneti and dello Stato* (1648), he celebrated Venetian painting and the superior role of color (above design), noting a painter's desire "to imitate the gracefulness of Parmegiano with the exquisite practice of coloring," that is, with the application of color.[41] A later elaboration of this idea can be found in Francesco Lana's discussion of color: "finally one must fill these [figures] with the appropriate lighting, with simple clearness or darkness; or even with colors, which make a much better effect because they imitate nature and give a vagueness and gracefulness to the design. In this, one should consider generally the manner of applying color and filling the surfaces with color."[42] In the visual arts, the semantic field for *leggiadria* included terms that described the artisan doing his work, manipulating materials and applying color to canvas. Artisans and connoisseurs in the period did not limit the term to a character of the object. With *leggiadria*, Croce cultivated a more suggestive vocabulary for the surgeon's skill, one that emphasized the vernacular traditions of art (and comportment) and elevated manual skill.

3.5 Conclusion

At the Renaissance university, the relationship between anatomy and natural philosophy was an increasingly developed and close one. The responses to anatomy and dissection, however, were multiple and various. The fine-grained analysis in this essay has taken, as a case study, the interaction between the texts of the surgeons, Croce and Fioravanti, in the second half of the sixteenth century. Whether we see Fioravanti as an interlocutor with Croce or as a mirror reflecting more pervasive

---

[40] Vasari 1966, preface. "allaquale mancava una leggiadria di fare svelte e graziose tutte le figure, e massimamente le femmine, e i putti con le membra naturali."

[41] Ridolfi 1648, 382. "ed in questa volle imitare la leggiadria del Parmegiano con esquisito colorire, si che paiono vive figure."

[42] Lana 1670, Chapter 3: Concepts pertaining to color, 150. "finalmente queste si devono riempire de'suoi proprii lumi, il che si fa o con semplice chiaro, e scuro; o pure con i colori, i quali fanno molto migliore effetto, perche piu imitano il naturale, e danno vaghezza, e leggiadria al disegno."

anxieties about anatomy and the surgeon's approach to the patient's body, his works suggest that the connections between anatomy and surgery were anxious ones; and perhaps because of this, they were intentionally deployed in arguments about a surgeon's expertise. Anatomy carried some negative connotations with it by the 1570s. Anatomical information had been circulating widely, and this is surely one cause for its more complicated reception. But there are at least two others. First, the medical marketplace was filled with more practitioners and healers, with more medicines, and with more information than ever before. In this setting, medical values were being debated, including the value of anatomical training. Second and more specific to Venice, for the three decades after the 1560s, the economic situation of skilled labor looked promising, perhaps promising enough for learned surgeons to reflect on their manual skill in new ways. This essay has argued that Croce responded to concerns about his anatomical training by shifting some of the terms of manual activity and the language of skill to those used by artisans and recognized as important by connoisseurs of art.

Given the interaction between Fioravanti and Croce and the momentary (conceptual and discursive) reticulation of surgery and the arts, we may wish to reconsider the role played by the sixteenth century in subsequent developments. Late seventeenth-century practitioners devoted themselves to the manipulation of bodies, making hygiene, beauty and well-being central to their practices. But the sixteenth century offers important preconditions for this devotion. In a marketplace where learned and non-learned practitioners encountered one another, the divisions between intellectual and manual labor could be partially overcome by a novel, robust language of skill—*ordine*, *arte*, *prudenza*, *leggiadria*—that both surgeons and their patients could use to understand or at least think about surgical care. Maybe they talked in these terms as well. Such developments suggest that for learned and non-learned practitioners, manual skill was in the process of acquiring a new cultural identity.

References

Alberti, Leon Battista. 1550. *L'architettura*. Florence: Lorenzo Torrentino.

Bernardi, F. 1826. *Elogio di Gio. Andrea dalla Croce, medico chirurgo ed anatomico veneziano del secolo xvi, letto nell'ateneo di Treviso*, 8 June, 1826. Venice: Giuseppe Picotti.

Braudel, Fernand. 1949. *La Méditerranée à l'époque de Philippe II*. Paris: Armand Colin.

Carlino, Andrea. 1994. *Books of the body: Anatomical ritual and renaissance learning*. Chicago: University of Chicago Press.

Carlino, Andrea. 1999. *Paper bodies: A catalogue of anatomical fugitive sheets, 1538–1687*. London: Wellcome Institute for the History of Medicine.

Carlino, Andrea. 2005. Petrarch and the early modern critics of medicine. *Journal of Medieval and Early Modern Studies* 35: 259–282.

Cassin, Barbara, Emily Apter, Jacques Lezra, and Michael Wood. Eds. 2014. *Dictionary of untranslatables: A philosophical lexicon*, 559–560. Princeton: Princeton University Press (This text was first published in French as Vocabulaire européen des philosophies: Dictionnaire des intraduisibles. Paris: Editions de Seuil, 2004).

Cherchi, Paolo, 1980. *Enciclopedismo e politica della riscrittura: Tommaso Garzoni*. Pisa: Pacini Editore.

Croce, Giovanni Andrea della. 1573. *Chirurgiae Ioannis Andreae a Cruce, Veneti medici libri septem, quamplurimis instrumentorum imaginibus Arti Chirurgicae opportunis suis locis exornati, Theoricam, Practicam, ac uerissimam experientiam continents*. Venice: Ziletti.

Croce, Giovanni Andrea della. 1574. *Cirugia universale e perfetta di tutte le parti pertinenti all'ottimo Chirurgo*. Venice: Giordano Ziletti.

Eamon, William. 1993. "With the Rules of Life and an Enema": Leonardo Fioravanti's medical primitivism. In *Renaissance and revolution: Humanists, scholars, craftsmen and natural philosophers in early modern Europe*, eds. Field and James, 29–44. Cambridge: Cambridge University Press.

Eamon, William. 1994. *Science and the secrets of nature: Books of secrets in medieval and early modern culture*. Princeton: Princeton University Press.

Eamon, William. 2010. *The professor of secrets: Mystery, medicine, and alchemy in renaissance Italy*. Washington, DC: National Geographic.

Farrington, Benjamin. 1932. The preface of Andreas Vesalius to *De humani corporis fabrica*—1543. *Proceedings of the Royal Society of Medicine* 25: 39–48.

Ferrara, Gabriele. 1596. *Nova selva di cirugia*. Venice: Bartolomeo Carampello.

Fioravanti, Leonardo. 1564. *Dello specchio di scientia universale in libri tre*. Venice: Valgrisi.

Fioravanti, Leonardo. 1570. *La cirugia distinta in tre libri*. Venice: Melchior Sessa.

Fioravanti, Leonardo. 1572. *Dello specchio di scientia universale in libri tre*. Venice: Marchiò Sessa.

Gentilcore, David. 1997. The organisation of medical practice in Malpighi's Italy. In *Marcello Malpighi, anatomist and physician*, ed. Bertoloni Meli et al., 75–110. Florence: Istituto e museo di storia della scienza, Olschki.

Gentilcore, David. 2006. *Medical charlatanism in early modern Italy*. Oxford: Oxford University Press.

Kusukawa, Sachiko. 2012. *Picturing the book of nature: Image, text, and argument in sixteenth-century anatomy and medical botany*. Chicago: University of Chicago Press.

Lana, Francesco. 1670. *Prodromo overo saggio di alcune inventioni nuove premesso all'arte maestro*. Brescia: Li Rizzardi.

Lawrence, Christopher. 1999. Medical minds, surgical bodies: Corporeality and the doctors. In *Science incarnate: Historical embodiments of natural knowledge*, 21–49. Chicago: University of Chicago Press.

Lewis, Rhodri. 2014. Francis Bacon and ingenuity. *Renaissance Quarterly* 67: 113–163.

Lomazzo, Giovanni Paolo. 1584. *Trattato dell'arte de la pittura di ...7 libri*. Milan.

Long, Pamela O. 2011. *Artisan/practitioners and the rise of the new sciences, 1400–1600*. Corvallis: Oregon State University Press.

Masciandaro, Nicola. 2007. *The voice of the hammer: The meaning of work in middle English literature*. Notre Dame: Notre Dame University Press.

McVaugh, Michael. 2006. *The rational surgery of the middle ages*. Florence: Sismel, Edizioni del Galluzzo.

Nutton, Vivian. 1985. Humanist surgery. In *The medical renaissance of the sixteenth century*, 75–99. Cambridge: Cambridge University Press.

Palmer, Richard. 1979. Physicians and surgeons in sixteenth-century Venice. *Medical History* 23: 451–460.

Park, Katharine. 1994. The criminal and the saintly body: Autopsy and dissection in renaissance Italy. *Renaissance Quarterly* 47: 1–33.

Park, Katharine. 1995. The life of the corpse: Division and dissection in late medieval Europe. *Journal of the History of Medicine and Allied Sciences* 50: 111–132.

Park. 2006. *Secrets of women: Gender, generation, and the origins of human dissection*. New York: Zone Books.

Pullan, Brian. 1964. Wage earners and the Venetian economy, 1550–1630. *Economic History Review* 16: 407–428.

Ridolfi, Carlo. 1648. *Le maraviglie dell'arte*. Venice: Gio. Battista Sgava.

Roberts, Lissa, Simon Schaffer, and Peter Dear. 2007. *The mindful hand: Inquiry and invention from the late renaissance to early industrialisation*. Amsterdam: Koninkliijke Nederlandse Akademie van Wetenschappen.

Rosand, David. 1988. *The meaning of the mark: Leonardo and Titian*. Lawrence: University of Kansas.

Roskill, Mark. 1968. *Dolce's Aretino and Venetian art theory of the Cinquecento*. New York: New York University Press.

Rostini, Pietro, and Lodovico Rostini. 1561. *Compendio di tutta la cirugia*. Venice: Lodovico Avanzo.

Siraisi, Nancy. 1990. *Late medieval and early renaissance medicine: An introduction to knowledge and practice*. Chicago: University of Chicago Press.

Smith, Pamela H. 2004. *The body of the Artisan: Art and experience in the scientific revolution*. Chicago: University of Chicago Press.

Summers, David. 1981. *Michelangelo and the language of art*. Princeton: Princeton University Press.

Summers, David. 1987. *The judgment of sense: Renaissance naturalism and the rise of aesthetics*. Cambridge: Cambridge University Press.

Vasari, Giorgio. 1966. In *Le vite de' più eccellenti pittori, scultori e architetti*, ed. Barocchi and Bettarini. Florence: Sansoni.

Vigo, Giovanni de. 1560. *La prattica universale in cirugia*. Venice: Francesco Sansovino.

Vocabolario della Accademia della Crusca. 1724. Venice.

von Staden, Heinrich. 2007. Physis and Techne in Greek Medicine. In *The artificial and the natural*, ed. Bensaude-Vincent and Newman, 21–49. Cambridge: MIT Press.

Whitney, Elspeth. 1990. Paradise restored: The mechanical arts from antiquity through the thirteenth century. *Transactions of the American Philosophical Society*, n.s. 80, pt. 1.

Chapter 4
Why All This Jelly? Jacopo Zabarella and Hieronymus Fabricius ab Aquapendente on the Usefulness of the Vitreous Humor

Tawrin Baker

Abstract At the end of the sixteenth century new anatomical knowledge led both empirically minded philosophers and philosophically minded anatomists to rethink theories of light, color, and vision in subtle but significant ways. In this paper I show how anatomy and philosophy conspired to understand the structure and the purpose of the parts of the eye in two important, but largely overlooked, works by professors at the University of Padua: the natural philosopher Jacopo Zabarella's *De visu* (first published in 1590) and the anatomist and physician Hieronymus Fabricius ab Aquapendente's *De visione* (1600). How they understood the roles of the various parts of the eye reveals much about the strategies different disciplines used to reconcile ancient authorities (particularly Galen and Aristotle) with new anatomical observations and experiments. Importantly, the two professors offer identical accounts of the size, shape, and clarity, as well as the *usus* (or Galeno-Aristotelian final cause), of the vitreous humor, the transparent gel that fills the space between the crystalline humor (or lens) and the retina. This account of the vitreous is at the center of a theory of vision that differs in crucial ways from previous perspectivists, natural philosophers, and anatomists. Given this striking similarity, I argue that the two must have interacted significantly at Padua. I also argue that (by way of a former student of Fabricius, the anatomist and physician Jan Jessenius) this theory of vision influenced Kepler's revolutionary account in his *Ad Vitellionem paralipomena* (1604) in certain respects.

Keywords Zabarella • Fabricius • Kepler • The eye • Vision

T. Baker (✉)
Department of History and Philosophy of Science,
Indiana University, Bloomington, IN, USA
e-mail: tawbaker@indiana.edu

P. Distelzweig et al. (eds.), *Early Modern Medicine and Natural Philosophy*,
History, Philosophy and Theory of the Life Sciences 14,
DOI 10.1007/978-94-017-7353-9_4

4.1 Introduction

An important element of the (purported) demise of Peripatetic natural philosophy in the seventeenth century concerns the development of new theories of sensation and sensible qualities, including theories of vision and color. Kepler's *Ad Vitellionem paralipomena* (1604) was among the more significant works in this transformation, and historians have stressed the role it played in the seventeenth-century mechanization of nature, the importance of the camera obscura as a model for vision, and Kepler's novel mathematical account of vision in which pictures are cast upon and sensed at the retina, displacing the crystalline humor (now called the crystalline lens) as the traditional site of sensation.[1] Kepler's scheme was also taken up by Scheiner, Plemp, and Descartes, and in many respects forms the basis for modern visual theory. Recently, attention has been drawn to developments in practical and mixed mathematics in the sixteenth century and Kepler's appropriation of this tradition in the service of natural philosophy, as well as the importance of courtly "experiments" with the camera obscura.[2] Accounts of vision by contemporary anatomists and Peripatetic natural philosophers have been largely overlooked.

Traces of earlier and now inadequate scientific revolution narratives still persist, and important early-modern works on vision by both natural philosophers and anatomists have yet to be carefully reexamined. To see what was truly novel about seventeenth century accounts of vision, and to understand reactions to them, it is necessary to reconstruct what was displaced. Not only treatises on mathematical optics (referred to as *perspectiva*, and its practitioners perspectivists or *perspectivae*), but also works of natural philosophy, anatomy, and medicine need to be reexamined. For example, after discussing Colombo, Bartisch, Estienne, Fabricius, Jessenius, Varolio, and Laurens on ocular anatomy, David Lindberg writes: "None of the post-Vesalian authors that I have mentioned made significant alterations in visual theory."[3] Although Hieronymus Fabricius ab Aquapendente's (1533–1619) *De visione* was first published in 1600, Lindberg does not seem to realize this and only cites the 1614 edition, so it is not surprising that he does not examine it carefully in a book aimed towards understanding Kepler's *Paralipomena*.[4] A. C. Crombie has also written on Fabricius, saying: "His visual theory was essentially a combination of the formulations of the problem by Aristotle and Galen with a version of the optical scheme with which Alhazen had prevented the reversal of the image as the visual cone passed through the transparent media."[5] Largely dismissing the anatomists, Crombie says: "It was the mathematicians who came to reform visual theory by proceeding through an optical analysis of ocular physiology, exploiting the models of eyeglasses and the camera obscura, and thus reformulating the problem

[1] See especially Crombie 1967; Straker 1970; Lindberg 1976.

[2] Dupré 2007, 2008, 2012; Shapiro 2008.

[3] Lindberg 1976, 175.

[4] *Ibid.*, 173.

[5] Crombie 1990, 629. This judgment also affects his analysis in Crombie 1991.

itself."[6] In the history of medicine Huldrych Koelbing sums up the predominant attitude towards Fabricius's *De visione*:

> Mais que fait-il de toutes ces observations? A peu pres rien! Fabrice a bien contribué à l'essor de l'anatomie du XVIe siècle, mais ses connaissances approfondies ne lui servent qu'à confirmer des doctrines anciennes, et plus particulièrement la théorie de la vision d'Aristote et d'Alhazen....[7]

These characterizations of Fabricius's work on vision have not been directly challenged, but they are incorrect. As I will show, Fabricius's model of vision differs in crucial ways from Alhazen's, particularly in the path of the visual rays in the eye.[8] This novel theory of vision was shared by his colleague at Padua, the logician and natural philosopher Jacopo Zabarella (1533–1589). They both provided detailed philosophical accounts of the nature of light, color, and vision, as well as anatomical accounts of the structure, action, and usefulness or purpose of the eye. (Zabarella uses the terms *officium* and *usus*, while Fabricius's preferred term is *utilitas*.)

Several historians have suggested that Zabarella influenced Fabricius in some way, particularly in the latter's Aristotelian scheme for his grand anatomical project,[9] but thus far evidence of their interaction has not been provided. A close reading of their texts, however, reveals a striking similarity in the account of vision that they give. In particular, both Zabarella and Fabricius reject Galen's widely accepted account of the usefulness (that is, the Galeno-Aristotelian teleological explanation[10]) of the part of the eye called the vitreous humor. Galen says that the vitreous exists to nourish the visually sensitive crystalline humor. Zabarella argues against this in his *De visu libri duo* first published in his philosophy textbook *De rebus naturalibus libri XXX* (Venice 1590), and Fabricius in his first anatomical publication *De visione, voce, auditu* (Venice 1600).[11] In its place they give the same account of the usefulness of this clear jelly that occupies the rear of the eye—in short, they view it as a large transparent chamber that dissipates the light passing through so that the color of the retina does not affect the crystalline humor. Despite much searching, I have found this account only in these two authors and those that have clearly been influenced by them. The uniqueness of this description of the vitreous humor, the striking details in their anatomical account, and the fact that Zabarella appeals to personal experience with ocular dissection at a time when Fabricius was the most

---

[6] *Ibid.*, 630.

[7] Koelbing 1990, 395. See also his incorrect assessment on page 365.

[8] Ibn al-Haytham (c. 965–c. 1040), most frequently called Alhacen in Latin before Friedrich Risner's 1572 printing, after which Alhazen became the dominant spelling. For simplicity I follow Fabricius and call him Alhazen throughout.

[9] Cunningham 1997, 170–174; Jardine 1997, 207. See also Bylebyl 1979; Cunningham 1985.

[10] This is a rich and important topic, but a full treatment is outside the scope of this paper. Excellent analyses can be found in Goldberg 2012, 90–104. Distelzweig 2014.

[11] Zabarella 1590; Fabricius 1600. *De visu* was also included (along with many other books in *De rebus* concerning the soul) in Zabarella's posthumous *De anima* commentary. All citations below are to the column number the 1617 Frankfurt edition. I also include the book number and chapter of *De visu* as: (*DV* book.chapter).

famous anatomist of his time (as well as the only person in Padua with permission to perform the annual public dissection) strongly suggest that the philosopher and the anatomist interacted in generating this novel theory of vision.

The first section of this paper is a brief outline of accounts of the eye from Galen until the end of the sixteenth century. I show that, as a result of sixteenth-century anatomical research as well as changes to the scope of anatomy as a discipline, the fact that there was so much of this jelly-like vitreous humor in the eye became a problem that no previous theory of vision could account for. After this I discuss Zabarella's theory of light, color, and transparency in *De visu*. Understanding these key aspects of natural philosophy allows us to make sense of Zabarella's criticism of Galen and his own account of the vitreous humor. I then examine Fabricius's *De visione*, and argue that the similarities between his and Zabarella's theory of vision strongly suggest that they interacted. Next I discuss the connection between this shared theory of vision and Kepler's revolutionary account given in his *Paralipomena*. I show that Kepler relied on knowledge about the size, shape, and refractive powers of the humors that was developed in Padua and conveyed to him by Fabricius's student Johann Jessenius. I also suggest that Kepler's theory of vision was perhaps influenced by certain qualitative elements about the path of rays in the eye given by Zabarella, Fabricius, and Jessenius.

4.2 The Three Ocular Humors in Sixteenth-Century Anatomy

The eye degrades rapidly, contains many fluid parts, and can differ greatly both across species as well as within an individual over time. In the Galeno-Aristotelian framework for vision every sensible aspect of the parts of the eye—color, clarity, texture, firmness, size, position, and connection—was significant for understanding vision, and all are unstable in a decaying body under dissection. Despite these challenges, during the sixteenth century some basic facts about the structure of the eye gained the consent of most anatomists. Two of these concern the three clear humors that make up the interior of the eye: the aqueous humor, located towards the front of eye, the crystalline humor, which is next, and the vitreous humor, located at the rear of the eye (respectively given by O, A, and C in Vesalius's diagram of the eye in Fig. 4.1). Galen's two main discussions of the eye are found in *On the Usefulness of the Parts* and *On the Doctrines of Hippocrates and Plato*, but Galen does not clearly describe either the relative size of the three humors or the location of the crystalline humor within the eye.[12] Mediaeval perspectivists as well as most pre-Vesalian anatomists placed the crystalline humor towards the front of the eye.[13] Rather than following his predecessors, Vesalius locates the crystalline in the direct center of the

[12] Galen 1968, 464–503; Galen 1980, 459.

[13] Carpi and Mondino 1521, 462r; Carpi 1959, 152; Benedetti and Ferrari 1998, 280.

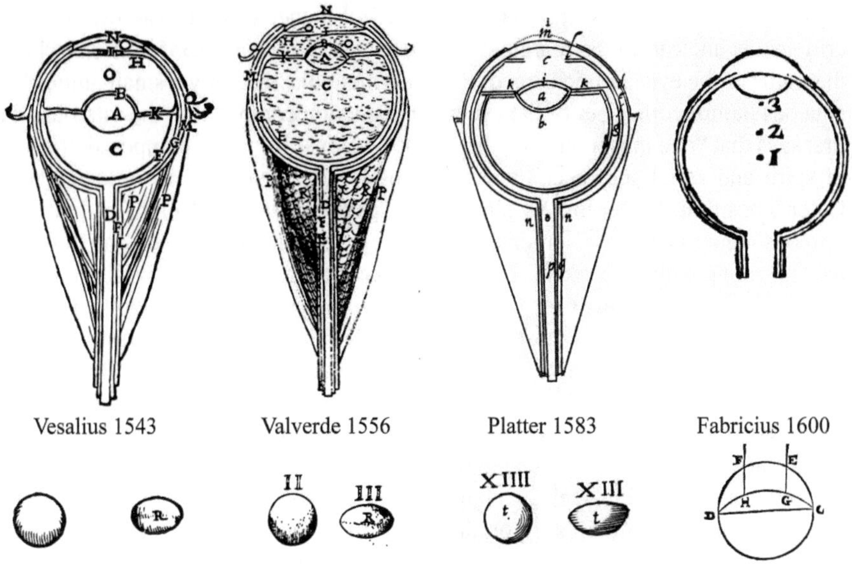

Fig. 4.1 Cross-sectional images of the human eye in four anatomical texts, showing the position of the crystalline humor and the relative sizes of the three humors, above, with accompanying representations of the crystalline humor itself. Note that Fabricius departs from the Vesalian pattern. Fabricius's crystalline humor (the lower image) consists of a half-sphere for the posterior part together with section of a sphere of larger radius for the anterior; compare to Vesalius's geometrical account of the crystalline in Fig. 4.2. Also note that Fabricius adds the centers of curvature of (*1*) the center of the eye, (*2*) the anterior of the crystalline, and (*3*) the cornea, making it incompatible with the theory of vision held by Alhacen and the mediaeval perspectivists

eye.[14] He cites Galen's *Usefulness* and *On the Doctrines* in his chapter on the eye, but historians often account for Vesalius's placement of the crystalline humor in the center by saying Vesalius was following a traditional medieval conception of the eye as a microcosm of the universe.[15] However, this traditional conception does not seem to have greatly influenced earlier anatomists. In any case, the surviving notes on Vesalius's 1540 dissection in Bologna of the eye suggest that he was far from careful. The student Hessler writes, "And he cut the eye through the middle with a razor, and he shook out into the hand the substance of the eye: the first humor, he said, is the albugineus one, the second is the vitreous and the third is the crystallinous humor, by which properly the vision occurs…."[16]

Vesalius's depiction of the humors quickly came under criticism, perhaps most famously by Realdo Colombo at Padua, and most subsequent anatomists placed the

[14] Vesalius 1555, 799–806.

[15] Saunders and O'Malley 1950, 200. They write that the view of the eye as a microcosm also hindered works in geometrical optics, but this is overly simplistic. Their opinion on this is cited in Vesalius 2014, 1301 n. 1.

[16] Heseler and Eriksson 1959, 290–291. Vesalius also says that "anyone can see for himself at home." One wonders if Vesalius ever did so.

crystalline humor towards the front of the eye.[17] Whether Vesalius was aware of this criticism is unclear. In his marginal annotations to his second 1555 edition (whose diagrams of the eye are unchanged), Vesalius mentions that only a small amount of aqueous humor comes out of the eye upon dissection compared to the vitreous. He first says that "one must conclude that it [the aqueous] is largely composed of a sort of spirit and aerial substance" which dissipates after death.[18] This is identical to Galen's position.[19] After this Vesalius writes "perhaps someone [might say] that the vitreous humor occupies a larger space in the eye than the rear portion and thus that the lens along with the vitreous humor [is placed] off-center in the front part of the eye."[20] However, Vesalius's language implies that he prefers the first explanation. If not endorsing it, he was at least raising the possibility that observations on a dead eye do not give us accurate knowledge of the structure of a living one. Regardless, a third edition to the *Fabrica* was never printed. It was due to anatomists such as Colombo and Valverde that the position of the crystalline humor, and as a consequence also the proportional volume of the three humors, underwent a shift. This was reflected both in changes to diagrams of the eye and frequently in the text itself. Eye diagrams were often based on the image from the *Fabrica*, and thus changes would be conspicuous, and descriptions of the position of the crystalline were often mentioned explicitly *contra* Vesalius. (Note that Colombo did not publish any images in his *De re anatomica*; his former student and collaborator Valverde did, although they were based on Vesalius's.[21])

Another change concerned the shape of the crystalline humor. Although anatomists such as Mondino and Benedetti described the crystalline humor in humans as having a more flattened anterior and protruding posterior,[22] nevertheless Vesalius described it as lenticular and symmetric.[23] In a marginal illustration and accompanying text, Vesalius says that its shape can be understood by removing a slice from the middle of a sphere, thus giving a geometrical account of its symmetry.[24] (See Fig. 4.2.) Many later anatomists, again in opposition to Vesalius, insisted that the

---

[17] Valverde 1556, 82–83; Colombo 1559, 220; Lindberg 1976, 173–174.

[18] Nutton 2012, 435.

[19] Galen 1968, 475–476.

[20] Nutton 2012, 435. Translation his. In a footnote Nutton mentions that a referee pointed out that it would be difficult to determine the place of the crystalline humor once the aqueous humor had leaked out and the bulbus collapsed, but in my experience this is not difficult if one is attending to this issue at all during dissection. Nutton also notes that earlier anatomists divided the eye into two equal cavities, which is not entirely correct. I would like to thank Gideon Manning for bringing this article to my attention.

[21] Klestinec 2005; San Juan 2008.

[22] Benedetti and Ferrari 1998, 280; Carpi and Mondino 1521, 462r. Alhazen and Witelo describe only the front of the crystalline as lenticular, although how to reconcile their characterizations of the shape of the crystalline humor seen through dissection with their geometrical account is not altogether clear. See Alhacen 2001a, lvii–lx, 12; Witelo 1991, 294–297.

[23] Galen 1968.

[24] Note that, in the annotations to his 1555 *Fabrica*, Vesalius gives no indication that his opinion had changed. See Nutton 2012.

·iormam.
dũ refert:
cus cōpri-
ecundùm
s globi par
teriora mi
odãmodo
recis nun-

Fig. 4.2 Marginal figure demonstrating the shape of the crystalline humor. Vesalius 1555, 801

crystalline humor was asymmetric, with a more flattened anterior and protruding or gibbous posterior, and this view became standard by the end of the century. As I will show, the growing consensus of these two points of anatomy had important ramifications for understanding not just the structure of the eye, but the visual process as a whole. Vesalius seems to be the catalyst for changes to seventeenth century ocular anatomy in two ways. Later anatomists formed their consensus of the parts of the eye by refuting Vesalius (and, to some extent, Galen) rather than building upon pre-Vesalian mediaeval and renaissance authors. Additionally, the importance of those shapes was considered in light of Vesalius's portrayal of anatomy: it is at once a *scientia* and an erudite activity that demands the analysis of ancient texts along with the body.[25]

4.3 Zabarella, Anatomical Experience, and the Usefulness of the Vitreous

Zabarella first published *De visu libri duo* in his natural philosophy textbook *De rebus naturalibus* in 1590, and *De visu* was reprinted many times throughout Europe in this text as well as in his *De anima* commentary.[26] Zabarella's works had considerable influence in Italy as well as northern Europe, and many references to his account of vision can be found in popular seventeenth-century natural philosophy textbooks, including those by Johannes Magirus (whose text was used in Cambridge during Newton's school days) and Daniel Sennert.[27]

Zabarella's theory of color, light, and vision is a culmination of the long, vigorous, and multifaceted Peripatetic tradition, and as such it differs significantly from what might be gathered from Aristotle's texts alone. Nevertheless, Zabarella's stated aim is to give a comprehensive account of natural philosophy *ad mentem Aristotelis*,[28] and his most important sources here are *De anima* and *De sensu*, although he refers

[25] Siraisi 1994, 65–66; Siraisi 1997; Vesalius 2013, 4.

[26] Edwards 1960, 368–373; Lohr 1982, 233.

[27] Magirus 1597; Sennert 1618; Kusukawa 2002; Maclean 2002.

[28] On unpacking this phrase, see Palmieri 2007.

to the *Meteorology*, book 5 of *On the Generation of Animals*, book 2 of *The Parts of Animals*, and the (Pseudo-Aristotelian) *Problems*. Note that the picture Aristotle gives in *Meteorology* III can be read as extramissionist, and indeed was taken to be evidence of such in Aristotle by some, perhaps most importantly Galen.[29]

According to Aristotle's account in *De anima* and *De sensu*, vision occurs when the colors of bodies cause a movement in a transparent medium which in turn reaches our eyes, causing our visual faculty to take on the forms of color present at the surfaces of bodies.[30] In order for color to pass through a transparent medium, however, the medium must be actualized, and thus the role of light for Aristotle is to turn a dark, potentially transparent medium (such as air or water) into a clear, actually transparent medium. Light can be thought of as, in some sense, switching on the transparency in the air, instantly removing the darkness that prevented the colors at the surfaces of bodies from issuing forth. There is little in Aristotle's writings to suggest that light plays a role in coaxing color itself into activity, although this was a common reading later on. Aristotle's account of the parts of the eye is also rudimentary, and he does not specify precisely where the seat of sensation lies.[31]

Aristotle gives a hylomorphic account of sensation. Sensitive beings consist of matter suitably organized for the task, and the crucial point for vision is that, in order for the eye to potentially receive any color, its matter must itself be uncolored, that is, transparent.[32] Aristotle concludes that all sensitive parts of the body must be homeomerous (a direct composition of the four elements) and that the eye must be predominantly watery.[33] Vision occurs when our visual faculty takes on the forms of the colors of a substance without that substance's underlying matter, and in *De sensu* Aristotle gives a definition of color as "the limit of the transparent in a determinately bounded body."[34] All bodies are said to be transparent to some degree, and color is just what you get when the transparency in a body ends. The precise mixture of the fundamental color-contraries white and black (or light and dark) determines the specific color of a body, with mixtures in simple ratios producing the most pleasing colors.[35] Thus how a body's transparency ceases somehow determines its color, but Aristotle does not say much about how this works.[36]

In many mediaeval and renaissance Peripatetic theories of vision, particularly under the influence of Averroes, density and rarity were the fundamental building blocks for theories of color and vision. Aristotle states that air and water are not transparent *qua* air or water, but because "each of them has contained in it a certain substance which is the same in both and is also found in the eternal upper body."[37]

---

[29] Galen 1988, 471–473.

[30] *DA* 419a1–21, 424a20. All quotations of Aristotle refer to the 1984 Barnes edition. Aristotle 1984.

[31] For Zabarella's discussion of this in Aristotle, see *DV* 1.8, Zabarella 1617, 874 C-D.

[32] *DA* 418b27.

[33] *PN* 438a13-438b6; *PA* 647a14.

[34] *PN*, 439b1–14.

[35] *PN*, 439b20–440a6.

[36] A more detailed account on color in Aristotle is Sorabji 2004.

[37] *DA*, 418b5.

Throughout his commentaries Averroes interprets this shared transparency in terms of the density and rarity of the underlying body, be it celestial and simple, or sublunar and thus a composition of the elements. He is most explicit in his commentary on *De coelo* and his *De substantia orbis*; in both of which he writes that water, air, and fire, as well as the celestial body, are rare and thus by default transparent. When condensed their transparency comes to an end, generating the color white, and in fire and the celestial body this also causes them to be luminous.[38] Earth, on the other hand, is naturally dense and black. Colors arise from a mixture of the darkness of the earth and the whiteness or transparency of the other elements, with earth playing a dual role as both one extreme on the color-contrary scale of colors as well as a condensing agent for transparent substances.

In *De visu*,Zabarella systematically builds from this foundation given by Averroes, and he carefully respects the scholastic distinction between *lux*, the property of a luminous body, and *lumen*, the effect of that body in transparent media.[39] After citing examples given by Avicenna—such as the fact that we can see color coming through a hole even in a completely dark cave or room—Zabarella concludes that vision requires not only that the medium be activated by *lumen*, but that the colored bodies themselves also be actualized by *lumen*.[40]*Qua* visibility, a body is a mixture of rare and dense matter, and *lumen* affects colored bodies insofar as they have some degree of transparency. Elemental earth is thus black because it has no degree of transparency and so does not admit *lumen* whatsoever. (This is in sharp contrast with modern notions according to which the color black is due to the absorption of light, rather than its failure to admit it.)

According to Zabarella*lumen* and color are joined at the limit of transparency, and so *lumen* both activates the colors at surfaces of bodies and also subsequently illuminates the transparent medium as it propagates together with the species of color. This is how the crystalline humor, which would otherwise be lying in the dark, becomes actually transparent and thus capable of receiving colors.[41] While color and light are ontologically distinct properties, any ray analysis would apply identically to both. Zabarella requires crystalline humor to have just the right degree of rarity: transparent enough to allow color and light to pass through the body of the crystalline, but just dense enough to capture the images as they pass through.[42] This is a requirement found at least as far back as Alhazen.[43] The sensitive soul then makes an active judgment in the eye and carries this judgment back from the web-like tunic or *aranea* (a thin membrane surrounding the crystalline humor, now called

[38] Aristotle and Averroes 1562, 125r F, 125v H–L, 127r A; Averroes 1984, 91–92. The intertwined histories of density and rarity and vision in pre-modern natural philosophy has been largely ignored, but there is no space to explore it here.

[39] Zabarella's lengthy discussion of *lux* and *lumen* is at *DV* 1.4–1.7, Zabarella 1617, 867–874.

[40] *DV* 1.9–11. *Ibid.*, 876–881.

[41] DV 1.10. *Ibid.*, 880 F.

[42] DV 2.5. *Ibid.*, 900 E. "…quia propter perspicuitatem recipiunt lumen introrsum, & propter densitatem retinent, atque uniunt."

[43] Alhacen 2001a, 88. For a translation, see Alhacen 2001b.

the lens capsule), then through the substance of the retina and the optic nerve, after which it is presented to the common sense. Thus it is crucial that the *aranea*, the retina, and the inner portion of the optic nerve are all connected and of the same substance. Before the adoption of a retinal theory of vision, anatomists for the most part agreed that they were in fact connected. Felix Platter denied their connection in his anatomical work, but he was an outlier on this point, albeit an important one.

In the first book Zabarella's task is to determine the correct Aristotelian theory of light, color, and vision, and to resolve related disputes arising within the Peripatetic tradition. In the second book of *De visu* Zabarella aims to vindicate the Peripatetic account over competing theories of vision, particularly Galenic-style extramission theories held by contemporary physicians, and this is where we find Zabarella's full account of the structure and usefulness of the parts of the eye. Zabarella is highly critical of Galen, and for instance writes, "Galen, whether writing on vision or on other things, is unable to distinguish the medical art from natural philosophy."[44] It is in the context of refuting Galen and, especially, contemporary Galenists on the subject of vision that Zabarella gives his own account of the usefulness of the vitreous humor.

The only purpose that Galen gives to the vitreous humor in *Usefulness* is to provide nutrition for the crystalline humor. He says that the color of the vitreous is somewhere in between that of blood and the perfect transparency of the crystalline, and nature has made the vitreous this way so that it doesn't mar the clarity that is essential for the action of the crystalline.[45] Zabarella says that, *pace* Galen in *Usefulness*, the vitreous is not only clear but in fact more transparent than the crystalline, and he appeals to his own experiences with anatomical observation to support this.[46] Furthermore, Zabarella says that if Galen's account of the usefulness of the vitreous were correct then its transparency would be superfluous.[47] Nature, Zabarella says, can turn blood into any number of substances—nerves, milk, and semen, for example—and yet the resulting color is always generated without difficulty. Indeed, if blood can be turned into the vitreous humor, which is more transparent than the crystalline, nothing prevents nature from using blood to nourish the crystalline.[48] Following Aristotle's statements on generation and corruption, Zabarella also says that, in the act of nutrition, the nourishing substance (such as blood) must be smaller in quantity than the nourished (such as flesh or, in this case,

[44] *DV* 1.8. Zabarella 1617, 874 B. "Galenus enim tum de visione, tum de plerisque aliis rebus scribens, nescivit artem medicam distinguere à naturali philosophia: quum enim plurima ad naturalem philosophum attinentia constituere potius, quam exquisitae tractare debuisset."

[45] Galen 1968, 464. Notably, Galen does not mention this in *On the Doctrines*. Galen 1980, 459.

[46] *DV* 1.8. Zabarella 1617, 875 B–C. His refutation of Galen is developed at length in *DV* 2.5. "nam ut ego iudicare videns potui, [vitreus] est fortasse quadruplo, vel etiam quintuplo maior [quam crystallinus]; sed maxime clarus, & albus, & in hoc manifestissimum est, deceptum esse Galenum… deceptus etiam in eo est Galenus, quod dixit huius humoris officium esse, ut ex eo crystallinus nutriatur."

[47] *DV* 2.5. *Ibid.*, 901 D–F.

[48] *DV* 2.5. *Ibid.*, 902 A.

the crystalline humor).[49] Zabarella says in two places that the vitreous is "four or perhaps five" times the size of the crystalline humor. Far from nourishing the crystalline, because of the overwhelming difference in bulk the vitreous would convert the crystalline into its own substance, not the reverse.[50] Finally, Zabarella argues that Galen's theory of vision, according to which a visual power is sent out to comprehend things, fails to make sense of the shape of the crystalline humor, in particular the gibbous posterior.

Zabarella's criticism of Galen's account of the vitreous humor is threefold: (1) if the usefulness of the vitreous humor was to provide nutrition for the crystalline, nature would be acting without purpose in making the vitreous clear; (2) nature would also be acting against its own ends by making far too much of the stuff; and finally (3) the crystalline humor would be shaped as it is for no reason. We can see here a thinly veiled accusation that, on Galen's account, nature would be acting without foresight. Not coincidentally, this was Galen's favorite argument against earlier anatomists, such as Erasistratus, whom Galen attacked for claiming (but failing) to follow Aristotle.[51] Zabarella also repeatedly says that the shape, size, and transparency of the humors is clear to sense, thus impugning not only Galen's reasoning but also his skill at anatomical observation.

For Zabarella, the *officium* or purpose of the vitreous humor is to put distance between the crystalline humor and the colored tunics at the rear of the eye, the retina and the uvea (or choroid). If the body immediately behind the crystalline were not transparent, the *lumen* passing through the crystalline would join with the color of this body. Both would reflect back to the crystalline, and the color and image of these tunics would be perpetually combined with images coming from outside.[52] But the final cause of vision is not to perceive the inside of our own eyes, and so not only is a certain space necessary, but the illumination progressing through the crystalline humor needs to be prevented from reflecting back. For this reason, Zabarella says, in addition to acting as the seat of sensation the crystalline humor is shaped so that *lumen* is refracted upon exiting, causing the luminous rays to unite at a point just behind it. Zabarella compares this with what "experience teaches" us about burning glasses: the *lumen* behind the glass is formed into a cone, the point of which can kindle a flame. Beyond this point the *lumen* is weakened, and Zabarella says

[49] For example, *GC* 321a30-322b1.

[50] *DV* 2.5. Zabarella 1617, 901 F–902 B. "quia mutatur in ea generatione color in colorem conveniente rei generandae, vel nutriendae, quod praestare sagacissima natura, quando ita expedit, facile potest. Est etiam absonum rationi, quod tanta moles, quanta est humor vitreus, crystallino ad eius nutritionem tradita sit; nam multo maior est crystallino vitreus, ac si sensui credimus, est quadruplo, & fortasse quintuplo maior, videmus autem in omnibus alimentum esse re nutrienda longe minoris quantitatis, idque omnino necessarium est: quia, quum alimentum in principio sit contrarium, & cum re alenda pugnet, si maius esset, opprimeret eam potius, quam aleret, tanquam validius, quoniam in maiori corpore vis maior inest; potius igitur natura crystallini in naturam vitrei mutaretur, quam è contrario."

[51] Von Staden 1997.

[52] *DV* 2.5. Zabarella 1617, 902 C.

that "the cone, extending to a peak, does not go past a certain determinate point."[53] Thus, the process of focusing *lumen* together into a point appears to weaken the *lumen* beyond that point (a notion that we will also see emphasized by Fabricius) and the true purpose of the vitreous humor is to facilitate this debilitation. Although space does not permit a full investigation of this curious idea, it is important that Zabarella's model for the inanimate inner-workings of the eye is a burning glass rather than a *camera obscura*. This makes sense if we keep in mind that, as Mark Smith succinctly puts it, "Alhacen and his medieval Latin followers were far more concerned with making sense of sight than with understanding light."[54] Before the seventeenth century optics was concerned with correcting mistakes in judgment about the location and shape of bodies due to refraction and reflection, and not with understanding what we would call real or projected images.[55] We must resist the urge to think of images being somehow projected onto the crystalline humor as if it were a screen. Insofar as mathematical optics did treat light itself it was in the context of burning mirrors and burning glasses, although this began to change in the sixteenth century in the context of practical rather than theoretical optics.[56] Thus placing a burning glass in the eye was an innovative conceptual move, but from the point of view of the history of vision it wasn't radical.

We can make some sense of Zabarella's account of what happens to *lumen* due to a burning glass by looking at his explanation for why *lumen* from the Sun heats the earth. When a ray of *lumen* strikes a body perpendicularly and reflects directly back, he says, there will be a ray ascending and a ray descending, and "from the collision of the two rays the air is thinned and made more hot."[57] It seems that the converging rays in the vitreous are debilitated through collision, but because they collide in water (which, unlike air, is not naturally receptive to heat) no flame is generated.[58] Indeed, this is a crucial reason why the vitreous is watery. Water and

---

[53] *DV* 2.5. *Ibid.*, 902 D–F. "hoc est absque dubio vitrei humoris oficium; nam experientia docet, lumen transiens per vitreum aliquod cavum uniri in illa cavitate, & permeans ultra vitrum in quadam certa ab eo distantia facere conum, in cuius extremetate intensissimum lumen apparet, sed minimae quantitas instar milii, nempe, si in illa certa distantia ponatur corpus aliquod solidum, in quod angulus impingat; nam si propinquius vitro corpus illud ponatur, maiore eius pars illuminabitur, & eo maior, quo sit propinquius vitro; at si paulatim removeatur, minuetur continue, donec ad minimam superficiei illuminatae quantitatem perveniat, ideo in illa minima quantitate ita est unitum & validum illud lumen, ut etiam accendat, & urat, quoniam ibi definit conus, & angulus a concursu radiorum productus; ideo si adhuc magis removeatur corpus illud, nullum amplius lumen ab illo vitreo ad ipsum pervenit, sed exinanitum, quia quum ad conum, & ad acumen tendat, non praetergreditur quoddam determinatum punctum."

[54] Smith 2004, 181.

[55] Dupré 2006, 2008.

[56] Dupré 2005.

[57] *De calore coelesti* chapter 10; Zabarella 1617, 574 F–575 B. "proiecti nanque radii Solis in terram resiliunt à terra refracti, & in aere duplicantur, nimirum descendentes, atque ascendentes, & ex radiorum inter se collisione extenuatur aer, & calidior fit: credendum quidem est radios Solis etiam rectà proiectos, ac simplices aliquid caloris efficere."

[58] *De calore coelesti* chapter 3. Zabarella 1617, 559–562. The relationship between heat and motion was an important and contested issue in the seventeenth century. Zabarella's treatment here is quite involved, and my short discussion necessarily omits a great deal.

fire are contraries—the former is cold and wet while the latter is hot and dry—and because of this water cannot become receptive to the form of fire by rarefaction alone.

Not only does Zabarella compare the crystalline humor to a burning glass, he mentions a dissection showing these properties of the eye.

> I saw the crystalline humor separated from the other humors in a dissection of the eye, which when placed near a small lit candle was made completely lucid, and gleamed just as if imbued with the *lumen* of the candle on account of its perspicuity. And the *lumen* traveled across the entire substance of the crystalline humor, and in the posterior part of the crystalline humor it turned into a cone, and into a peak not much beyond the bulge of the crystalline humor, so that that peak and the running together of the lines stood apart very little from the crystalline humor—indeed almost seemed to touch [the crystalline humor] itself. Therefore it is certain that the peak of that cone is exhausted (*exinaniri*) in the vitreous humor, which has a great depth, and thus is not able to reach the posterior tunics.[59]

Zabarella stresses the shape of the crystalline humor, in particular the gibbous back end that facilitates refraction. His novel account of the *usus* of the vitreous humor also hinges on it being less refractive than the crystalline humor. Although he does not analyze vision mathematically, Zabarella's account is in some respects consistent with Alhazen and his Latin followers. But whether or not he was influenced by mediaeval optics, Zabarella's account becomes incompatible with this tradition once the rays pass through the crystalline humor. Alhazen, Witelo, and Pecham all require the visual image to remain upright as it is funneled through a supposed hole in the optical nerve and carried to the common sense. In the perspectivist tradition the visual cone formed from incoming rays needed to be suitably refracted upon passing from the crystalline humor to the vitreous humor (See Fig. 4.3).

Alhazen's theory of vision cannot be explained at length here, but some discussion is necessary. It is assumed that every part of a body emits rays in all directions, but in the visual field only one ray from each point meets the spherical cornea at a right angle. Arguing that oblique rays are weakened by refraction, Alhazen and his mediaeval followers constructed a pointwise one-to-one map between the thing seen and the eye by positing that the visual faculty somehow distinguishes between orthogonal and oblique rays.[60] Alhazen also requires the rays that primarily affect sight to enter the crystalline humor orthogonally, and so the surfaces of the cornea and the crystalline must form portions of concentric spheres. After entering the crystalline, the resulting image is sent upright through an aperture in the optic nerve,

[59] *DV* 2.5. Zabarella 1617, 903. A–B. "Ego igitur in oculorum sectione vidi crystallinum ab aliis humoribus separatum, cui quum accensa candelula apponeretur, totius fiebat lucidus, & splendens tanquam candelae lumine imbutus ob suam perspicuitatem, & trans totam crystallini substantiam meabat lumen, & in posteriore crystallini parte transibat in conum, & in acumen, non multo post intimam crystallini gibbositatem, ita ut acumen illud, & linearum concursus parum distaret à crystallino, imo ipsum fere attingere videretur; ideo certum est, illius coni acumen exinaniri in humore vitreo, qui magnam habet profunditatem, ideoque ad posteriores tunicas pervenire non posse."

[60] Alhacen 2001a, lx–lxi, 26–43; Lindberg 1976, 71–80. Note that for Alhazen the *glacialis* and the vitreous are not always treated as distinct humors, but sometimes separate regions of a single humor. For the edition that Zabarella and Fabricius would have examined, which differs in some important ways from Latin mediaeval manuscripts, see Alhazen and Witelo 1572.

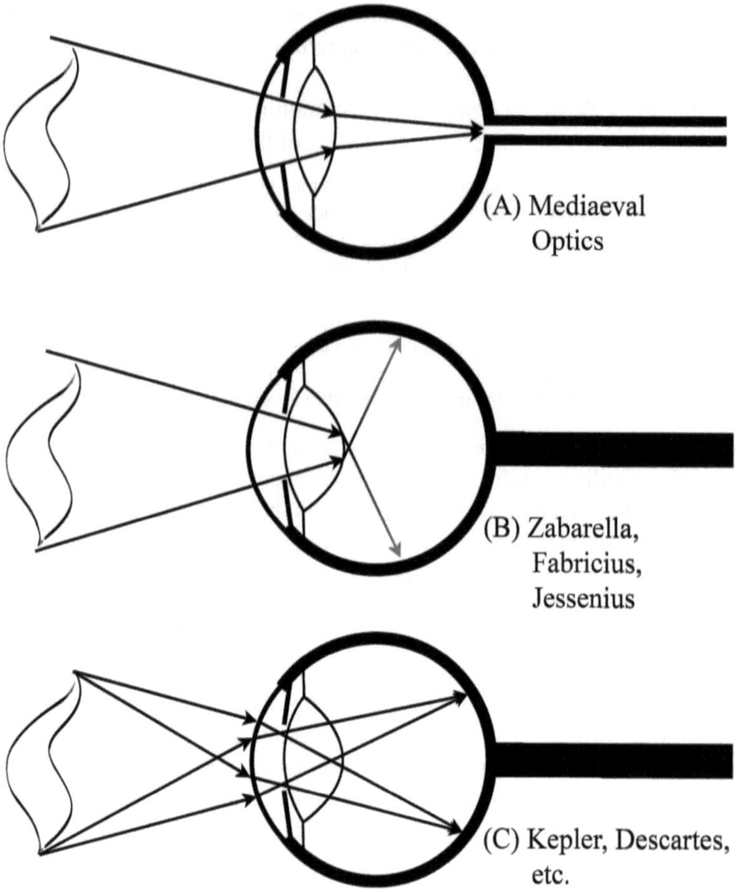

Fig. 4.3 Simplified representation of three main theories of the refraction of rays in the eye: (**a**) upright image (here of a candle flame) entering the aperture of the optical nerve, (**b**) weakening and dispersion of light in the vitreous, and (**c**) pencils of rays converging on the retina. No attempt was made to accurately portray the shape or location of humors in (**a**) (that is, the visual theory of Alhazen, Witelo, Pecham, and their followers) as this varied considerably depending on the manuscript or printed text. The multiple refraction of rays due to the various humors in (**b**) and (**c**) are also ignored

after which the image is carried along the twisting path of the nerve by the visual spirits, whose powers allow for non-rectilinear path propagation of rays.[61] According to the discussion of refraction that Alhazen himself presents, for the rays to be appropriately refracted after exiting the crystalline humor the vitreous would have to be denser than the crystalline. Yet Alhazen never seems to say whether the crystalline or the vitreous is more dense,[62] and indeed he says that the humors of the

[61] Alhacen 2001a, lxi, 83; Lindberg 1976, 80–85.

[62] Lindberg 1976, 244 n. 106.

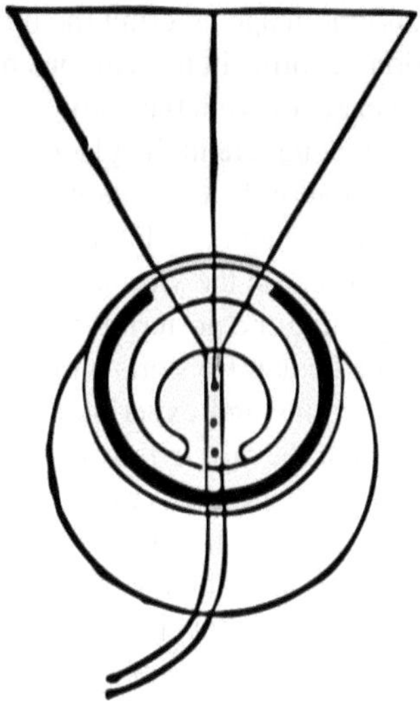

Fig. 4.4 The visual cone in some sixteenth-century editions of Pecham's *Perspectiva*. Note the refraction of the rays funneling the image into the hole in the optical nerve, typical of the perspectivist account of perception. The points, from top to bottom, represent the centers of: (*1*) the cornea and crystalline, (*2*) the vitreous, and (*3*) the consolidativa (the outer tunic). From Peckham 1592, 13v. The same diagram is found in Peckham 1504. Note that the Nurenburg edition gives a biconvex crystalline, as in Fig. 4.3 above. See Peckham 1542

eye refract light differently than other transparent media because of a "receptivity of sense," a capacity that differs between the humors.[63] Thus, on Alhazen's account the refraction that occurs within a living eye cannot be discovered through experiments on the parts of an anatomized eye. A dead eye has no visual spirits and no sensitive power, and so would refract color and light differently compared to a living one. Zabarella, on the other hand, has the visual faculty make an active judgment at the crystalline humor itself, and this judgment (rather than an optical image) is carried back through the aranea, retina, the optical nerve, and to the commons sense. At no point does the visual faculty alter the path of rays. This is a crucial difference compared to Zabarella's—and Fabricius's—understanding of the properties of the humors.

Mediaeval perspectivists writing in Latin followed Alhazen's scheme for the refraction within the eye.[64] As Lindberg says of Pecham, "Once inside the vitreous humor, the rays are no longer bound by the laws of transparency".[65] Some (but not all) printed editions of Pecham's *Perspectiva* depicted an eye that is at odds with the eye revealed through dissection (see Fig. 4.4), but even in these diagrams the required refraction demands an intervention of the sense faculty.[66] Likewise, for

[63] Alhacen 2001a, lxii, 51–52, 79–97; Lindberg 1976, 82–3.

[64] Peckham 1970, 118–119.

[65] *Ibid.*, 38.

[66] How to interpret Pecham on this point is not always clear. See *ibid.*, 118–119.

Witelo vision involves an upright image entering the hole in the optical nerve, and he says "the kind of reception of forms in the vitreous humor along refracted lines is due to its difference in transparency from the body of the glacial and to the quality of sensible reception that is not complete in the glacial humor."[67]

For the *perspectivae*, the action of the eye follows from the demands of a visual theory in which an upright image enters a hole in the optical nerve. Zabarella reasons in the opposite direction: because the vitreous humor is manifestly less dense than the crystalline, the rays exiting the posterior of the crystalline humor must be refracted away from the perpendicular. He starts with the material properties of the eye, determined through dissections and experiments that he has witnessed, and from this he constructs the path of the rays in the eye. In order to fit this path of rays into his theory of vision, and to refute Galen's extramissionist account, he says that the true purpose of the vitreous humor is to act as sort of light dampener. Zabarella's *usus* accounts for not only the great size of the vitreous humor compared to the other humors, but also the shape of both the flattened front and the gibbous posterior of the crystalline and the relative optical density of the humors. Indeed, he accounts for almost every sensible aspect of the humors, observed personally but in alignment with the most up-to-date anatomical accounts at the time. Furthermore, his account of the *usus* of the vitreous humor does not appeal to any properties of the humors that depend on the faculty of vision, properties that would necessarily be unobservable in dissection (and vivisection in this case would be highly impractical). For the purposes of ray analysis—which is qualitative—the eye is equivalent to a dead eye. Although his model for vision is not the *camera obscura* as a device that projects a picture or real image, Zabarella does have a different dark room in mind drawn from experiences listed by Avicenna. The crystalline humor is conceived of as a semi-transparent body with omnidirectional sensitivity placed in a dark room or cave: light and color penetrate the darkness, but if too much of the interior space is illuminated the image entering from outside would be continuously merged with the colors of the walls. To deal with the problem of excess illumination another sort of optical device, a burning glass, is appealed to.

4.4 Fabricius, Dissection, and Visual Theory

The same account of the purpose of the vitreous humor given by Zabarella was also advanced by Fabricius ab Aquapendente, who was his exact contemporary. Born in 1533, his career at Padua lasted over 50 years, and for at least 25 years his time at the university overlapped with Zabarella's. Fabricius's *De visione, voce, auditu* was first published in Venice in 1600 as the first part of his ambitious anatomical project, the theatre of the whole animal, which was to be published in many installments.[68]

[67] Witello 1992, 319. On transparency of the humors in Witelo, see *ibid*, 105, 128–129. See also Unguru's assessment at *ibid.*, 216 n. 8.

[68] Cunningham 1985.

His goal for this long-term project was in part to surpass Vesalius in accuracy, quality and number of images, and depth of philosophical treatment. Indeed, the degree to which Fabricius integrates anatomy, natural philosophy, and mathematical optics in his *De visione* was unprecedented in any genre, and his other books published together with *De visione*, on hearing and the voice, are similarly ambitious. Yet as much as Fabricius aims to expand the scope and prestige of the anatomical genre, these three books are still ultimately anatomical treatises, and each is divided, according to a Galeno-Aristotelian scheme, into three sections: *historia*, *actio*, and *utilitas*.

Scattered throughout *De visione* are numerous references to a theory of color and light consistent with Zabarella's.[69] There are, however, a few notable differences. Fabricius explicitly says that he does not make the distinction between *lux* and *lumen*,[70] and he treats color as an affection that light picks up from bodies rather than a separate quality propagated from the surface of bodies.[71] Fabricius's conception of light also has Neoplatonic influences, and he cites approvingly Plotinus's Ennead 4, book 5, chapter 6 on the nature of light.[72] However, given that *lumen* and color are joined and propagate together according to Zabarella, these differences do not result in a structurally different account of vision. Furthermore, Fabricius follows Aristotle's account in *De sensu* on the relationship between transparency and color.[73] As with Zabarella, density and rarity underlie accounts of luminosity, transparency, and color, but as a practicing physician he harmonizes this theory of the origin of colors with the more practical concerns of diagnosis through temperament, with theoretical notions of the color and consistency of the humors, as well as with some important issues related to animal generation.[74]

In his section on the *historia* of the eyeFabricius mentions that the vitreous humor is more than four times the magnitude of the crystalline, a figure that he repeats in the *utilitas* section—the same figure that Zabarella uses.[75] To my knowledge the only other writers before Fabricius, apart from Zabarella, to give quantitative estimates of the vitreous humor are Colombo and his student Valverde. They give

---

[69] See, for example, his discussion of the color of the crystalline humor due to age or boiling and its connection to the elements at Fabricius 1600, 12.

[70] *Ibid.*, 48.

[71] "Atque haec sola lux existit, quae si colorem corporis attingat, coloratur;" *Ibid.*, 40. The collapse of light and color in the seventeenth century was a radical conceptual change in the history of vision and demands far more analysis than has hitherto been given, but there is no space for anything like an adequate discussion here.

[72] *Ibid.*, 41.

[73] *Ibid.*, 44–45.

[74] For this last point, see his discussion of the parts of the egg in Fabricius and Adelmann 1967, 215, 220–221.

[75] *Ibid.*, 109. "vitreus enim quadruplo, & amplius Crystallino est copiosior, quae res omninò Crystallinum à tunicis quàm maximè distare facit, ut scilicet omninò lux in tanta vitrei amplitudine prius evanescat & obumbretur, quàm ad tunicas pertingere, atque ab ipsis reflecti possit: atque hoc ita evenire, si Crystallinum & vitreum adversae luci opponas, facilè conspicies. See also page 13, where the vitreous "ferè quadruplo crystalloidem exsuperans."

different figures, however, and they also do not compare the size of the vitreous to the crystalline, but rather to the interior chamber as a whole.[76] Furthermore the ratio Zabarella and Fabricius give is incorrect: in humans it is closer to 16:1, and although it varies from species to species, for the animals Fabricius is most concerned with (human, sheep, cow) the estimate of four or five to one is far too low.

In part III, chapter 10 Fabricius discusses the *utilitas* of the vitreous at length. He rejects the account of Alhazen and Witelo, in which vision is perfected only once the image enters the supposed aperture in the optical nerve, as "obscure," implying philosophical as well as sensible, anatomical problems with the account.[77] He rejects Galen's belief that the vitreous humor merely nourishes the crystalline, and as with Zabarella this is because he observes that the vitreous is the most pellucid of all the parts.[78] Fabricius says that one *officium* of the retina is to prohibit contact between the choroid tunic (or uvea) and the vitreous humor so that the choroid does not spoil the "the exceedingly pure vitreous substance" (*purissima vitrei substantia*) with its dark color.[79]

Fabricius's opinion on the usefulness of the vitreous humor and his reasons justifying it are also identical to Zabarella's. He writes, "my opinion can be easily followed, if first it is imagined that the vitreous humor, or something diaphanous, were not next to [the crystalline humor]."[80] If this were the case, something colored would be there and it would be like shining a light against a colored wall: the light would return to the crystalline humor tinged with the color of the tunic. Nature, therefore, had to put some distance between the posterior tunics and the crystalline humor so that light passing through the crystalline can "disperse and disappear, and be prevented from reflecting back."[81] This is why the vitreous humor is so large, and the crystalline humor so far away from the retina. Additionally, the reason the posterior of the crystalline humor protrudes is, as with Zabarella, so that light can be united in the vitreous directly behind the crystalline humor, thereby debilitating the light.[82] He writes:

[76] Valverde and Colombo both say that the vitreous occupies three-fourths of the interior. Valverde 1556, 82: "Este umor llamaron los Griegos Udatoydes, los Latinos Vitreo, el qual occupa las tres partes del huesco del ojo." Colombo 1559, 219: "Neque ibi solum sed anterioris quoque non exiguam portionem, ita ut ex quatuor oculi partibus tres occupet hialoides [i.e., the vitreous]."

[77] Fabricius 1600, 107.

[78] *Ibid.*, 105.

[79] *Ibid.*, 106.

[80] *Ibid.*, 107. "Id autem mea sententia facile assequemur, si primo vitreum, aut eius diaphanum non adesse imaginemur. Quod si diaphanum post crystallinum non esset positum, necessario opacum collocari corpus oporteret [...] igitur retinam, & Choroidem, crystallinum attingere necessarium esset: indeque lux crystallinum transgressa, & has tunicas, quasi coloratum parietem pertingens, pertundensque ac tunicarum coloribus, ob contactum affecta, foedataque denique retro ad crystalloidem reflexa crystallinum potius tunicarum nativis coloribus afficeret, quam extrinsecus assumptis, sine ulla sensus videndi utilitate."

[81] *Ibid.*, 107. "Neque hoc loco illud est astruendum in crystallino, lucem dispergi, & evanescere, & ita reflexum prohiberi."

[82] *Ibid.*, 110. "Crystallini postica extuberantia, quae lucis unionem in vitreo prope Crystallinum finiri cogit."

> What, therefore, will be the proposed usefulness of the roundness [of the posterior of the crystalline humor]? It is surely, in my opinion, so that the light carried past the crystalline should be united into itself and not progress very far from the crystalline, but cease in the vitreous, and in a certain way perish.[83]

Thus we see that this curious notion that being brought into a cone debilitates light is also found in *De visione*, and Fabricius also has the crystalline act like a burning glass. Zabarella, we have seen, makes the argument that the collision of rays of lumen produces heat in air but not water because the former is hot by nature, the latter cold. Likewise, for Fabricius the primary *utilitas* of the watery nature of the eye is precisely to prevent the kindling of flame in the eye.[84]

Zabarella and Fabricius stayed in Padua almost their entire lives and they certainly interacted in some capacity. Given that Zabarella is not known to have travelled beyond Venice, and even there only rarely, it is certain that he observed his dissections in Padua, and because Fabricius was the public lecturer in anatomy from 1566 onwards it is likely that Zabarella attended one or several of Fabricius's public demonstrations. As Cynthia Klestinec points out, in these public dissections Fabricius focused on natural philosophy rather than the art of dissection.[85] Furthermore Fabricius's interest in visual theory, and certainly his heavy citation of the *perspectivae*, was rare for an anatomist of his time. The common points in their theory of vision that I have been unable to find in any previous author are as follows: (1) that the vitreous is four or perhaps five times the size of the crystalline (which is in fact incorrect); (2) that the rays of light need to refract into a cone just behind the crystalline humor, which is why nature has made the rear gibbous; (3) that the vitreous is more transparent and thus less optically dense than the crystalline in order to cause this cone; (4) that, contra Galen, the vitreous is clear because otherwise the color of the vitreous would reflect back and interfere with the crystalline humor; (5) that the uniting of rays into a cone debilitates and exhausts light; (6) that there is such a great quantity of vitreous humor in order to provide large chamber for this cone to occur, and so that the colored retina and uvea are far from the crystalline; (7) that there a sort of burning glass in the eye, and thus the primary *utilitas* of the watery nature of the vitreous is to prevent burning within the eye. Furthermore, while not necessarily unique to them, both have the visual spirits make a judgment at the aranea that is carried back through the retina and optical nerve.

It is possible that one developed a theory of vision independently and the other copied it without attribution. On the other hand they may have developed the theory together (even if not consciously), perhaps with the involvement with others present in Padua during discussion and public disputation. Disputation, it should be emphasized, was an expected part of public anatomical demonstrations. Their theory of

---

[83] *Ibid.*, 102–103. "Quae igitur erit propositae rotundus utilitas? Ea certe, mea sententia, ut lux crystallino transvecta, tum in seipsa uniatur, tum longius à crystallino non progediatur, sed in vitreo cesset, ac quodammodo commoriatur."

[84] *Ibid.*, 60. "Aqueus deinde est non aereus; ut intus facilè contineri vicissimque continere visilium formas possit, neque lux vehemens intus in oculo ignem accendere valeat."

[85] Klestinec 2011. For student records of Fabricius's ocular dissections, see Favaro 1911–1912, vol. 1, 227; vol. 2, 32.

vision could have been formed between them privately as well. Finally, it is possible that there is some earlier source that both Zabarella and Fabricius are drawing upon and which neither acknowledged. Absent further evidence it is by far most likely that the theory arose out of their interaction: Zabarella clearly attended many different anatomical demonstrations, and these were most likely performed by Fabricius. Indeed, as a fellow professor Zabarella would have been granted front row access to the annual dissections. Any common source for their theory would have to have been a contemporary who did not discuss their theory in print, and whose anatomical knowledge was as up-to-date as theirs. Zabarella, as we have seen, says that many physicians followed Galen's extramission theory, and so the pool of potential sources excludes many physicians and anatomists. In all likelihood this was a Paduan theory, developed within a culture of anatomical demonstration and public disputation (which doesn't exclude the informal exchange of ideas), and perhaps held by other professors as well. Nevertheless, Zabarella and Fabricius have the best claim to it. Although Zabarella revealed it in print ten years before Fabricius, what Fabricius was teaching before 1590 has not been investigated, and so absent further evidence it seems most appropriate to consider it a shared theory. Indeed, they both had a hand in disseminating it throughout Europe via their texts as well as their students, although Fabricius seems to have been more influential in this regard. The Jesuit François de Aguilón cites Fabricius several times in his *Opticorum libri sex*, and his account of the structure and *utilitas* of the eye and its parts is essentially a paraphrase of *De visione*. Aguilón also gives the same account of shape of the crystalline and the size vitreous causing the rays of light cross and perish in the vitreous so that "the stain is not reflected from the opaque and colored body of the retina to the crystalline."[86] Another follower of the theory was Jan Jessenius, one of Fabricius's students who began his studies in Padua in 1588 and moved to Prague in 1600.

4.5 Kepler and Paduan Anatomy and Visual Theory

In his brief *Anatomia Pragensis* of 1601 Jessenius draws heavily on his teacher. Indeed, his section on the eye functions as a paraphrase Fabricius's *De visione*, from the description of the shape and clarity of the humors, to the usefulness of the vitreous humor, and even the statement that the vitreous humor is four times the size of the crystalline.[87] Given how closely together they were published, is quite possible that Jessenius's account of the eye is based on notes from Fabricius's anatomical

[86] Aguilón 1613, 6. "Vitreum autem post crystalloidem natura collocavit, ut si quid luminis crystallinum praetergressum fuerit in eo hebetetur, ne, ut iam antè dictum est, ab opaco coloratoque retinae corpore foedatum ad crystallinum reflectatur." He also says that crystalline is dense and protrudes in the rear "ut lux in ipso commoriatur, ne longiùs progressa vitreumque praetervecta, ad retinam redeat, ab eaque ad crystallinum resiliens nova affectione visum perturbat."

[87] Jessenius 1601, 113r-126v. An interesting recent paper concerning Zabarella and Fabricius on the body, which also contains a short (but unfortunately rather superficial) discussion of Kepler in this context, is De Angelis 2008.

demonstrations rather than Fabricius's recently published text. We know Kepler did not have access to a copy, and it is likely that Jessenius did not have one in Prague either.

Jessenius and Kepler knew each other well, and in his *Paralipomena* Kepler mentions that he gathered all his knowledge of ocular anatomy from Felix Platter's *De corporis humani structura et usu* and the *Anatomia Pragensis* of his friend Jessenius. Kepler refers to the latter in part because of Jessenius's own anatomical efforts, but also because "he professed chiefly to follow Aquapendente."[88] The shapes of parts of the eye, the relative optical densities of the two humors, and the crossing of rays in the vitreous humor are all critical for Kepler's theory of vision, yet Kepler claimed that he "never before had been either spectator or performer" at an anatomical dissection of the eye, trusting instead in the expertise of Platter and Jessenius.[89] Much of what Kepler takes for granted is absent from Platter's work, and so the visual theory of Zabarella and Fabricius had some influence on Kepler's revolutionary treatise, contrary to the frequent dismissal of the role played by anatomists and scholastic natural philosophers regarding visual theory.[90]

Comparing the theory of vision just given to Kepler's reveals important continuities and some crucial differences. Kepler provides the physical and causal foundation for his theory in the first chapter of the *Paralipomena* and an appendix to it; there he develops his own novel theory of color and light and refutes Aristotle on this. Importantly, he refutes Aristotle himself, and does not seriously engage with the much more complex Aristotelian accounts given by contemporary natural philosophers like Zabarella. Kepler's keen attention to the shape of the posterior of the crystalline humor, and the importance of the fact that the vitreous is less dense and thus more transparent than the crystalline, was clearly derived from Jessenius (whose account is entirely from Fabricius) rather than Witelo or Platter.[91] Jessenius's *Anatomia Pragensis* contains no images, and so Kepler famously reproduced images from Platter. However, besides these anatomical plates the latter contains only four (folio) pages of text spread out in diagrammatic form, and thus relatively little detail about either the action or *utilitas* of the parts of eye. Jessenius has far more text on the eye, and we can observe Kepler following Jessenius's much more detailed account of the shape of the humors, their translucency, and their refractive powers. Concerning the posterior of the crystalline humor, Kepler writes, "In fact Jessenius reports that [the posterior of the crystalline] is not spherical, as Platter

[88] Kepler 2000, 171–172.

[89] Translation from Kepler 2000, 171–2; see also Kepler 1604, 158–159.

[90] To give one more recent example Lefèvre 2007, 55–56: "No anatomical discoveries fed into this [i.e., Kepler's] model: a seventeenth century anatomist's knowledge of this organ did not differ significantly from that of a fifteenth century artist-anatomist like Leonardo da Vinci (1452–1519)." See also notes 3–7 above.

[91] In Witelo the shape of the rear of the crystalline is unclear, while Platter says that the posterior of the crystalline is "sphaericus," which Kepler does not follow. Furthermore, Platter says that the vitreous is "aequè splendidus ac crystallinus, sed mollior", though he never explicitly refers to the density/rarity or thickness/thinness of the vitreous, which terms typically denote refractive power. Platter 1583, 187.

asserts, but that it protrudes greatly (*valdè protuberare*), and is made oblong, as if rising into a cone".[92] Kepler uses this to argue (without having seen a dissection) that the posterior is a hyperbolic conoid, a shape that satisfies the requirements of his mathematical account. In the *Anatomia Pragensis* Jessenius, speaking of the *usus* of the crystalline, does indeed say that it "protrudes greatly (*valdè protuberat*)", adding "so that the light transmitted through the crystalline should be united into itself and not progress very far from the crystalline, but disappear at the crystalline, and in a certain way perish."[93] This is exactly what we read in Fabricius's *De visione*, and thus we can see that the very text that Kepler cites on the shape of the crystalline humor is appropriated from Fabricius and conveys an important and novel aspect of his and Zabarella's theory of vision. Certainly, Kepler's combination of experience with optical devices, his Neoplatonic concept of light, and his mathematical acumen led him to reject the notion that the light "perishes" just behind the crystalline, from which it follows that light and color would be cast upon the retina—the very thing that Zabarella, Fabricius, and Jessenius wished to avoid. However, among all of Kepler's sources only Jessenius describes rays of light forming a cone within the vitreous due to the difference in refractive power of the humors. Kepler makes much of the 1589 edition of della Porta's *Magia naturalis*, where both the *camera obscura* and vision are separately discussed, but he laments della Porta's failure to connect the two.[94] In fact, his *De refractione* (of which Kepler says he was unable to find a copy) della Porta assumes that vision takes place at the anterior of the crystalline humor, and he ignores what happens beyond.[95]

It is well accepted that Platter's comment that the retina is visually sensitive influenced Kepler's theory of vision. The shapes and sizes of the humors, and that their differing refractive powers cause incoming rays from different points in the object to cross in the eye, are crucial features in Kepler's theory of vision, but for this information he relies entirely on his anatomical authorities. Apart from the shape of the crystalline (about which Kepler explicitly follows Jessenius), Platter says nothing about these things, and so Kepler's source for this empirical information is Jessenius—and thus ultimately Fabricius.

Importantly, Zabarella and Fabricius (and following them Jessenius) believe that the refraction that occurs within a living eye can be exactly demonstrated by doing experiments on a dead eye. This is a significant break from nearly all past writers. As we have seen, for the *perspectivae* the visual faculty actively refracts the light beyond the crystalline. For Galen and other extramissionists a luminous pneuma

[92] Kepler 1604, 167. "Sic enim refert Jessenius, non sphaericum esse, quod Platterus aiebat, sed valdè protuberare, & oblongum fieri, quasi in conum assurgat: anteriore verò facie, depresse esse rotunditate."

[93] Jessenius 1601, 117v. "ita posteriùs valdè protuberat, eo fine ut lux crystallino transmissa, cùm in seipsa uniatur, tum longius à crystallino non progressa, illicò in crystallino evanesceret, & quasi commoreretur." This is repeated and emphasized at 124v–125r.

[94] Kepler 2000, 224–226; Kepler 1604, 209–211.

[95] Porta 1593, 83–86; Frangenberg 1991, 153. For the account given by Francesco Maurolico in his *Photismi*, see Frangenberg 1991,147–150.

fills the front of the eye, but this immediately dissipates upon death. Alan Shapiro emphasizes the fact that "Kepler treated the eye as an optical instrument without any active powers—a 'dead' eye,"[96] but we can see Zabarella and Fabricius doing the same. However, in contrast to the refraction of rays coming to a point in the vitreous in the manner of a burning glass, for Kepler this refraction of the rays results in the projection of an inverted *pictura* upon the retina.[97]

Most analyses of Kepler's *Paralipomena* stress the analogy of the eye to a *camera obscura*, and this is no doubt justified. Using this analogy and applying his expertise in mathematics Kepler was able to account for not only a single ray originating from each point in the object, as had past perspectivists, but an innumerable quantity of rays issuing forth from every point on the visible body into the eye in the form of a cone. These cones (or pencils of rays) then combine again through refraction into a single point on the retina. This is Kepler's double cone model, taken up by Scheiner and Descartes, and it is the basis for modern visual theory. Furthermore, Kepler's experience with the *camera obscura* led him to make a distinction between a *pictura*, or a real projected image, and an *imago*, a perceived image that is the product of the imagination (e.g., the image seen in a mirror or crystal ball).[98] However, in placing the seat of vision at the retina Kepler had a problem: the retina is, he says, colored.[99] Without the elaborate theory of light and color developed in Kepler's first chapter—a theory that is finely tuned to solve this problem—the possibility that a colored body could be the seat of visual sensation would have been ridiculous. As it is, however, Kepler's solution appears *ad hoc*. He writes that the retina "is said to resemble the substance of the cerebrum, but to be more mucous and reddish (bluish, according to Jessenius), whence one concludes that it seems to be above all a diluted white tinged with redness or blueness."[100] After describing how each point of an object is resolved into a single point on the retina, he refers his reader to chapter 1. There Kepler writes:

> There follows hence a kind of corollary to Props. 30 and 31: that the rays that have flowed to black surfaces are perceived most distinctly, and to white ones most evidently; and if a surface be a mean between black and white, such as blue, white washed with red, and the like, they will stand about equally in rendering both the individual colors and their differences.[101]

---

[96] Crombie 1967, 54–55. Shapiro 2008, 310.

[97] On this crucial point, Sven Dupré writes that, prior to the sixteenth century, the "*punctum inversionis* was not used in the perspectivist tradition of optics. Rather, this point was regarded as either the point of inversion or the point of combustion, but it fell outside the conceptual framework of perspectivist optics that this point could possibly be the locus of both." Dupré 2012, 515. For more on the conceptual framework of mediaeval and renaissance perspectivists and their failure to treat image projection, see Smith 2005.

[98] Shapiro 2008; Dupré 2012.

[99] Kepler 2000, 185; Kepler 1604, 175.

[100] Kepler 2000, 178; Kepler 1604, 166.

[101] Kepler 2000, 38–39; Kepler 1604, 25.

What would otherwise be an embarrassing lack of cooperation by nature as revealed through anatomical observation is reframed four chapters earlier, in the foundational material of his work, by positing a theory of light and color in which surfaces tinted either red or blue are ideally suited to image projection. If the retina turns out to be reddish, as Platter says, or bluish, as according Jessenius, no matter—either color will do just fine. It is also convenient that Kepler chose Platter's position that the aranea and the retina are not connected. The "entire opinion [that the crystalline is the seat of sensation]… is knocked down when the crystalline is cut off from the nerve and from the retina, and joined with the uvea, as was shown from Platter."[102] Kepler was no authority on this matter, and so he ignored authorities that did not support his theory.

Kepler conceived of the activity of the objects of vision—i.e., light and color—as points that have become super-rarefied into two-dimensional surfaces, propagating outwards from luminous and colored bodies in all directions instantaneously.[103] These sense objects are not received by a three dimensional body, but ultimately by super-rarefied spirits at a two dimensional boundary.[104] In addition to solving the problem of how the colored surface of the retina can be the seat of sensation, Kepler claimed that his characterization of light as two-dimensional also gave a better causal understanding of reflection and refraction.[105] What Kepler is demanding of his readers, however, is to abandon their previous understanding of transparency, light, and color; to discard the philosophically well-grounded notion that two-dimensional surfaces are mathematical abstractions, not physical entities, and thus to accept that two-dimensional beings can, somehow, interact with our three dimensional bodies; and finally, to abandon the causal principle at the heart of previous theories of visual perception—that color, the object of vision, can only affect that which is potentially colored, i.e., uncolored, i.e., transparent. Kepler's remedy for the maladies affecting the visual theory of his contemporaries must have appeared attractive, but just how many, at least initially, believed that Kepler's cure was worth its side effects has not been sufficiently considered.[106]

[102] Kepler 2000, 219; Kepler 1604, 204.

[103] Kepler 1604, 6–25; Lindberg 1986.

[104] Kepler 1604, 170, 204, 220–221.

[105] Kepler 1604, 13–21; Straker 1970, 503–506, 509–520.

[106] This attack is merely rhetorical. My aim here is to present potential contemporary objections to his theory in order to imagine a time in which it was not at all clear that Kepler's theory of vision would win out. Good historical work documenting actual objections or reservations towards Kepler's theory is the proper corrective, but this is a significant project that has yet to be undertaken by scholars in detail.

4.6 Conclusion

To borrow an image from Francis Bacon, the activity of Zabarella and Fabricius neither resembles the behavior of ants, piling up natural histories and experiments without purpose, nor that of spiders, spinning webs from themselves alone. Whether they might reflect Bacon's ideal, the bee—gathering material from nature and digesting it into a philosophical honey—perhaps depends on whether one considers their ruminations to be proper digestion.

A great deal has been written on Zabarella's *regressus* method in connection to Galileo and the development of the so-called modern scientific method, and I don't wish to delve too far into the issue here.[107] Rather than focusing on the abstract notion of scientific method or the slippery distinction between experience and experiment, I have presented some specific ways that Zabarella applied his experiences with anatomy, and I hope to have shown that the theory of vision he endorsed, and likely helped to create, had some influence. Putting aside Zabarella's opinion on the proper method of demonstration and the influence of his *regressus* method on later figures, we can see that he was clearly connected to important empirical work on the animal body being carried out in Padua. He relied on this new knowledge to argue for specific philosophical positions on light, color, and sensation (in *De visu* book 1), to argue against theories of vision that were in competition with Aristotelian ones (in *De visu* book 2), and to formulate a new theory compatible with both Aristotle and his experiences with dissection.

As a member of the arts faculty at Europe's most prestigious medical school, Zabarella argued that Aristotelian natural philosophy and logic are the true foundations for medicine. As an anatomist and physician, Fabricius's methods of investigation and argumentation were of a different sort: his Aristotle was not so much that of the *Posterior Analytics*, but of the *History of Animals*, the *De anima*, and the *Parts of Animals* understood alongside Galen's *Anatomical Procedures*, *On the Natural Faculties*, and *On the Usefulness of the Parts*. These three pairs of texts were packaged, respectively, into the framework of *historia*, *actio*, and *utilitas*.[108] Fabricius's Aristotle was in substantial harmony with Galen, whereas Zabarella's Aristotle was irreconcilable with Galen on nearly every issue: on questions of logic and demonstration,[109] on a theory of vision and everything that goes into it, and on foundational anatomical and physiological issues. Yet behind this disciplinary divide Zabarella and Fabricius presented the same theory of vision. They gave identical accounts of the structure of the eye. Although their theory of light differed, they gave the same account of the action of the eye—that is, where vision takes place and by what means, including the notion that the visual faculty makes an

[107] In my view the most succinct and accurate assessment in English is Poppi 2004. For the thesis that Zabarella contributed significantly to the development of the modern scientific method, see Cassirer 1922, 136–143; Randall 1940; Edwards 1960, 323–353; Wallace 1988. For its refutation, see Schmitt 1969; Jardine 1976; Palmieri 2007.

[108] See the introduction to *De voce* in Fabricius 1600.

[109] Not addressed here, but see Edwards 1960.

active judgment at the *aranea*. Finally, they centered their theory of vision around the same things and for the same reasons: a novel account of the usefulness of the vitreous humor, an identical (qualitative) analyses of the path of light in the eye, and the presence of a burning glass in the eye. Their treatises on vision are not well accounted for by recent historiography on the role of experience and experiment and the rise of the experimental method in the sixteenth and seventeenth centuries, a historiography which has tended privilege the exact sciences.[110] More importantly, their works reveals some limitations of this approach. Discussions by historical actors about the proper form of demonstration, the precise relationship between sensory experience and universal knowledge, and the admissibility of singular events and contrived experiments into natural philosophy all ought to be taken into account. But this approach should not be at the expense of detailed historical investigations into the content of neglected written works, or to other methods (such as historical replication) to investigate past practices.

Throughout I have stressed the similarities between Zabarella and Fabricius on vision and the eye, but one item of note is unique to Fabricius. Along with a geometrical diagram of a human eye, Fabricius also gives one for a sheep's eye, and notably the position of the centers of curvature occur in different places within the eyes (although the order is the same). On these diagrams he writes:

> But so that those who produce works of optical science can accurately observe the diverse progression of rays, which are called visual, while they cross over from one humor into another; and [so that] they can accurately measure off the angles of refraction, and thence grasp the innumerable *utilitates* of the parts: we provide, with the most exact care, human and sheep eyes divided through the middle. And the whole magnitude and that of the individual parts, including their situations and figures, are described, and the place that each of their centers occupy is revealed, and everything is outlined in tables below. Diligent investigators of the works of nature will have much to contemplate, where they are able.[111]

Throughout his treatise Fabricius's discussion of rays is merely qualitative, but here we see the expert at investigating animal bodies handing over his results to experts in optics. What Fabricius provides, however, is not merely the scheme for one individual eye, or even one kind of eye, but two kinds of animal eyes. He poses the problem of solving, geometrically, the question of vision in two different kinds of animal, each with different sized humors whose surfaces that have different centers of curvature. At play here is the problem of vision in animals as a whole.

[110] See especially Dear 1987; Dear 2006. The former is particularly relevant as it looks at the development of experiment in mathematical optics. A comparison between the use of experience and experiment in works on vision written by anatomists and physicians with those written by mathematicians would be particularly fruitful. I see little reason to privilege the latter over the former, as has thus far been the case.

[111] Fabricius, *De visione*, p. 105. "Ut autem qui Opticae scientiae operam dant, accuratè obervare possint, progressum varium radiorum, quos visuales appellant, dum ab uno in alium humorem transeunt; atque angulos refractionis dimetiri, & inde innumeras utilitates partium excepere: curavimus exactissima diligentia, oculum humanum & ovilem per medium secari, & magnitudinem totius, ac singularum partium, nec non earundem situs, & figuras describi, & loca qua eorum centra obtinent inveniri, & omnia in subiecta tabella delineari. Habebunt enim curiosi indagatores operum naturae, ubi multa contemplari possint."

Notably, because the centers of curvature of the cornea and the crystalline are not identical, and also are not in the same place in the two animals, the visual theory of Alhazen and the rest of the perspectivists becomes impossible on empirical grounds. Jessenius did not include corresponding images or any descriptions of them in his text, and Kepler approached the problem quite differently by extrapolating from the refraction of a sphere and applying these results to the eye.[112] However, at least two important writers on optics did follow Fabricius's approach. As we have seen, François de Aguilón relied almost entirely on Fabricius for his anatomy of the eye in his 1613 *Optica*. He uses the knowledge that the centers of curvature of the cornea and the anterior crystalline are in different places to argue against Alhazen and Witelo and to generate his own crystalline-centered theory.[113] And in his *Oculus* of 1619 the Jesuit mathematician and natural philosopher Christoph Scheiner quoted this very passage, expressed his delight in reading it, and took up the challenge in his retinal theory of vision.[114]

References

Aguilón, François de. 1613. *Francisci Agvilonii E Societate Iesv Opticorvm Libri Sex*. Antwerp: Ex officina Plantiniana.

Alhacen. 2001a. Alhacen's theory of visual perception: A critical edition, with English translation and commentary, of the first three books of Alhacen's 'De Aspectibus', the Medieval Latin Version of Ibn Al-Haytham's 'Kitāb Al-Manāzir': Volume one., ed. A. Mark Smith. *Transactions of the American Philosophical Society* 91(4): i–337.

Alhacen. 2001b. Alhacen's theory of visual perception: A critical edition, with English translation and commentary, of the first three books of Alhacen's 'De Aspectibus', the Medieval Latin Version of Ibn Al-Haytham's 'Kitāb Al-Manāzir': Volume Two., ed. A. Mark Smith. *Transactions of the American Philosophical Society* 91(5): 339–819.

Alhazen, and Witelo. 1572. In *Opticae thesaurus*, ed. Risner Friedrich. Basel: Per Episcopios.

Aristotle. 1984. *The complete works of Aristotle: The revised Oxford translation*, 2 vols., ed. Jonathan Barnes. Princeton, N.J.: Princeton University Press.

Aristotle, and Averroes. 1562. *Aristotelis Stagiritae De coelo; De generatione et corruptione; Meteorologicorum; De plantis libri cum Averroes Cordubensis in variis eosdem comentariis*. Venice: Apud Iuntas.

Averroes. 1986. In *Averroes on the substance of the celestial sphere: Critical edition of the Hebrew text with English translation and commentary*, ed. Arthur Hyman. Cambridge: Medieval Academy of America.

Benedetti, Alessandro, and Giovanna Ferrari. 1998. *Historia corporis humani, sive Anatomice*. Firenze: Giunti.

Bylebyl, Jerome. 1979. The school of Padua: Humanistic medicine in the sixteenth century. In *Health, medicine, and mortality in the sixteenth century*, ed. Charles Webster, 335–370. Cambridge: Cambridge University Press.

[112] Shapiro 2008.

[113] Aguilón 1613, 11–12, 119–125. Note that he nevertheless reverses the order of the centers of the cornea and crystalline compared to Fabricius, which has a significant effect on his analysis of vision.

[114] Scheiner 1619, 20.

Carpi, Jacopo Berengario da. 1959. *A short introduction to anatomy: (Isagogae Breves)*. Chicago: University of Chicago Press.

Carpi, Jacopo Berengario da., and Mondino. 1521. *Carpi Commentaria cum additionibus super anatomia Mundini*. Bologna: Impressum per Hieronymum de Benedictis.

Cassirer, Ernst. 1922. *Das Erkenntnisproblem in Der Philosophie Und Wissenschaft Der Neuren Zeit*. Vol. 1. Berlin: B. Cassirer.

Colombo, Realdo. 1559. *De re anatomica libri XV*. Venice: ex typographia Nicolai Builacquae.

Crombie, A.C. 1967. The mechanistic hypothesis and the scientific study of vision. *Proceedings of the Royal Microscopical Society* 2: 1–112.

Crombie, A.C. 1990. Expectation, modelling, and assent in the history of optics: Part I. Alhazen and the medieval tradition. *Studies in History and Philosophy of Science* 21(4): 605–632.

Crombie, A.C. 1991. Expectation, modelling, and assent in the history of optics—II. Kepler and Descartes. *Studies in History and Philosophy of Science* 22(1): 89–115.

Cunningham, Andrew. 1985. Fabricius and the 'Aristotle Project' in anatomical teaching and research at Padua. In *The medical renaissance of the sixteenth century*, ed. Andrew Wear, Roger Kenneth French, and Iain M. Lonie, 195–222. New York: Cambridge University Press.

Cunningham, Andrew. 1997. *The anatomical renaissance: The resurrection of the anatomical projects of the ancients*. Brookfield, VT: Ashgate.

De Angelis, Simone. 2008. From text to the body: Commentaries on De Anima, anatomical practice and authority around 1600. In *Scholarly knowledge: Textbooks in early modern Europe*, 205–228. Genève: Librairie Droz.

Dear, Peter. 1987. Jesuit mathematical science and the reconstitution of experience in the early seventeenth century. *Studies in History and Philosophy of Science Part A* 18(2): 133–175.

Dear, Peter. 2006. The meanings of experience. In *The Cambridge history of science: Early modern period*, vol. 3, ed. Lorraine Daston and Katherine Park, 106–131. Cambridge: Cambridge University Press.

Distelzweig, Peter. 2014. Fabricius's Galeno-Aristotelian teleomechanics of muscle. In *The life sciences in early modern philosophy*, ed. Ohad Nachtomy and Justin E.H. Smith, 65–84. New York: Oxford University Press.

Dupré, Sven. 2005. Ausonio's Mirrors and Galileo's lenses: The telescope and sixteenth-century practical optical knowledge. *Galilaeana* 2: 145–180.

Dupré, Sven. 2006. Optica Est Ars Bene Videndi: From Gemma's radius to Galileo's telescope. In *Astronomy as a model for the sciences in early modern times*, ed. M. Folkerts and A. Kühne, 355–368. Ausburg: Rauner.

Dupré, Sven. 2007. Images in the air: Optical games, magic and imagination. In *Spirits unseen: The representation of subtle bodies in early modern European culture*, ed. C. Göttler and W. Neuber, 71–92. Leiden: Brill Academic Publishers.

Dupré, Sven. 2008. Inside the camera obscura: Kepler's experiment and theory of optical imagery. *Early Science and Medicine* 13(3): 219–244.

Dupré, Sven. 2012. Kepler's optics without hypotheses. *Synthese* 185(3): 501–525.

Edwards, William F. 1960. The logic of Iacopo Zabarella. PhD dissertation, Columbia University.

Fabricius, Hieronymus. 1600. *De Visione, Voce, Auditu*. Venice: Bolzetta.

Fabricius, Hieronymus, and Howard Bernhardt Adelmann. 1967. *The embryological treatises of Hieronymus Fabricius of aquadenpente*. Ithaca: Cornell University Press.

Favaro, Antonio. 1911–1912. *Atti Della Nazione Germanica Nello Studio Di Padova*. 2 vols. Venice: Venezia Società.

Frangenberg, Thomas. 1991. Perspectivist aristotelianism: Three case-studies of cinquecento visual theory. *Journal of the Warburg and Courtauld Institutes* 54: 137–158.

Galen. 1968. *Galen on the usefulness of the parts of the body. De Usu Partium. Translated from the Greek with an introduction and commentary by Margaret Tallmadge May.* Trans. Margaret Tallmadge. 1st ed. Ithaca, N.Y.: Cornell University Press.

Galen. 1980. *On the doctrines of Hippocrates and Plato: Books VI – IX*. Trans. Phillip De Lacy. Berlin: Akademie-Verlag.

Goldberg, Benjamin Isaac. 2012. William Harvey, Soul searcher: Teleology and philosophical anatomy. PhD dissertation, University of Pittsburgh, Pittsburgh.

Heseler, Baldasar, and Ruben Eriksson. 1959. *Andreas Vesalius' first public anatomy at Bologna: 1540*. Uppsala: Almqvist & Wiksell.

Jardine, N. 1976. Galileo's road to truth and the demonstrative regress. *Studies in History and Philosophy of Science Part A* 7(4): 277–318.

Jardine, Nicholas. 1997. Keeping order in the school of Padua: Jacopo Zabarella and Francesco Piccolomini on the offices of philosophy. In *Method and order in renaissance philosophy of nature*, ed. Daniel A. Di Liscia, Eckhard Kessler, and Charlotte Methuen, 183–209. Brookfield, VT: Ashgate.

Jessenius, Johannis. 1601. *Iohannis Jessenii a Iessen, Anatomiae, Pragae, anno M.D.C. abs se solenniter administratae historia: accessit eiusdem de ossibus tractatus*. Wittenberg: excudebat Laurentius Seuberlich.

Kepler, Johannes. 1604. *Ad Vitellionem Paralipomena, Quibus Astronomiae Pars Optica Traditure.* Frankfurt: Apud Claudium Marnium & haeredes Ioannis Aubrii.

Kepler, Johannes. 2000. In *Optics: Paralipomena to Witelo & optical part of astronomy*, ed. William H. Donahue. Santa Fe: Green Lion Press.

Klestinec, Cynthia. 2005. Juan Valverde de (H) Amusco and print culture. In *Zergliedungeren: Anatomie Und Wahrnehmung in Der Frühen Neuzeit*, ed. Albert Schirrmeister, 78–96. Zeitsprünge: Forschungen Zur Frühen Neuzeit. Frankfurt: Vittorio Klostermann.

Klestinec, Cynthia. 2011. *Theaters of anatomy: Students, teachers, and traditions of dissection in renaissance Venice*. Baltimore: Johns Hopkins University Press.

Koelbing, Huldrych M. 1990. Anatomie De L'œil Et Perception Visuelle, De Vésale à Kepler. In *Le Corps à La Renaissance. Actes Du XXXe Colloque De Tours 1987*, 389–398. Paris: Aux amateurs de livres.

Kusukawa, Sachiko. 2002. Meditations of Zabarella in Northern Europe: The preface of Johann Ludwig Hawenreuter. In *La Presenza dell'Aristotelismo Padovano Nella Filosofia Della Prima Modernità*, ed. Gregorio Piaia, 199–214. Roma-Padova: Antenore.

Lefèvre, Wolfgang. 2007. Exposing the seventeenth-century optical camera obscura. *Endeavour* 31(2, June): 54–58.

Lindberg, David C. 1976. *Theories of vision from al-Kindi to Kepler*. Chicago: University of Chicago Press.

Lindberg, David C. 1986. The genesis of Kepler's theory of light: Light metaphysics from Plotinus to Kepler. *Osiris* 2(January 1): 4–42.

Lohr, Charles H. 1982. Renaissance latin Aristotle commentaries: Authors So--Z. *Renaissance Quarterly* 35(2, July 1): 164–256.

Maclean, Ian. 2002. Meditations of Zabarella in Northern German, 1586–1623. In *La Presenza dell'Aristotelismo Padovano Nella Filosofia Della Prima Modernità*, ed. Gregorio Piaia, 173–198. Roma-Padova: Antenore.

Magirus, Johann. 1597. *Physica peripatetica ex Aristotele, eiusque interpretibus collecta, et in sex libros distincta: in usum Academiae Marpurgensis Studio & opera Johannis Magiri Doctoris Medici & Physiologiae ...: acceßit tum capitum, tum verborum ac rerum Index geminus*. Frankfurt: Palthenius.

Nutton, Vivian. 2012. Vesalius revised. His annotations to the 1555 Fabrica. *Medical History* 56(4, October): 415–443.

Palmieri, Paolo. 2007. Science and authority in Giacomo Zabarella. *History of Science* 14: 404–427.

Peckham, Johannes. 1504. *Perspectiva communis*. Venice Per Io. Baptistam Sessam.

Peckham, Johannes. 1542. In *Perspectiva communis*, ed. George Hartman. Nurenburg: Apud Iohan. Petreium.

Peckham, Johannes. 1592. *Perspectivae Communis Libris Tres*. Cologne: In officina Berkmannica, sumptibus Arnoldi Mylii.

Peckham, John. 1970. In *John Pecham and the science of optics: Perspectiva Communis*, ed. David C. Lindberg. Madison: University of Wisconsin Press.

Platter, Felix. 1583. *De corporis humani structura et vsu Felicis Plateri... libri III...* Basel: ex Officina Frobeniana, per Ambrosium Frob.

Poppi, Antonino. 2004. Zabarella, or Aristotelianism as a rigorous science. In *The impact of Aristotelianism on modern philosophy*, Studies in philosophy and the history of philosophy 39, ed. Riccardo Pozzo, 35–63. Washington, DC: The Catholic University of America Press.

Porta, Giambattista della. 1593. *De refractione optices parte: libri novem ...* Naples: Apud Io. Iacobum Carlinum & Antonium Pacem.

Randall, John Herman. 1940. The development of scientific method in the school of Padua. *Journal of the History of Ideas* 1(2): 177–206.

San Juan, Rose Marie. 2008. Restoration and translation in Juan de Valverde's Historia de La composition Del Cuerpo Humano. In *The virtual tourist in renaissance Rome: Printing and collecting the Speculum Romanae Magnificentiae*, ed. Rebecca Zorach. Chicago: University of Chicago Library.

Saunders, John B. de C. M., and Charles D. O'Malley. 1950. *The illustrations from the works of Andreas Vesalius of Brussels: With annotations a. Translations, a discussion of the plates a. Their background, authorship a. Influence, and a biographical sketch of Vesalius*. Cleveland: World Publishing Company.

Scheiner, Christoph. 1619. *Oculus, Hoc Est, Fundamentum Opticum*. Oeniponti: Apud Danielem Agricolam.

Schmitt, Charles B. 1969. Experience and experiment: A comparison of Zabarella's view with Galileo's in De Motu. *Studies in the Renaissance* 16(January): 80–138.

Sennert, Daniel. 1618. *Epitome naturalis scientiae*. C. Heiden: Wittenberg.

Shapiro, Alan E. 2008. Images: Real and virtual, projected and perceived, from Kepler to Dechales. *Early Science and Medicine* 13(3): 270–312.

Siraisi, Nancy G. 1994. Vesalius and human diversity in De Humani Corporis Fabrica. *Journal of the Warburg and Courtauld Institutes* 57(January): 60–88.

Siraisi, Nancy G. 1997. Vesalius and the reading of Galen's teleology. *Renaissance Quarterly* 50(1): 1–37.

Smith, A. Mark. 2004. What is the history of medieval optics really about? *Proceedings of the American Philosophical Society* 148(2): 180–194.

Smith, A. Mark. 2005. Reflections on the Hockney-Falco thesis: Optical theory and artistic practice in the fifteenth and sixteenth centuries. *Early Science and Medicine* 10(2): 163–186.

Sorabji, Richard. 2004. Aristotle on colour, light and imperceptibles. *Bulletin of the Institute of Classical Studies* 47: 129–140.

Straker, Stephen. 1970. Kepler's optics: A study in the development of seventeenth-century natural philosophy. PhD dissertation, Indiana University, Bloomington.

Valverde, Juan de. 1556. *Historia de la composicion del cuerpo humano*. Roma: Impressa por Antonio Salamanca y Antonio Lafrerij.

Vesalius, Andreas. 1555. *Andreae Vesalii... De humani corporis fabrica libri septem...* Basel: per Ioannem Oporinum.

Vesalius, Andreas. 2013. *The fabric of the human body: An annotated translation of the 1543 and 1555 editions of "De Humani Corporis Fabrica Libri Septem"*. 2 vols. Basel: S Karger Ag.

Von Staden, Heinrich. 1997. Teleology and mechanism: Aristotelian biology and early hellenistic medicine. In *Aristotelische Biologie*, ed. W. Kullmann and S. Föllinger, 183–208. Stuttgart: Franz Steiner.

Wallace, William A. 1988. Randall Redivivus: Galileo and the Paduan Aristotelians. *Journal of the History of Ideas* 49(1): 133–149.

Witelo. 1991. In *Witelonis Perspectivae Liber Secundus et Liber Tertius*, Studia Copernicana, XXVIII, ed. Sabetai Unguru. Wrocław: Ossolineum.

Zabarella, Giacomo. 1590. *De Rebus Naturalibus Libri XXX*. Venice: Apud Paulum Meietum Bibliopolam Patauinum.

Zabarella, Giacomo. 1617. *De Rebus Naturalibus Libri XXX*. Frankfurt: Lazari Zetzneri.

Part II
Life and Mechanism

Chapter 5
Machines of the Body in the Seventeenth Century

Domenico Bertoloni Meli

Abstract This essay discusses the role of new mechanical devices put forward in the seventeenth century in anatomy and pathology, showing how several of those devices were promptly deployed in anatomical investigations. I also discuss the role of dead bodies as boundary objects between living bodies and machines, highlighting their problematic status in experimentation and vivisection.

Keywords Mechanism • Mechanistic anatomy • Experiment • Pathology

5.1 Introduction

The seventeenth century – especially the second half – was arguably the golden age of mechanistic anatomy, of the attempts to explain an increasing number of operations of the human and animal body, and also of plants, in mechanistic terms. In a now classic paper dating from almost half a century ago, Italian medical historian Luigi Belloni offered an insightful analysis of this area, which in recent years has attracted a large number of studies. The topic has become so rich as to require a book-length study rather than a short paper; the work by René Descartes alone, the most prominent figure among those who defended a machine-like explanation of bodily operations, would require an extensive analysis.[1] Thus my aim in this essay is to focus on a small number of themes, highlighting some especially intriguing or thought-provoking aspects, rather than attempting a broad or comprehensive survey. I seek more to raise questions than to provide answers.

This essay was conceived as a stimulus for discussion at an oral presentation and retains its relatively informal character even in its present form. I am grateful to all those who offered comments and suggestions, to the anonymous referees, who forced me to expand and clarify my views, and to Joshua Ewigleben for his insightful comment on Hooke's experiment and organ playing.

[1] Belloni 1963. Des Chene 2001. Aucante 2006. Manning 2012.

D. Bertoloni Meli (✉)
Department of History and Philosophy of Science and Medicine,
Indiana University, Bloomington, IN 47405, USA
e-mail: dbmeli@indiana.edu

P. Distelzweig et al. (eds.), *Early Modern Medicine and Natural Philosophy*,
History, Philosophy and Theory of the Life Sciences 14,
DOI 10.1007/978-94-017-7353-9_5

I am going to consider three aspects: the first involves the usage in anatomy of devices or machines that were novel to the seventeenth century, highlighting how the growing field of mechanical (in a broad sense) tools affected the conceptualization of animal and plant anatomy and physiology; the second focuses on the usage of mechanical devices to conceptualize an often neglected area, namely diseased states; lastly, I discuss dead bodies and body parts as boundary objects between the world of living organisms and that of artificial or man-made machines.

Before embarking on my brief excursus, a few observations are in order. The philosophical implications of mechanistic anatomy and the set of views opposed to it, what we may call its contrast class, were highly specific to the time when they were formulated. Those seventeenth-century anatomists opposed to a mechanistic understanding of the body were by and large defending the role of the faculties of the soul or of nature, which could not be reproduced artificially by a machine. The notion of vitalism, for example, developed mainly in the eighteenth century and has to be studied in its specific temporal and conceptual context. At a later time still, in the nineteenth century, mechanism was often contrasted to teleology; however, many seventeenth-century mechanists – though by no means all, Descartes being a notable exception – had no particular objection to teleology; anatomists such as Malpighi and Steno saw the body as a divinely organized and planned machine. Their teleology, however, concerned God's plan for the creation and was not immanent to individual living bodies, acting like an internal principle guiding their operations. Thus in this respect most seventeenth-century mechanists differed profoundly from nineteen-century mechanists, despite the fact that they are all grouped together under the same category.[2]

5.2 New Devices

Defining what a machine is and which machines were deemed relevant to understanding the body in the seventeenth century are not straightforward matters; in fact, several historians have debated this issue, at times taking the notion of machine to include not only purely mechanical devices, such as clocks, mills, fountains, and pneumatic devices, but also hybrid chemico-mechanical ones involving processes like fermentation, distillation, and even explosion, as in a gun. Traditional devices dating from antiquity include the lever and the filter, for example. The seventeenth century, however, was the time when a large number of new machines or devices and notions were introduced; thus the tool-kit for understanding how the body works mechanically was not fixed but in a state of flux at the time, and perhaps it

[2] Duchesneau 1998. Von Staden 1997. Lennox 1992. Lenoir 1989. Manning 2012. Bertoloni Meli 2011, 12–16.

Fig. 5.1 de Caus, pump used by an early fire brigade, 1624

always is, with the introduction mostly after our period of feed-back mechanisms, self-regulating devices, or even electrical and cybernetic machines.[3]

Historian Gweneth Whitteridge, for example, has shown that two machines discussed in relation to the early seventeenth-century reflect the contrasting views about the heartbeat put forward by William Harvey and René Descartes: the first is a fire-pump acting by spurting water when the men press the piston, thus one could say in systole or contraction (Fig. 5.1); the other is a water-engine ejecting water when the heat makes water boil over, thus one could say in diastole or relaxation (Fig. 5.2). These two examples highlight an obvious point that is still worth making explicit: analogies between machines and anatomical structures or processes could be established by mechanists and non-mechanists alike. Descartes had to provide a mechanistic understanding of all the structures and operations of the animal body; by contrast, Harvey opposed the mechanistic worldview though he could still adopt mechanical analogies in limited domains. Discussing the valves in the veins, for example, he compared them to "the sluice gates which check the flow of streams" ("valvularum, quibus cursus fluminum inhibentur, in morem)".[4]

Many mechanical devices that we take for granted today were introduced and conceptualized in the seventeenth century; the pendulum, the barometer or Torricellian tube, and the spring, for example, though familiar to us, were introduced in a meaningful way at the time of Galileo and Mersenne, whilst their behavior was

[3] Gaillard et al. 2013, especially the essay by Roux (2013). Bertoloni Meli 2006, 14–6. Machamer et al. 2000. Keller 2010. Craver and Darden 2013.

[4] Whitteridge 1971, 169–72, plates IV and VII. De Caus 1624, 4 and problem XX. O'Rourke Boyle 2008. Harvey 19933, 65. Harvey 1957, 186.

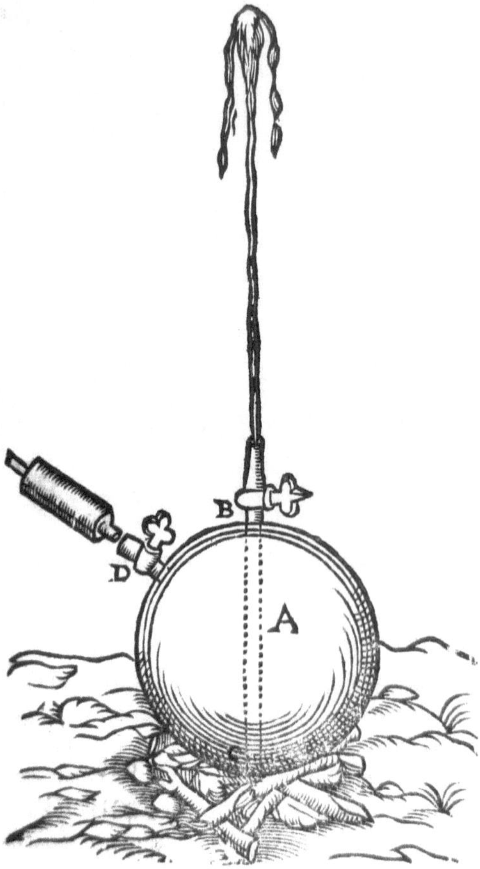

Fig. 5.2 de Caus, expulsion of water from a heated sphere, 1624

understood in a more refined way later in the century. The pendulum and its basic rule whereby the period is proportional to the square root of the length became prominent in the early 1630s in Galileo's *Dialogue Concerning the Two Chief World Systems* and in Mersenne's edition of Galileo's *Mechaniche*. The barometer was first used in the 1640s, when several experiments were performed especially in Italy and in France; lastly, the mathematical law of the spring was formulated by Robert Hooke in 1678, though elastic phenomena were known and discussed well before then, by Galileo's disciple Benedetto Castelli and Descartes, for example.[5] What impact did these new devices have on the understanding of the body?

While studying insects, Robert Hooke saw under "the curious *Mechanism* of the wings" (Hooke's term and italics) of the fly some structures that he identified as "pendulums"; relying on a relatively novel tool of investigation – the microscope, dating from the beginning of the century – Hooke saw a new structure and identified or conceptualized it in terms of a novel device. He aptly compared it in shape to

[5] Bertoloni Meli 2006, chapters 3, 4, 6, 8.

Fig. 5.3 Hooke, *Micrographia*, Plate XXVI, figure 2, 1665

"a long hanging drop of some transparent viscous liquor"; although the halteres – as we call them today – are solid, Hooke's description is quite convincing because they do look exactly like drops of a viscous fluid (Fig. 5.3). He further observed that they are set in motion just before the wings begin to move and speculated that they may serve to regulate those motions. He also proposed alternative explanations, such as a possible use in respiration, whereby the "pendulums may be somewhat like the staff of a pump" – another mechanical analogy – but then considered this second explanation as less plausible. Either way, here Hooke was not gesturing towards abstract mechanical explanations; rather, he was trying to interpret the role of a moving device he had identified with the microscope in an animal. Although Harvey's and Hooke's philosophical outlooks are vastly different, in the specific instances we have seen, both compared anatomical structures (valves in the veins and "halteres" in flies) to mechanical devices, such as sluices and pendulums.[6]

Today halteres are understood to serve as gyroscopic or balancing organs helping to stabilize flight; without them, the insect could still fly but its flying would be erratic and the insect would be likely to bump against objects.[7] But regardless of how consonant Hooke's views may be or not with our own, his was undoubtedly a pioneering attempt to use the pendulum in an anatomical context.

Italian physico-mathematician Giovanni Alfonso Borelli too refers to the pendulum in his investigation of animal motion, though his usage was often more based on analogy in behavior than anatomy proper, as when he compared in general terms the beating of the heart with the oscillation of a pendulum, for example, or oscillations pertinent to respiration. In his analogies Borelli highlighted an important difference among machines: in some, such as a balance, when one alters the equilibrium conditions, the lighter side rises and the heavier one descends, and they remain in such a position. In a pendulum, however, and also in a spring, by altering the equilibrium conditions, the machine is set in motion and this motion is at least in principle perpetual, thus such machines are especially useful in thinking about and conceptualizing regular motions in the body, such as the heartbeat. In other words, often the significance of pendulums and springs for Borelli is that they instantiate some mechanical behavior capable of mimicking in some crucial respects an automaton or a self-moving device, a key conceptual tool of mechanistic anatomy. His analogies were considerably more abstract and less precise than those established by Harvey and Hooke: the Italian physico-mathematician was satisfied with capturing some significant physiological features gesturing towards the possibility of a mechanical explanation without necessarily identifying the precise structural and anatomical elements involved.[8]

Often Borelli established mechanical analogies in some respects and then proceeded to question them in others, highlighting why they should be ultimately rejected. One may think that his way of proceeding resembles Galen's, who also rejected parallels between animal processes and machines. For example, in *On the Natural Faculties* Galen argued that growth – which is one of the natural faculties – is unlike any process that could be imitated by humans, such as weaving, for example, because a small liver that will grow larger is still a liver even when small, whereas a basket does not become a basket until the weaving is complete. Another

[6] Hooke 1665, quotation at 173. On the early history of the pendulum see Büttner 2008; Bertoloni Meli 2006, index. On microscopy see Wilson, Catherine 1995. Ruestow 1996. Fournier 1996.

[7] I wish to thank my colleague Armin Moczek for informing me about halteres and their role in the flight of "Diptera", the two-winged insects (as opposed to all other winged insects, which have 4 wings).

[8] Borelli 1989, 318, 185, 283. Borelli is a complex thinker; I am not even confident that he could license for publication the posthumous *De motu animalium*, especially the second book. Therefore my comments refer to the specific passages cited and should not be taken as representative of his general views.

example was inspired by a game played by children, who used pig bladders – the air balloons of the second century of the Common Era. The children blow into the bladders and heat them in warm ashes, and also sing melodies to them to make them grow. Alas, says Galen, this is not real growth, because what the bladders gain in surface they lose in thickness; real growth, argues Galen, can be performed only by nature. Expansion and compression, however, are interesting phenomena that attracted a great deal of attention and we will discuss them again shortly.[9]

Borelli's concerns often are of a different nature: generally for him there is no fundamental difference in kind between physiological and mechanical processes, contrasts and comparisons do not highlight the role of the faculties that cannot be imitated by art but rather survey a range of possible mechanical explanations and their similarities and differences from the machines of the body; the issue for him is to identify the most appropriate machine actually employed by nature among many possible options, or even a combination of machines.[10]

The next set of devices relies on the incompressibility of water, and the pressure, compressibility, and elasticity of air. The 1640s was the golden age of such pieces of apparatus; in 1648 Raffaello Magiotti invented a device called "ludione" consisting of a glass tube filled with water in which glass globules partially filled with air float. The tube is closed at one end and has either a tiny opening or a flexible membrane at the other end. By pressing on the tiny opening – or the membrane – the air in the globules is compressed, water raises inside them and they descend; this happens because water cannot be compressed, whereas air can. Removing the pressure of the hand, the air in the globules expands again and they rise inside the tube (Fig. 5.4). Other similar devices would work relying on temperature variations rather than pressure: with a higher temperature air would expand and the globules would ascend; a lower temperature would make them descend.[11]

Here I would like to mention an anatomical application of Magiotti's device, one known in Italian as "diavoletto di Cartesio" or "ludione". In 1665 Bologna anatomist Carlo Fracassati argued that nervous transmission would occur this way: the nerve would be the glass tube and external stimuli on the sense organs would work like the hand, exerting pressure and making the globules move up or down, toward the brain or away from it. Fracassati's account seems rather crude at several levels: its implicit reliance on gravity makes its application to sense perception problematic; moreover, it is hard to envisage how any specific sensation might be transmitted this way. However, the ludione opens a window onto an attempt to rely on a novel device in order to account for nervous transmission, long before the idea that such a process would have to do with electric impulses was first put forward.[12]

Luigi Belloni has shown how similar experiments were used in the seventeenth century to account for the floating of fish, whose air bladder would behave similarly to the globules in Magiotti's tube; more than that, air bladders – like the pig bladders

[9] Galen 1916, I.7 and II.3.

[10] Des Chene 2005, especially 251–4, discusses several examples.

[11] The treatise by Magiotti is discussed in Belloni 1963, and is reprinted there at 271–82.

[12] Bertoloni Meli 2011, 95–6. On the rise of animal electricity see Piccolino and Bresadola 2013.

Fig. 5.4 Magiotti, *Renitenza certissima dell'acqua alla compressione*, "ludione", 1648

mentioned by Galen – would behave differently from glass because they could expand and contract. Members of the Accademia del Cimento in Florence performed several experiments on the floating of fish and Borelli too discussed them in his posthumous work, *De motu animalium*, in which he compared a fish to a floating device as in the hydrostatics by Archimedes (Fig. 5.5). Borelli used knowledge from mechanical devices to understand animals in water, and, reciprocally, knowledge from animals in water to develop new mechanical devices, such as a submarine. Borelli was not the first to think of a new mechanical device inspired by what he had seen in nature. Hooke had realized that the "beards" of oats consist of two microscopic filaments twisted together; since the filaments react differently to moisture, they move in reaction to different environments. Their behavior enabled Hooke to construct a new instrument for measuring humidity, the hygroscope, which reacted to "a little breath of moist or dry Air."[13]

My next device – the barometer – is seldom considered in anatomy; yet in 1651 French anatomist Jean Pecquet reported in his groundbreaking *Experimenta nova*

[13] Belloni 1963. Middleton 1971, 105–66, especially 159–66. Borelli 1989, 183–202, especially 197–202. Hooke 1665, 147–52, quotation at 152.

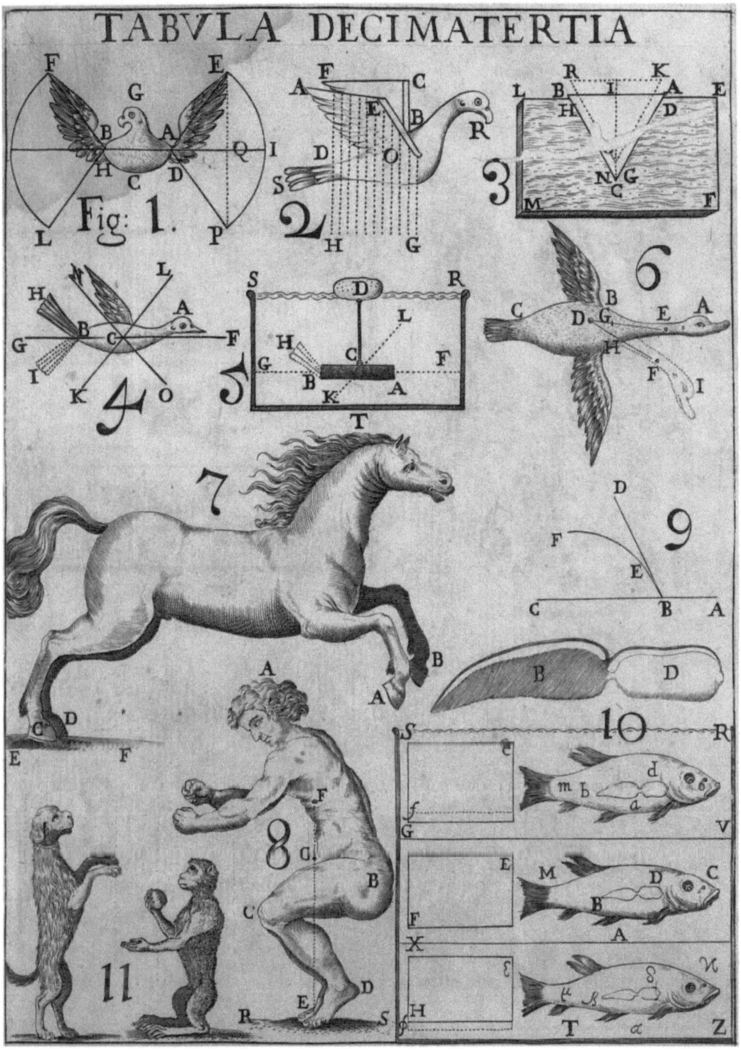

Fig. 5.5 Borelli, *De motu animalium*, air bladder of fish, figure 10, 1680

anatomica a number of barometric experiments, a cutting-edge area of research in the physico-mathematical sciences of those years. Although it had long been known that water could not be raised higher than approximately ten meters, the reason for this phenomenon was unclear. It was in the early 1640s that a number of scholars between Florence and Rome conceived and performed experiments with mercury rather than water. We can gain a feel for the nature of the experiments performed immediately later by looking at one due to the mathematician Gilles Personne de Roberval and reported by Pecquet. The experiment consisted in inserting the emptied air bladder of a fish in an evacuated Torricellian tube; once it is in the empty

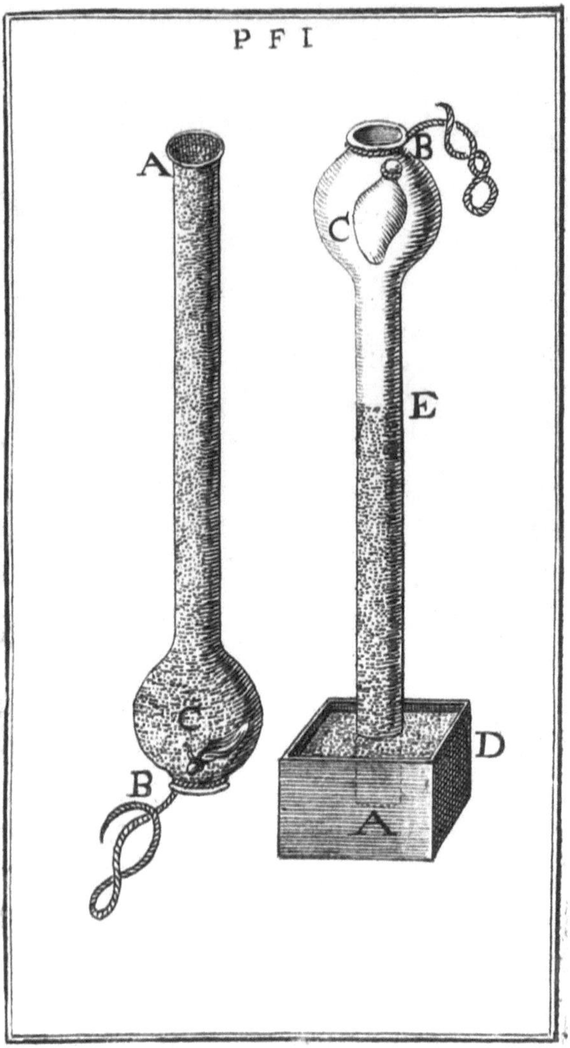

Fig. 5.6 Pecquet, *Experimenta nova anatomica*, Roberval's experiment, 1651

space at the top of the tube, the bladder – working like a tiny air balloon – inflates because of the air's elasticity or "elatery", as it is called in the 1653 English translation of his work – "elastrum" is the original Latin. Since the top of the tube is almost empty, the tiny amount of air left inside the bladder expands and fills it (Fig. 5.6).[14]

Pecquet had two main reasons for having recourse to such experiments: the first was to explain the motion of chyle – or digested food – in a mechanistic way inside

[14] I cite from the contemporary English translation: Pecquet 1653, 89–92. Bertoloni Meli 2008, 670–7.

the body without having recourse to attraction, purely as a result of elasticity or "elatery" and pressure, from respiration for example. Thus elasticity is used here qualitatively, not quantitatively in terms of pressure and volume. Pecquet provides a physiological application of a recent physico-mathematical experiment.[15]

Pecquet extended his reflections on the "elatery" to other areas as well; one very brief passage, predictably, deals with respiration, which involves not only compressible air but also the "elatery" of the lungs, which are distended and contracted; thus elasticity affects respiration in a double way, for the air and the organ. Curiously, in the case of the carp bladder Pecquet focused on the elasticity of air but did not mention the elasticity of the membrane.[16] Another process involving "elatery" is digestion, in which the fibers of both the stomach and intestines would expand and contract "like into an Elatery" – notice here the interesting specific reference to fibers.[17] Yet another relevant area concerns the "elatery" of blood vessels, both arteries and veins. Pecquet argued that immediately after the cardiac systole, arteries distend; the same would happen to veins when blood enters them.[18]

Pecquet's account modifies Harvey's, who had devoted his attention to this issue opposing Galen's views: Galen had argued that the arteries move because of a faculty transmitted to them by the heart, "faculty" being here a *terminus technicus* related to his philosophical stance. Harvey, by contrast, had argued that the arteries fill because of the impulsion of blood, adding as a clarification that they fill like leather gloves one blows into, rather than like bellows; the difference here is that leather containers are purely passive whereas bellows fill as a result of an action, that of the hands operating them, just as arteries would fill by expanding because of the faculty transmitted by the heart. Pecquet's analysis modified this dichotomy because it attributed a more active role to the walls of arteries and veins: they were no longer purely passive containers but contributed through their "elatery" to the motion of blood, helping with their expansion and contraction the heart's action.[19]

Pecquet may have been the first to introduce the notion of elasticity in anatomy but he was not the last in the seventeenth century; others followed suit in different forms, some of which resonate with our own current views, some do not. Borelli, for example, attributed a key role to elasticity in respiration: he argued that air particles consist of spiral machines or springs and that once they are mixed with blood they oscillate and keep the blood in a state of constant internal motion. Thus for him respiration would be a mechanical process of oscillation requiring air particles to be mixed with blood in order to keep it in a state of internal motion.[20]

About the same time Nehemiah Grew investigated in a much more concrete way why the plant Coded Arsmart "ejaculates" its seeds – as he put it. Much like Borelli, Grew too attributed the behavior of a living organism to coiled springs;

---

[15] Pecquet 1653, 141–8.

[16] Pecquet 1653, 149–50.

[17] Pecquet 1653, 144.

[18] Pecquet 1653, 135–6.

[19] Harvey 1993, 10–1 and 112–4.

[20] Borelli 1989, Part II, chapter 8, especially Propositions 115–6, pages 318–20.

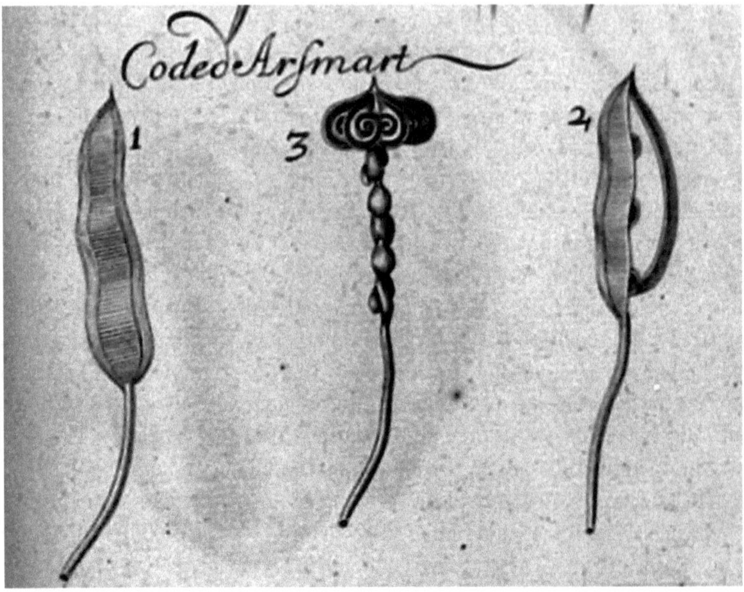

Fig. 5.7 Grew, *Anatomy of Plants*, discharge mechanism of Coded Arsmart. Figure 3 in the middle. 1682

unlike Borelli, however, he could actually see them and explain that the coiled membrane constitutes a discharge "mechanism" – a term used by Grew to mean the mechanical arrangement of the parts enabling them to perform the task of violently discharging the seed – that projects the seed away from the plant by unfolding (Fig. 5.7). In his own words (italics in the original)[21]:

> From this *Mechanism*, the manner of that violent and surprising *Ejaculation* of the *Seeds*, is intelligible. Which is not a motion originally in the *Seeds* themselves; but contrived by the *Structure* of the *Case*. For the *Seeds* hanging very loose, and not on the *Sides* of the *Case*, as sometimes, but on the *Pole*, in the *Centre*, with their thicker end downward, they stand ready for a discharge: and the *Sides* of the *Case* being lined with a strong and Tensed *Membrane*, they hereby perform the office of so many little *Bows:* which, remaining fast at the *Top*, and (contrary to what we see in other *Plants*) opening or being *lett off* at the *Bottom*, forceably curle upward, and so drive all the *Seeds* before them.

Just a few years before, Grew's friend and collaborator Hooke had published a treatise on the spring, *De potentia restitutiva*, which provided the quantitative formulation bearing his name, Hooke's law, stating that the force is proportional to the displacement (Fig. 5.8). Notice the striking visual similarity between Hooke's coiled spring and the coiled membrane seen by Grew in Coded Arsmart.[22]

[21] Grew 1682, 188–9. Bertoloni Meli 2011, 266–7. Dear 2006, chapter 1. Craver and Darden 2013, 15–20.

[22] Hooke 1678, 1. Bertoloni Meli 2006, 242–6.

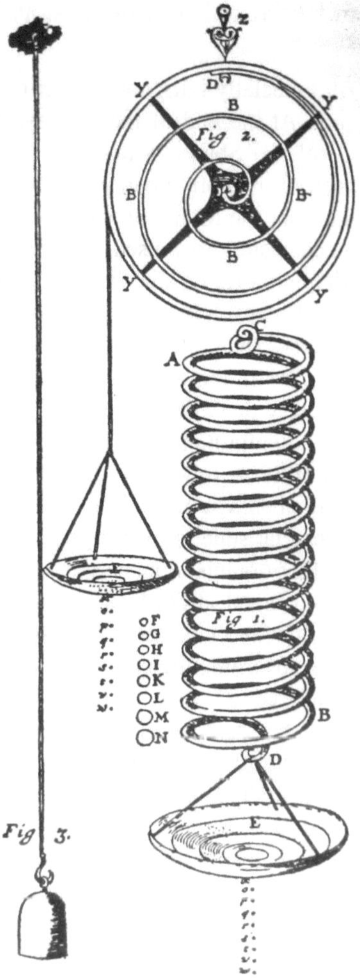

Fig. 5.8 Hooke, *De potentia restitutiva*, examples of springs. Figure 2 top left, 1678

Thus Pecquet relied on elasticity to account for the motion of chyle and the behavior of the lungs, stomach, intestine, and blood vessels; he did not attempt to provide a microscopic explanation of the action of these body parts, but offered what could be called a phenomenological description of those actions. Borelli, by contrast, tried to provide a microscopic account of the internal motion of blood by hypothesizing the existence in air of tiny spring-like oscillating particles that mix with blood. Lastly, Grew offered a much more limited but more concrete contribution by visually identifying in Coded Arsmart small coiled membranes attached to the top of a pole that unfold from the bottom, thus enabling the discharge mechanism of the seeds and making it intelligible.

5.3 Machines and Disease

An especially intriguing aspect of mechanistic anatomy concerns the study of disease. At the end of the century, the professor of medicine at Bologna Giovanni Gerolamo Sbaraglia attacked his colleague Marcello Malpighi and mechanistic anatomists more broadly for their inability to tackle disease in a new way. Thus in his response Malpighi was forced to spell out in unusual detail how machines and mechanisms were used both to understand disease and to devise effective cures. Malpighi offered some reflections on the role of machines, often very simple artificial devices, and provided a list of those that had been used in the course of the century. The philosophical underpinnings of his views are quite complex: they can be found in the belief that disease consists in the structural alterations of the body parts and that therefore the task of the physician is to correct those alterations, rather than attempting to cure the alleged faculties of the soul – or perhaps the *archeus*, if one were to follow Jan Baptiste van Helmont. Moreover, Malpighi believed that nature behaves in a uniform manner both in health and disease; hence disease is not a peculiar state following laws of its own, but rather it follows the same ones as the healthy organism, namely mechanical ones. By laws Malpighi here did not mean only general laws of matter, such as the law of inertia, for example, but also the specific laws governing processes in the living organism, such as those mechanical laws governing the process of growth in plants and animals, for example, which he compared to a weaving process.[23]

In his defense of mechanistic anatomy against the attacks of his colleague, Malpighi provided a list of devices with a pathological significance; many of them had actually been built and were used for teaching and research. For example, in a medical consultation of 1687 for a case of gout, Malpighi sought to explain what happens inside the body, the cause of the disease being an excess of acids in the chyle. He also stated, "All this can be seen in proportion also mechanically mixing spirit of vitriol, or another acid that is especially austere, with different fluids." Here Malpighi used the term "mechanically" rather broadly, in conjunction with the act of mixing and a chymical operation reproducing in vitro processes occurring inside the body in order to investigate disease and to devise suitable therapies.[24]

Malpighi mentioned that the camera obscura could serve to understand sight and its problems, whereby, as he put it: "the way of seeing and its lesions are demonstrated by means of the cognition of the man-made machine analogous to the eye." Indeed, we know that he experimented on the eye and the properties of its parts with

[23] Malpighi 1967, 491–631, especially 512–16. Bertoloni Meli 2001, especially 517–20. For a recent philosophical analysis of health and disease in Descartes see Manning 2012; for broader reflections on diseases see Canguilhem 1978; Wilson 2000. Giglioni 2000, 97–142.

[24] I have treated some of these views in Bertoloni Meli 2013a. Malpighi 1975, 3:1268–69, Malpighi to Tarantino, 29 March 1687.

his colleague Giandomenico Cassini, the celebrated astronomer and professor of mathematics at Bologna.[25]

Another instance concerns the model of an artery, one closely resembling that mentioned by Harvey in the second reply to Jean Riolan, which would enable us to study blood circulation and its diseases, by which Malpighi presumably meant aneurysms and extravasations. Harvey states[26]:

> If you take what length you will of the inflated and dried intestines of a dog or wolf (such a preparation as you find in an apothecary's shop), cut if off and fill it with water, and tie it at both ends to make a sort of sausage, you will be able with a finger-tap to strike one end of it and set it a-tremble, and by applying fingers (in the way that we usually feel the pulse over the wrist artery) at the other end to feel clearly every knock and difference of movement. And in this way (as also in every swollen vein in the living or dead body) anyone will be able to teach students, by demonstration and verbal instruction, all the differences occurring in the amplitude, rate, strength, and rhythm of the pulse. For just as in a long full bladder and an oblong drum every blow to one end is felt simultaneously at the other, so in dropsy of the belly, as also in every abscess filled with liquid matter, we are accustomed to distinguish anasarca from tympanites.

Here Harvey joins investigations of the healthy and diseased body. He suggests a use of the intestine sausage going even beyond the normal operations and diseases of the circulatory system. Here he considers anasarca, or the swelling up of the entire body, and tympanites, a distension of the abdomen; notice also the reference to experiments on dead bodies – a topic we shall return to.

In addition, in his study of heart polyps, Malpighi repeatedly compared the circulatory system and aqueducts, explaining the phenomena occurring in the heart and blood vessels as analogues to obstructions and sedimentations due to mineral deposits in water pipes. In both cases time would play a key role in enabling the processes to occur, especially because the inside of the heart is not smooth but full of cavities and fibers. Heart polyps are formations found postmortem in the heart and nearby vessels that most seventeenth-century anatomists and physicians believed slowly formed while the patient was alive. It is worth adding here that in the following century Malpighi's pupil Ippolito Francesco Albertini used the simile of the water mill in his study of heart disease and especially aneurisms: "As for bloodletting, I have prescribed it with a beneficial result at the beginning of the disease, just as if we were to turn away water from a water mill that is beginning to work badly and to leak." Here a machine, the water mill, suggests a therapy.[27]

Malpighi further referred to the operations of the lungs, which he compared to a machine exploiting by expanding the weight and "elatery" of air. Malpighi then argued that an artificial thorax would serve to study what happens when the lungs fill with fluid or solid bodies and therefore "helps to uncover a priori nature's way

---

[25] Malpighi 1967, 513. On Cassini and Malpighi see Bertoloni Meli 2008, 692–3. Bertoloni Meli 2011, 317–8.

[26] Malpighi 1967, 513. Harvey 1993, 124–25.

[27] Malpighi 1967, 189–216, especially 203, 215; 514–6. Malpighi 1995, especially 485, 492. Belloni 1956, at 28–35. Bertoloni Meli 2001, 519–20. Albertini in Jarcho 1980, quotation at 332.

Fig. 5.9 Swammerdam, *De respiratione*, 1667

of operating and the phenomena in the diseased states of respiration." Probably he was thinking along the lines of what Dutch anatomist Jan Swammerdam had done in *De respiratione*, when he had used a mechanical apparatus consisting of a bladder attached to a tube inside a glass phial (Fig. 5.9) to understand the punctured thorax; Swammerdam's example also had a pathological significance in showing instances when respiration is hindered.[28]

Lastly, in his response to Sbaraglia and defense of mechanistic anatomy in the study of disease, Malpighi mentioned the articulations of bones with threads attached to them, thus a hybrid device combining different components. Although he did not refer to specific pathological conditions, such a device would have had immediate surgical applications. The role of tendons in moving muscles was well known since antiquity and had been singled out by Vesalius at his 1540 Bologna anatomical demonstrations, when he warned barbers of the dangers of accidentally damaging the tendons during venesection, for example.[29] This last example

[28] Malpighi 1967, 513–14. Swammerdam 1667, 30, 36–37. Schierbeek 1974, 67–71.

[29] Malpighi 1967, 513. Eriksson 1959, 252–55.

involving a machine partly skeleton and partly artificial seems especially suited to introduce my next topic, the role of dead bodies and body parts as boundary objects in understanding the living body.

5.4 Dead Bodies as Boundary Objects

There is another dimension that I would like to explore briefly, one crossing the boundaries from diseased to dead bodies. We may take it for granted that anatomy deals with dead bodies but there are some conceptual issues at stake here: both Aristotle and Galen, for example, believed that quite a lot could be learnt from dissecting dead bodies and both relied extensively on this practice, albeit of animals rather than human. Still, a dead body differs from a live one and if the operations of the body depend on the faculties (as both Aristotle and Galen believed, whether of the soul or of nature), they cannot all be investigated effectively through anatomy. In *Parts of Animals*, for example, Aristotle, argued that a cadaver is no longer a human being, just as much as the hand or the eye of a dead man are no longer a hand or an eye; although they may be shaped in the right way and made of the right materials – at least for a while –, they are a hand and an eye in name alone because they have lost their ability to operate as a hand or an eye. In addition, at the moment of death the hand and eye begin to lose their organization in a process of decay that advances inexorably with time. As Tawrin Baker has recently shown, according to Galen dissecting a dead eye would fail to reveal its mode of operation because some of its physical characteristics were so ephemeral that they vanished with death.[30]

This is the reason why the study of the structure and material composition of bodies goes only up to some point for Aristotle and Galen, it provides important structural data on the basis of which to explain some aspects of how the body works, though other aspects remain beyond the domain of dissection of dead bodies in that they cannot be explained on structural grounds alone but need an investigation of the soul and the faculties; it is because of the faculties that not only operations like motion and sensations occur, but also those related to generation, growth, and nutrition. Of course, investigating or dissecting the live animal may prove helpful in some cases, though there are problems of a different sort associated with tampering with a live organism as well.[31]

In the sixteenth century Paracelsus held an extreme position, questioning whether "dead anatomy", as he called it, would teach us anything useful at all about the body, which functions when it is alive and whose mode of operation is irremediably lost with death. It seems to me that Paracelsus implied here that the examination of a cadaver might uncover mechanical processes, but these would be of secondary importance at best in understanding how the body works. Admittedly Paracelsus

---

[30] Aristotle 1937, I.1, 640b34–641a21. Baker 2014.

[31] Aristotle 1937, I.1, 641a17–36. Bertoloni Meli 2013b. On boundary objects the classic paper is Leigh Star and Griesemer 1989.

may have been a rather extreme case, and others were a lot more accommodating, though his views highlight a widely perceived problem. William Harvey, for example, relied extensively on dissections of dead bodies, yet he too highlighted important differences between the dead and the living body. The 52nd exercise of his treatise on the generation of animals, for example, deals with blood and addresses precisely this point[32]:

> For *blood*, as it is a Natural body, being an Heterogeneous, or Dissimilar substance, is compounded of those *parts*, or *juyces*. But as it lives, and i[s] the chief Animal part, compounded of a body and soul. But when that soul, by reason of the expiration of the native heat, doth vanish, and its native substance is presently corrupted, and is dissolved into those parts, of which it was formerly made: namely, first into a Watry Blood, next into Red, and White parts: and the Red parts, which are uppermost, are most florid: but those that sinck downwards grow dark, and black. Now some of the parts also are *fibrous*, and thicker, as being the tye, and connexion of the rest; others are *ichorous* and *serous*, upon which the coagulated *lump* useth to float. And into this *Serum* almost all the *blood* degenerates. Now these parts are not in the live *blood*, but onely when it is now corrupted and dissolved by death.

What I would like to discuss here is the usage of dead bodies or body parts for anatomical – or, as we would rather say, also physiological – experimentation. These experiments raise the question of what are the key features associated to life and what is the status of a dead body at the junction between, on the one hand, a structure profoundly similar to that of a living organism, and, on the other, one that is no longer alive and that in at least some respects can be compared to a machine. The practice of dealing with dead bodies not just for dissection but also for experimentation may sound rather macabre but it was not entirely unusual: in fact, surgeons routinely practiced their skills not only on live animals but also on cadavers. Animals were generally alive, thus alerting the surgeon to the danger of operating on a living body, risking damage to the vital parts; cadavers were human bodies, thus offering unparalleled opportunities for a more realistic and effective training. In his celebrated experiments in the eighteenth century also Austrian physician Leopold Auenbrugger injected fluids into the lungs of cadavers in order to investigate changes in their acoustic properties through percussion.[33]

There are several references in the literature to experiments performed on dead bodies or body parts in order to understand the living body. Such practices were not novel to the seventeenth century: Berengario da Carpi, for example, performed an experiment by injecting water into the bladder of a fetus in order to investigate the passage for urine; it may be no accident that Berengario was a surgeon, since his training would have involved practicing on cadavers.[34]

Whether surgical practices led to experiments on cadavers or not, the motivation for such experiments in the seventeenth century came from anatomical (and physiological, as we would call them) questions, mostly related to the circulation of the blood and respiration. Let me start with some of these intriguing experiments.

---

[32] Wear 1995, 315. Harvey 1653, 292.

[33] Jones 1960, 113. Keel 2001, 186–254.

[34] French 1985, 51.

In 1628 Harvey put forward the thesis that blood circulates and, specifically, that venous blood moves towards the heart, not away from it. In *De motu cordis* he did experiment on dead bodies by pushing a probe into a vein, showing that valves offer no resistance towards the heart but prevent motion in the opposite direction. It is in this context, as we have seen above, that he also compared the valves to "the sluice gates which check the flow of streams," a mechanical analogy relying on specific technological devices Harvey would have seen in operation.[35] There are other experiments on body parts in *De motu cordis;* Harvey argued that the heart of an eel continues to beat even after it is removed from the animal and even after it has been chopped into pieces, which then beat in unison. He noticed that even the flesh of eels goes on moving after the animal has been skinned, disemboweled, and cut into pieces. In another experiment Harvey could revive the motion of the heart that had been extracted from a dove even after it had stopped moving, by moistening it with warm saliva. Such examples posed a problem to mechanist and non-mechanist views alike: if the soul is essential to the motion of an animal, how can parts move not only after death but even after they have been removed from the body of the animal and chopped into pieces? The motion of the heart is a problem from a mechanist standpoint too; which mechanism can explain the motion of portions of the heart? Harvey suggested that such phenomena may be peculiar to those animals that cling more to life. Borelli suggested that the heartbeat may be due to a nerve working like a leaky tap, where an irritant drips out inducing contractions; such contractions may occur for some time even after the heart has been removed from the animal, if some irritant fluid remains in the nerves – another intriguing mechanical analogy.[36]

In the late 1630s Dutch physician and anatomist Johannes Walaeus became a strong advocate of Harvey's circulation of the blood and performed experiments supporting Harvey's views. The one especially relevant to our discussion concerns a specific aspect of the circulation, namely the passage of blood from arteries to veins; it was unclear whether blood always remained inside blood vessels, passed through the flesh, or collected in some small pools at some stage and then exited through veins. This was a grey area in Harvey's work, one for which he could not provide the ocular demonstrations that were so crucial to his approach. Walaeus addressed this point by means of an experiment on a dead dog; his purpose was to infer the existence of inosculations or anastomoses between arteries and veins, allowing blood to pass from one to the other. To this end Walaeus laid bare an artery and a vein in a leg of the dog; he emptied and ligated the crural vein, then after ligation of the main vessels, both arteries and veins, he was able to press blood from the artery, which thus emptied, into the crural vein, which became filled, thus supporting his claim that blood could pass from arteries to veins through inosculations. Walaeus's work was published several times in the 1640s in prominent anatomical works.[37]

---

[35] Harvey 1993, 65. O'Rourke Boyle 2008.

[36] Harvey 1993, 26–7. Borelli 1989, 283–5.

[37] Schouten 1974, 262, 271.

In 1650, in the aftermath of Walaeus's letters, William Harvey too performed an experiment on the cadaver of a throttled man: as he wrote to the Hamburg physician Paul Marquard Schlegel, he wished to refute the denial of the pulmonary transit by the renowned Paris physician Jean Riolan the Younger, a debated point since the time of Realdo Colombo. Colombo had denied Galen's view that venous blood seeped through pores in the septum of the heart and argued instead that blood moved from the right ventricle to the lungs, and then back to the left ventricle. Having ligated the *vena arteriosa* (pulmonary artery), the *arteria venosa* (pulmonary vein), and the aorta, Harvey fastened an ox bladder to a tube, as was usually done in clysters, and injected warm water into the vena cava; the usage of warm water was presumably a precaution against the objection that the pores in the septum of the heart had closed because of lack of heat, thus warm water would be, in a very prosaic way, emulating the live animal, almost replacing the role of the soul. While the right ventricle filled with water, not a drop reached the left ventricle, thus showing that there were no pores in the septum. Having released those ligatures, Harvey inserted the tube into the *vena arteriosa* and ligated it between the tube and the heart, to prevent water from returning to the right ventricle. On pressing the bladder, this time water came out from the left ventricle of the heart, thus revealing an easy passage through the lungs to the *arteria venosa*.[38] Harvey's second experiment in particular presents strong similarities to Walaeus's: both tried to investigate the passage of blood through capillary blood vessels in dead animals.

The last example I am going to consider is probably the best known and also the most elaborate. It was performed in London in the circles around the Royal Society by physician and anatomist Richard Lower, relying also on techniques devised by Robert Hooke, curator of experiments at the Royal Society and Professor of Geometry at Gresham College. They would have been certainly aware not only of Walaeus's experiment, but in all probability of Harvey's too, because Harvey had performed his experiment in front of several colleagues and his letter to Schlegel was known to the physician and anatomist George Ent, a friend of Harvey's who remained active at the Royal Society for several decades. Despite its apparent simplicity, the experiment must have involved at least two people, besides a dead dog and several pieces of equipment. The experiment was part of an elaborate series devoted to investigating several features of respiration. In the specific experiment on the "strangled dog, after sensation and life had completely deserted it," its lungs were kept inflated with the two pairs of bellows, a technique devised by Hooke in vivisection experiments intended to keep the lungs still. In all probability this technique stemmed from organ playing; organs rely on a constant flow of air forced through small and large pipes, producing sounds. At the time the air flow was produced by two or more sets of bellows, as shown in the woodcut from *Spiegel der Orgelmacher* (Speyer 1511), by early renaissance German organist and composer Arnolt Schlick, for example (Fig. 5.10); the man at top right is operating two sets

[38] Harvey 1993, pp. 140–5, at 140–1, from the letter to Schlegel dated London, 26 March 1651. French 1994, pp. 279–85. Cole 1917–1921, 2:290–1. French 1985, p. 54, deals with injections using a syringe in order to study the passageways of fetal anatomy.

Fig. 5.10 Arnolt Schlick, *Spiegel der Orgelmacher*, organ concert, 1511

contemporaneously. Interestingly, Hooke played the organ and was therefore familiar with its operations. Pipe organs, such as those found in churches, were invoked by Descartes to explain bodily operations, with animal spirits moving through the body much like air moves through the pipes of the organ, while external stimuli play the role of the organist's hands pressing this or that key and forcing air in the appropriate openings. Here, however, Hooke used bellows not to represent the operations of the body, but as devices to experiment on the body.[39]

By having not one but two pair of bellows blowing air into the lungs, it was possible to keep them still during respiration; the air would escape through some holes at the bottom of the lungs. The other experimenter injected venous blood into the

[39] Pugliese 2004. Descartes 1972, 71–2. Descartes's relevant passage is discussed by several authors; see for example Gaukroger 1995, 279–81.

vena cava – as we have seen above, Harvey had used warm water, possibly because of his belief in the degeneration of blood. Blood was thus propelled into the right ventricle, the pulmonary artery (*vena arteriosa*), and the lungs, coming out from the pulmonary vein (*arteria venosa*) bright red, as if it had been drawn from the artery of a living animal, says Lower. Thus color change in blood from dark to bright red occurred not in the heart because of the heart's heat—or indeed of any vital flame or property, since the animal was dead—but in the lungs purely as a result of fresh air. This experiment resolved several controversial points at the time concerning the role of the motion of the lungs and the site where blood changes color. Respiration had been traditionally investigated through vivisection since antiquity because it was associated with life and motion: Lower enacted respiration in a dead animal by blowing air through its lungs, thus showing that one of the key operations associated with life involved only chemical and mechanical processes. As Lower put it at the end of his analysis of the matter in *Tractatus de corde*: "Wherever, in a word, a fire can burn sufficiently well, there we can equally well breathe," establishing a striking parallel between processes in living and non-living subjects.[40]

In the examples we have seen the dead body works like a boundary object; on the one hand, one could argue that it is akin to a machine in that there is no life in it, no faculty of the soul, no heat due to a vital flame – only the injection of warm water keeps its pores open. On the other hand, a recently deceased body retains to a large extent the structure of a live one and it is not a man-made machine, like the bellows used to inflate the lungs. It is relatively easy to show in a cadaver the action of tendons moving the muscles, for example, as in the celebrated *Anatomy Lesson of Dr Nicolaas Tulp*, where Tulp shows with his left arm the same movements controlled by the tendons in the cadaver, in a striking visual interplay so effectively captured by Rembrandt. The experiments carried out by Berengario, Walaeus, or Harvey investigated the arrangement of the vessels or the "plumbing" of the body, assuming that there were no substantial differences between a living and a recently deceased body. Respiration, however, was a process quintessentially related to life that Lower showed could be enacted in a dead body. One may well ask, which other processes could be enacted in dead bodies and what could we learn from them?

5.5 Concluding Reflections

In the enormously rich and complex field of mechanistic anatomy in the seventeenth century I have selected three areas and even in these limited areas my analysis has been partial and limited. Even so, I hope to have shown that it is useful to look at the new devices invented at the time and immediately deployed by anatomists to understand how the body works – regardless of whether their explanations are still accepted today or not. The mechanistic understanding of the body in the seventeenth century is best seen as a project growing intellectually, philosophically,

[40] Frank 1980, 213–7, at 214. Lower 1669, 163–71, quotations at 165, 171.

experimentally, and also in conjunction with the growing number of material tools and technological devices being developed at the time; this will come as no surprise to all those familiar with Descartes's references to the machines in the King's grottoes. The pendulum, the spring, the barometer, but also locks and sluices, played an important role in conceptualizing the body, together with notions associated with them, such as periodic motion, elasticity, pressure, and unidirectional flow. More generally, the mechanistic understanding of the body is constantly changing because our empirical and experimental knowledge of the body is changing, and so are both our general notion of machine and our knowledge of specific machines; the two changes act in a fruitful interaction. The examples of Hooke's hygroscope inspired by the "beards" of oats and Borelli's submarine inspired by the air bladder of fishes highlight a peculiar aspect of this reciprocal relation.[41]

Nor should we look only at the anatomy or physiology of the healthy body. Since the laws at work in health and disease are the same, pathology or the diseased body too can offer a rich set of examples relevant to mechanistic anatomy; while mechanistic anatomy can help understand disease, pathology, in its turn, can enrich our grasp of mechanistic representations of the body.

Lastly, I have raised the issue of dead bodies at the intersection between man-made devices and living organisms; the ambiguous nature of cadavers poses intriguing questions to the philosopher and the experimentalist alike: how much can we understand by dissecting dead bodies? Which operations that we associate to the live animal, such as respiration, can be replicated in a cadaver?

Seen together, these issues highlight the complexity of the problems raised by mechanistic anatomy in the 17th; far from being limited to levers and tubes, scholars at the time raised original questions based on the latest technological and mechanical devices, challenging traditional views and philosophies. Thus the history of science, of medicine, of philosophy, and of technology come together in a fruitful way and provide us with the tools for exploring this area in an inter-related way.

References

Albertini, Ippolito Francesco. 1980. Animadversiones super quibusdam difficilis respirationis vitiis a laesa cordis et praecordiorum structura pendentibus. In *The Concept of Heart Failure from Avicenna to Albertini*, Trans. Saul Jarcho, 324–336. Cambridge, MA: Harvard University Press.

Aristotle. 1937. *Parts of animals*. Cambridge, MA: Harvard University Press.

Aucante, Vincent. 2006. *La philosophie médicale de Descartes*. Paris: Presses Universitaires de France.

Baker, Tawrin. 2014. *Color, Cosmos, Oculus*: Color, vision, and the eye in the works of Jacopo Zabarella and Hieronymus Fabricius ab aquapendente. PhD thesis, Indiana University.

Belloni, Luigi. 1956. Introduzione storica alla patologia dell'aterosclerosi. In *Aterosclerosi umana e sperimentale*, ed. Società italiana di patologia, 25–66. Milan: Ambrosiana.

Belloni, Luigi. 1963. Schemi e modelli della macchina vivente nel Seicento. *Physis* 5: 259–298.

[41] Des Chene 2001, 2005. Craver and Darden 2013. Dear 1995, chapter 6.

Bertoloni Meli, Domenico. 2001. Blood, monsters, and necessity in Malpighi's *De polypo cordis*. *Medical History* 45: 511–522.

Bertoloni Meli, Domenico. 2006. *Thinking with objects. The transformation of mechanics in the seventeenth century*. Baltimore: Johns Hopkins University Press.

Bertoloni Meli, Domenico. 2008. The collaboration between anatomists and mathematicians in the mid-seventeenth century with a study of images as experiments and Galileo's role in Steno's *Myology*. *Early Science and Medicine* 13: 665–709.

Bertoloni Meli, Domenico. 2011. *Mechanism, experiment, disease. Marcello Malpighi and seventeenth-century anatomy*. Baltimore: Johns Hopkins University Press.

Bertoloni Meli, Domenico. 2013a. Machines of the body between anatomy and pathology. In *L'automate*, ed. Gaillard et al., 53–68.

Bertoloni Meli, Domenico. 2013b. Early modern experimentation on live animals. *Journal of the History of Biology* 46: 199–226.

Borelli, Giovanni Alfonso. 1989. *On the Movement of Animals*. Berlin: Springer. English translation by Paul Maquet of *De motu animalium*, 2 vols. Rome: Angeli Bernabò, 1680–1681.

Büttner, Jochen. 2008. The pendulum as a challenging object in early-modern mechanics. In *Mechanics and natural philosophy before the scientific revolution*, ed. Walter Roy Laird and Sophie Roux, 223–237. Dordrecht: Springer.

Canguilhem, Georges. 1978. *On the normal and the pathological*. With an introduction by Michel Foucault. Dordrecht: Reidel.

Caus, Salomon de. 1624. *Les raisons des forces mouvantes, avec diverses machines tant utiles que plaisantes*. Paris: Hierosme Droüart. First published in 1615.

Cole, Francis Joseph. 1917–1921. The history of anatomical injections. In *Studies in the history and method of science*, 2 vols., ed. Charles Joseph Singer, 285–343. Oxford: Clarendon Press.

Craver, Carl F., and Lindley Darden. 2013. *In search of mechanisms. Discoveries across the life sciences*. Chicago: University of Chicago Press.

Dear, Peter. 1995. *Discipline and experience: The mathematical way in the scientific revolution*. Chicago: University of Chicago Press.

Dear, Peter. 2006. *The intelligibility of nature: How science makes sense of the world*. Chicago: University of Chicago Press.

Des Chene, Dennis. 2001. *Spirits and clocks. Machine and organism in Descartes*. Ithaca: Cornell University Press.

Des Chene, Dennis. 2005. Mechanisms of life in the seventeenth century: Borelli, Perrault, Régis. *Studies in History and Philosophy of Biological and Biomedical Sciences* 36: 245–260.

Descartes, René. 1972 (originally published in French in 1664). *Treatise on Man*. French text with translation and commentary by Thomas Steele Hall. Cambridge, MA: Harvard University Press.

Duchesneau, François. 1998. *Les modèles du vivant de Descartes à Leibniz*. Paris: Vrin.

Eriksson, Ruben. 1959. *Andreas Vesalius' First Public Anatomy at Bologna. 1540. An eyewitness report by Baldasar Heseler*. Uppsala/Stockholm: Almqvist & Wiksell.

Fournier, Marian. 1996. *The fabric of life. Microscopy in the seventeenth century*. Baltimore: Johns Hopkins University Press.

Frank Jr., Robert G. 1980. *Harvey and the Oxford physiologists*. Berkeley: University of California Press.

French, Roger K. 1985. Berengario da Carpi and the use of commentary in anatomical teaching. In *The medical renaissance of the sixteenth century*, ed. Andrew Wear, Roger K. French, and Iain M. Lonie, 42–74. New York: Cambridge University Press.

French, Roger K. 1994. *William Harvey's natural philosophy*. Cambridge: Cambridge University Press.

Gaillard, Aurélia, Jean-Yves Goffi, Bernard Roukhomovsky, and Sophie Roux (eds.). 2013. *L'automate: Modèle Métaphore Machine Merveille*. Bordeaux: Presses Universitaires de Bordeaux.

Galen. 1916. *On the Natural Faculties*. Trans. Arthur John Brock. Cambridge, MA: Harvard University Press.

Gaukroger, Stephen. 1995. *Descartes. An intellectual biography*. Oxford: Oxford University Press.

Giglioni, Guido. 2000. *Immaginazione e malattia. Saggio su Jan Baptiste van Helmont*. Milano: Franco Angeli.

Grew, Nehemiah. 1682. *The anatomy of plants*. London: W. Rawlings.

Harvey, William. 1653. *Anatomical Exercitations Concerning the Generation of Living Creatures*. London: James Young. English translation of *Exercitationes de generatione animalium*. London: Typis Du-Gardianis, 1651.

Harvey, William. 1957. *De motu cordis. Movement of the heart and blood in animals*. Springfield: Charles C. Thomas.

Harvey, William. 1993. *The Circulation of the Blood and Other Writings*. London: Everyman. English translation by Kenneth J. Franklin of William Harvey. 1628. *Exercitatio anatomica de motu cordis et sanguinis in animalibus*. Frankfurt: Wilhelm Fitzer; and William Harvey. 1649. *Exercitatio anatomica de circulatione sanguinis*. Cambridge: Roger Daniels, 1649.

Hooke, Robert. 1665. *Micrographia*. London: John Martyn and James Allestry.

Hooke, Robert. 1678. *Lectures de potentia restitutiva*. London: John Martyn.

Jones, Ellis W.P. 1960. The life and works of Guilhelmus Fabricius Hildanus (1560–1634). *Medical History* 4: 112–134 and 196–209.

Keel, Othmar. 2001. *L'avènement de la médicine clinique moderne en Europe, 1750–1815*. Montreal: Presses de l'Université de Montreal.

Keller, Vera. 2010. Drebbel's living instruments, Hartmann's microcosm, and Libavius's thelesmos: Epistemic machines before Descartes. *History of Science* 48: 39–74.

Leigh Star, Susan, and James R. Griesemer. 1989. Institutional ecology, 'Translation', and boundary objects: Amateurs and professionals in Berkeley's museum of vertebrate zoology, 1907–1939. *Social Studies of Science* 19: 387–420.

Lennox, Jim. 1992. Teleology. In *Keywords in evolutionary biology*, ed. Evelyn Fox Keller and Elisabeth A. Lloyd, 324–333. Cambridge, MA: Harvard University Press.

Lenoir, Timothy. 1989. *The strategy of life: Teleology and mechanism in nineteenth-century German biology*. Chicago: University of Chicago Press.

Lower, Richard. 1669. *Tractatus de corde, item de motu et colore sanguinis et chyli in eum transitu*. London: Typis Jo. Redmayne impensis Jacobi Allestry. Facsimile edition with introduction and translation by Kenneth J. Franklin, in *Early Science in Oxford*, vol. 9, ed. Robert T. Gunther. Oxford: Oxford University Press, 1932.

Machamer, Peter, Lindley Darden, and Carl F. Craver. 2000. Thinking about mechanisms. *Philosophy of Science* 67: 1–25.

Malpighi, Marcello. 1967. *Opere scelte*, ed. Luigi Belloni. Turin: UTET.

Malpighi, Marcello. 1975. *Correspondence*, 5 vols., ed. Howard B. Adelmann. Ithaca: Cornell University Press.

Malpighi, Marcello. 1995. "De polypo cordis. An Annotated Translation", by John M. Forrester. *Medical History* 39: 477–492.

Manning, Gideon. 2012. Descartes' healthy machines and the human exception. In *The mechanization of natural philosophy*, ed. Sophie Roux and Daniel Garber, 237–262. New York: Kluwer.

Middleton, W.E. Knowles. 1971. *The experimenters. A study of the Accademia del Cimento*. Baltimore: Johns Hopkins University Press.

O'Rourke Boyle, Marjorie. 2008. Harvey in the Sluice: From hydraulic engineering to human physiology. *History and Technology* 24: 1–22.

Pecquet, Jean. 1653. *New Anatomical Experiments*. London: Octavian Pulleyn. English translation of *Nova experimenta anatomica*. Paris: Sebastian and Gabriel Cramoisy, 1651.

Piccolino, Marco, and Marco Bresadola. 2013. *Shocking frogs: Galvani, Volta, and the electric origins of neuroscience*. Oxford: Oxford University Press. Originally published in Italian, 2003.

Pugliese, Patri J. 2004. Robert Hooke. In *Oxford dictionary of national biography*. Oxford: Oxford University Press. http://www.oxforddnb.com/view/article/13693. Accessed 31 Mar 2015.
Roux, Sophie. 2013. Quelles machines pour quels animaux? Jacques Rohault, Claude Perrault, Giovanni Alfonso Borelli. In *L'automate*, ed. Gaillard et al., 69–113.
Ruestow, Edward G. 1996. *The microscope in the Dutch Republic*. Cambridge: Cambridge University Press.
Schierbeek, Abraham. 1974. *Jan Swammerdam. His life and works*. Amsterdam: Sweets & Zeitlinger.
Schlick, Arnolt. 1511. *Spiegel der Orgelmacher und Organisten*. Speyer: Peter Drach.
Schouten, Jan. 1974. Johannes Walaeus (1604–1649) and his experiments on the circulation of the blood. *Journal of the History of Medicine* 29: 259–279.
Swammerdam, Jan. 1667. *Tractatus physico-anatomico-medicus de respiratione usuque pulmonum*. Leiden: Gaasbeeck.
Von Staden, Heinrich. 1997. Teleology and mechanism: Aristotelian biology and Hellenistic medicine. In *Aristotelische Biologie*, ed. Wolfgang Kullman and Sabine Föllinger, 183–208. Stuttgart: Franz Steiner.
Wear, Andrew. 1995. Medicine in early modern Europe, 1500–1700. In *The Western medical tradition, 800 B.C. to A.D. 1800*, ed. Lawrence I. Conrad et al., 215–361. Cambridge: Cambridge University Press.
Whitteridge, Gweneth. 1971. *William Harvey and the circulation of the blood*. London/New York: Macdonald/American Elsevier.
Wilson, Catherine. 1995. *The invisible world. Early modern philosophy and the invention of the microscope*. Princeton: Princeton University Press.
Wilson, Adrian. 2000. On the history of disease-concepts: The case of pleurisy. *History of Science* 38: 271–319.

Chapter 6
"Mechanics" and Mechanism in William Harvey's Anatomy: Varieties and Limits

Peter Distelzweig

Abstract English anatomist William Harvey (1578–1657), and especially his *De motu cordis* (1628), played a prominent role in the rise of mechanical and experimental approaches to natural philosophy in the seventeenth century. Famously, he compares the expansion of the arteries to the inflation of a glove or the expansion of a bladder; the motion of the heart to that of interlocking gears and the firing mechanism of a gun; and the heart to a pump. Less well known, in unpublished notes he compares the digestive organs to chemical apparatus and devotes an entire section to the *artificium mechanicum* of the muscles. It is perhaps surprising, then, that Harvey's was a self-consciously Aristotelian and Galenic approach to anatomy. He understood the goal of anatomy to be final causal Aristotelian *scientia* of the parts of animals articulated using the Galenic notions of the "actions" and "uses" of the parts. Furthermore, he was critical of Descartes' mechanistic theory of the heart and, more generally, of the corpuscularianism associated with (e.g.) Descartes, Gassendi, and Boyle. He even criticizes his one-time teacher Hieronymus Fabricius ab Aquapendente (who was no mechanical philosopher!) for being overly influenced by the "petty reasoning of mechanics." In this chapter, I explore the complex and varied uses of mechanics/mechanical in Harvey's works. I argue that, despite the apparent diversity, Harvey's attitude toward mechanism is consistent, stable, and creative, reflecting the seventeenth-century semantic ambiguities of "mechanics" and the "mechanical," as well as his own Galeno-Aristotelian understanding of anatomy.

Keywords William Harvey • Mechanism • Mechanical philosophy • Teleology

P. Distelzweig (✉)
Department of Philosophy, University of St. Thomas, St. Paul, MN, USA
e-mail: peter.distelzweig@stthomas.edu

P. Distelzweig et al. (eds.), *Early Modern Medicine and Natural Philosophy*, History, Philosophy and Theory of the Life Sciences 14,
DOI 10.1007/978-94-017-7353-9_6

6.1 Introduction

In the preface to his 1655 *De Corpore*, arch-mechanist Thomas Hobbes identified William Harvey as the first to discover and demonstrate (*detexit & demonstravit*) the science of the human body (*Scientiam Humani Corporis*), and set him alongside Copernicus and Galileo as a founder of genuine natural science (*Physica*).[1] Nor was Hobbes alone among "mechanical" philosophers in his high opinion of Harvey's work, the foundational significance of his discoveries, the effectiveness of his arguments, or the excellence of his anatomical research. Although he disagrees with Harvey on the motion of the heart, René Descartes (somewhat uncharacteristically) acknowledges and credits Harvey for his discovery of the circulation of the blood in his 1637 *Discourse on Method* and again in his 1649 *Passions of the Soul*.[2] Robert Boyle, too, was clearly impressed by Harvey's work.[3] William Harvey—and especially his *De motu cordis* (1628)—played a prominent role in the rise of mechanical and experimental approaches to natural philosophy in the seventeenth century.[4]

It is perhaps surprising, then, that Harvey's was a self-consciously Aristotelian and Galenic approach to anatomy. He understood the goal of anatomy to be final causal Aristotelian *scientia* of the parts of animals articulated using the Galenic notions of the "actions" and "uses" of the parts. Furthermore, Harvey defended the existence of a non-mechanical pulsific "force" or "faculty" in the heart. He was critical of Descartes' mechanistic theory of the heart and, more generally, of the corpuscularianism associated with (e.g.) Descartes, Gassendi, Hobbes, and Boyle. In his work on animal generation he even criticizes his one-time teacher Hieronymus Fabricius ab Aquapendente (who was no mechanical philosopher) for being overly influenced by the "petty reasoning of mechanics." At the same time, in the *De motu cordis* Harvey compares the passive expansion of the arteries to the inflation of a glove and the expansion of a bladder or wineskin. There, too, he compares the motion of the heart to that of the gears of a machine and of the components of the mechanical contrivance used to fire a gun. In the *De circulatione*, published in 1649, Harvey compares the forceful, rhythmic exit of blood from an opened artery to that of water from a pump or syringe.[5] Furthermore, in his anatomy lecture notes Harvey

[1] Hobbes 1655, *Epistola Dedicatoria*.

[2] Harvey is mentioned by name in the margins of the first edition of the *Discourse*; in the text he is referred to as a *médecin d'Angleterre* (AT VI 50). In *Passions of the Soul* (Article 7) he is credited and referred to by name in the text (AT XI 332). Descartes also credits Harvey for the discovery of the circulation in his posthumously published *Description of the Human Body* (AT XI 239).

[3] Though Boyle was, of course, familiar with Harvey's discovery of the circulation of the blood, he mentions and praises Harvey most frequently in the context of animal generation and Harvey's later *De generatione animalium*. See Hunter and Macalpine 1958 for a catalogue and discussion of Boyle's references to his much older fellow Englishman.

[4] See Frank 1980; French 1994, Ch. 11.

[5] Harvey uses the word *sypho*, which could, it seems, mean a spout (any artificial tube-like passage from which water is forcefully ejected), a pump, a syringe, or a pump-driven fire-engine. Harvey uses the word four times in the text (Harvey 1649, 13, 51, 72, and 108). I follow the 1653 English

compares various digestive organs to chemical apparatus, and in his working notes on muscles, he devotes an entire section to the mechanical construction (*artificium mechanicum*) of the muscles and considers a multi-step process leading to muscle contraction under the heading *ratio mechanica*. What is this but the mechanization of the animal that is championed by Descartes and others?

Clearly, Harvey's attitude towards "mechanics" and the "mechanical" is a complex one. This should be no surprise, because the nature and meaning of "mechanics" and "mechanical" in the seventeenth century is itself a complex and multi-faceted issue. In this paper I explore the complex and varied uses of the mechanics/mechanical in Harvey's works. I argue that, despite the apparent diversity, Harvey's attitude toward mechanism is consistent and stable, reflecting both the contemporary semantic ambiguities of "mechanics" and the "mechanical" and his own Galeno-Aristotelian understanding of anatomy. Before turning to Harvey's texts (Sect. 6.4), I first clarify more precisely my goal and method in this chapter (Sect. 6.2), and articulate the relevant semantic ambiguities of "mechanical" and "mechanics" in the seventeenth century (Sect. 6.3).

6.2 William Harvey: Influence vs. Influences

It is important for my purposes to distinguish between Harvey's place in the subsequent development of iatromechanism, on the one hand, and Harvey's own view of "mechanics" and its place in anatomy, on the other. The former concerns Harvey's reception, how he was read, perceived, and even portrayed by others. The latter concerns Harvey's own intentions, what he wrote and what he meant by what he wrote. Of course, these two are interrelated. Harvey didn't write in isolation nor use a private language. He wrote to be understood and through much of his reception. Still, the two can and should be distinguished, and my concern here is only with the latter. For this reason I will give significant attention to characterizing Harvey's larger project in anatomy and how he understood that project.

I will also give significant attention to Harvey's influences. Hieronymus Fabricius ab Aquapendente will, for this reason, feature prominently below. Fabricius taught Harvey at Padua and was one of the signatories of his medical degree. More important, though, is Harvey's later, sustained interaction with Fabricius's publications. In the *Praefatio* of his 1651 *Exercitationes de generatione animalium*, William Harvey famously writes, "But in chief, of all the *Ancients*, I follow *Aristotle*; and of the later Writers, *Hieronymus Fabricius ab Aquapendente*. Him [i.e., Aristotle] as my *General*, and This [i.e., Fabricius] as my *Guide*."[6] This is not mere lip service; much

translation in interpreting Harvey as referring variously to an unspecified spout (twice), a syringe, and a specific kind of pump-driven fire engine.

[6] "Praecaeteris autem, *Aristotelem* ex antiquis; ex recentioribus verò Hieronymum *Fabricium ab Aquapendente*, sequor; illum, tanquam *Ducem*; hunc, ut *Praemonstratorem*."(Harvey 1651, 36). The translation is taken from Harvey 1653.

of the rest of this work is structured around critically examining the relevant views of Aristotle and Fabricius in light of Harvey's own research.[7] However much Harvey meant his comment to be a claim particularly about his *Exercitationes de generatione animalium*, the point has wider validity. Harvey's earlier work also clearly shows the influence of a sustained engagement with the texts of his own teacher at Padua (and of Aristotle). The three (very different) sources we have for Harvey's earlier anatomical research reflect a general, critical appropriation of Fabricius's methods and views. In *De motu cordis*, Fabricius's influence is apparent. His work on the valves in the veins appears in Chapter 13 and plays a prominent role in Harvey's argument for the circulation of the blood. In addition, Harvey refers to Fabricius's work on the organs of respiration in the *Prooemium* (he is the first anatomist mentioned, save Galen) and says in Chapter 1 that he was motivated, in part, to work on the heart because Fabricius did not publish on it. Harvey also frequently refers to Fabricius's published views in his *Prelectiones anatomiae universalis* and even more frequently in his working notes for a work on muscle anatomy.[8] Harvey's careful reading of Fabricius is also evident from his own copy of Fabricius's *Opera Physica Anatomica* (1625), a posthumous collection of some of his works.[9] In it we see marginalia and underlining in Harvey's hand sprinkled throughout the two embryological texts included in the collection. At places, we find underlining and marginalia in three distinct pens (but, it appears, all in Harvey's hand), suggesting that Harvey read and reread the work multiple times. If we aim to understand Harvey's approach to anatomy, we must appreciate Fabricius's own project and Harvey's interaction with it.[10]

[7] In this way Harvey "follows" Aristotle and Fabricius in more than simply nomenclature (the immediate point of his comment).

[8] In his lecture notes on anatomy Harvey refers to Fabricius's work numerous times (Harvey 1964 76, 106, 120, 164, 216, 222, 230, 234, 238, 252, and 334). Even more conspicuously, Fabricius's works on muscles and joints appear in Harvey's working notes on muscle anatomy (Harvey 1959 42, 54, 68, 72, 74, 76, 78, 80, 82, 86, 88, 90, 106, 110, 112, 114, 116, 132, 134, and 136.).

[9] This copy is held by the Lilly Library at Indiana University, Bloomington. I am grateful to the Lilly Library for a Helm Visiting Fellowship which funded a visit to examine the book.

[10] Roger French (1994) and Andrew Cunningham (2006) both appreciate Fabricius's importance for understanding Harvey. Andrew Wear (1983), in his effort to place Harvey in a specifically anatomical and Galenic context, chooses Andreas Laurentius (1600) as representative. Despite the prominence of Laurentius's work, this seems an odd choice, given the relative prominence in Harvey's work of references to Fabricius and scarcity of references to Laurentius. Perhaps under the influence of Cunningham's emphasis on the Aristotelian and natural philosophical aspects of Fabricius's anatomical project, Wear supposes that one has to look beyond Fabricius to find a distinctly Galenic and anatomical influence on Harvey. This is unnecessary and unlikely. Fabricius is unquestionably an anatomist and deeply influenced by Galen. Harvey, too, bears an unmistakable Galenic mark, but there is no reason to think this reflects in some special way a non-Fabrician influence.

6.3 Mechanics, Mechanical, Mechanism

In order to disentangle Harvey's various statements about mechanics, it is helpful also to make explicit some of the semantic complexity of the term (and its cognates) in the seventeenth century.

In calling something in the seventeenth century "mechanical" or "mechanics," one could have any of at least six things in mind. First, [1.] one could mean the mechanical or manual arts. In connection with this meaning, one could mean lowly, coarse, or undignified (one thinks of the "Rude Mechanicals" of Shakespeare's *A Midsummer Night's Dream*). One could also mean [2.] the mathematical science of mechanics, which was already by the turn of seventeenth century firmly established as a theoretical, mathematical science of machines, typically located within the intellectual landscape as an Aristotelian subordinate science. Here, one could also mean to pick out especially the conceptual developments and progress made in this science through the assimilation and expansion of works by (e.g.) Archimedes or Pappus. Or, again, closely related, but distinct still, is [3.] the seventeenth century transformation of mathematical mechanics into what many called "physico-mathematics." Thus, as Alan Gabbey nicely points out,[11] Boyle can write in 1671 of "Mechanics" in a broader sense:

> …I do not here take the term mechanicks in that stricter and more proper sense wherein it is wont to be taken, when it is used only to signify the doctrine about the moving powers (as the beam, the lever, the screws, and the wedge) and of framing engines to multiply force; but I here understand the word mechanicks in a larger sense, for those disciplines that consist of the applications of pure mathematicks to produce or modify motion in inferior bodies; so that in this sense they comprize not only the vulgar staticks, but divers other disciplines, such as the centrobaricks, hydraulicks, pneumaticks, hydrostaticks, balisticks, &c. the etymology of whose names may inform you about what subjects they are conversant.[12]

Another distinct sense [4.] is given the term "mechanical" in the context of the "mechanical philosophy." Here, unlike in the case of physico-mathematics, the successful harnessing of mathematical tools is not of the essence. Rather, the guiding idea is that proper (true or promising or excellent) natural philosophy will invoke only the small set of properties typically employed in our understanding of machines: shapes, size, motion, contact forces, etc. In close connection with this commitment is the employment of sub-visible bodies (corpuscles or atoms) and structure to explain macroscopic phenomena. Such explanations were quite often entirely qualitative and devoid of *mathematical* inference (be it geometrical, arithmetical, or algebraic). As Domenico Bertoloni Meli has recently stressed, the relevant contrast to

---

[11] Gabbey 2004. My discussion of varieties of mechanics in the seventeenth century has benefited greatly from Gabbey's work.

[12] Boyle 1772, vol. 3, 435. In our context, one thinks particularly of Borelli's 1680–1681 *De motu animalium*. In the Dedicatory to Queen Christina Borelli invokes the Platonic idea that God geometrizes and insists that since animals are bodies and their operations are or require motions they are subject to geometrical study.

such mechanism in the context of seventeenth century medicine was not teleology so much as appeal to the activity of soul or Galenic faculties.[13] In calling a seventeenth century thinker mechanical one could also mean [5.] that he privileges the use of machine analogies—be it in heuristic, explanatory, or rhetorical contexts. Of course, typically machine analogies were invoked by supporters of a "mechanical philosophy;" but making precisely this point requires distinguishing these two senses and reflects the state of seventeenth-century technology.

Finally, in the wake of the seminal paper by Machamer et al. (2000), the term mechanism has become an important concept in contemporary philosophy of science. Taking a cue from that original paper, and abstracting from a host of subtleties and controversies spawned by it, we can understand a thinker to be mechanistic, in this sense [6.], if he champions or primarily employs a particular kind of explanation: explanation by the description of a "mechanism."[14] Machamer, Darden, and Craver define "mechanism" in the following way:

> Mechanisms are entities and activities organized such that they are productive of regular changes from start or set-up to finish or termination conditions.[15]

To call a seventeenth century thinker "mechanistic," in this sense, would be to suggest that they privilege a particular mode of explanation—a particular approach to rendering the natural world intelligible: explanation by the description of this kind of "mechanism."

Thus, when considering the question of the mechanical in Harvey, we can distinguish six distinct, if variously related, senses of the term.

1. The manual arts (house building, etc.); and so, perhaps, lowly, mean, or coarse ("Rude Mechanicals")
2. Mathematical Mechanics (varieties of which depend more or less on Jardanus, the Pseudo-Aristotelian *Quaestiones Mechanicae*, Archimedes, Pappus, and other rediscovered ancient sources)
3. Physico-Mathematics (of which Galileo's first science in the *Two New Sciences* is an early instance)
4. The Mechanical Philosophy (as coined and conceived of by Boyle, who took Descartes as a paradigmatic example)
5. Privileging heuristic, explanatory, rhetorical, or pedagogical use of machine analogies
6. Explanation by description of a mechanism (Machamer, Darden, and Craver)

[13] Bertoloni Meli 2011, 12–16.

[14] See the opening line of their paper: "In many fields of science what is taken to be a satisfactory explanation requires providing a description of a mechanism." (Machamer et al. 2000, 1)

[15] Machamer et al. 2000, 3.

6.4 Texts and Contexts

With these senses distinguished, I now turn to Harvey's texts. I am concerned here primarily to understand Harvey's attitude in these various texts toward the "mechanical" in the first five senses. But I also consider the place in Harvey's anatomical project of "mechanism" in the final sense. This last I think is important because of the way it helps us appreciate the precise explanatory project of Harvey's anatomy.

6.4.1 Anatomia maechanica *and Anatomical Method*

I will begin at the beginning: the first folio of the *Prelectiones Anatomiae Unversalis*, Harvey's lecture notes started around 1616, after being appointed Lumleian Lecturer.[16] This set of notes is particularly helpful, because they open with an explicitly methodological discussion of the definition, divisions, and goals of anatomy. This *accessus* of sorts is a rich resource—particularly so, if it can be connected with Harvey's practice as reflected in his discussions of individual organs throughout the *Prelectiones* and in his other works.[17]

In the *Prelectiones* presentation of several different ways of dividing anatomy, one is of particular interest. Harvey divides anatomy into philosophical, medical, and mechanical (*maechanica*). (See Fig. 6.1; this division is in the lower, left hand corner of the transcription.[18]) Although it may be tempting to connect this use of "*maechanica*" with one of the other senses of mechanical, it is most likely that Harvey employs it here in the first sense (the mechanical or manual arts). *Anatomia maechanica*, understood in this way, is the manual or craft-like skill involved in anatomy: the technological know-how and hand-eye coordination required to successfully dissect. This mechanical aspect of anatomy appears in Harvey's list of five *capita* of anatomy on this same folio. The last of these is "know-how and skill at dissection and the preparation of the preserved cadaver."[19] It also appears later in the

[16] For an introduction to these notes see Whitteridge's introduction in Harvey 1964 and Keynes 1966, 84–111.

[17] Whitteridge shows convincingly that in the anatomy proper, after this methodological introduction, Harvey depends heavily on Caspar Bauhin's *Theatrum anatomicum* (Bauhin 1605). However, Benjamin Goldberg argues that Harvey's use of Bauhin is more creative than Whitteridge seems to imply (Goldberg 2012). Regardless, Harvey is more straightforwardly responsible for the content and structure of the methodological introduction. On the anatomical *accessus* in the middle ages see French 1979.

[18] 1v. This and all transcriptions from the *Prelectiones* are my own. Transcriptions are made from the images of the manuscript provided in the 1886 transcription and reproduction (Harvey 1886). In making my transcriptions I have benefited greatly from consulting both the transcription provided in this edition and Whitteridge's transcription (Harvey 1964). I provide the folio number for the quotations (e.g. "1v" signifies folio 1 *verso* and "3" signifies folio 3 *recto*). Translation is my own (I have consulted and benefited from Whitteridge's translation).

[19] 1v. Peritia aut divisionis dexteritas et praeparatio cadaveris conditio.

Anatomia est facultas quae occulari
inspectione et sectione partium usus et actiones.

Anatomiae ad 5 Capita
- Historia
- usus action utilitates propter quid
- observatio eorum quae raro et morbi de
- problemata ex autoribus resolvere
- peritia aut divisionis dexteritas
- et praeparatio cadaveris conditio

Anatomia
- philosophica
- medica
- maechanica

Anatomia alia
- popularis
- quae hic libro iij ventrium
- curiosa et philosophica medica partes

1 de partibus externis, physiognomie
2 ossium sceleton
3 musculorum ligamentorum
4 de organis sensetivis et vocis
5 de vasis venis arteriis nervis
6 de partibus similaribus
7 de genitalibus embrione mammis

Fig. 6.1 Transcription of part of folio 1v, Harvey's *Prelectiones*

fourth of his *Canones Anatomiae Generalis*: "Cut up as much as may be in present so that know-how is learned along with *historia*."[20] At the top of folio 1v, Harvey inserts a general definition of anatomy: "Anatomy is the faculty that by ocular inspection and dissection [grasps] the *usus* and *actiones* of the parts."[21] I will return below to the first part of his definition, characterizing the knowledge at which anatomy aims. Here it is important to stress that Harvey understands anatomy to have a characteristic *method* by which it attains this knowledge: dissection and ocular inspection. For this reason, anatomy includes a "mechanical"—i.e., manual—component.

When we turn to Harvey's criticism of Fabricius for employing the "petty reasoning borrowed from mechanics," we see that here too he uses the term in the sense of the manual arts. This is clear from the context of Harvey's critique in the *Exercitationes de generatione animalium.* Harvey is disagreeing with Fabricius's account of the order of the formation of the parts of the chick during its development in the egg:

> But when he asserts that the bones are made before the muscles, the heart, liver, lungs and all the praecordia, and maintains that all the inward parts must exist before the outward, he relies on probable arguments [*rationibus probabilibus*] rather than on ocular inspection, and laying aside the judgment of the senses which is grounded upon dissections, *he flies to petty reasonings borrowed from mechanics* [*ratiunculas e mechanicis*], a thing which is very unbeseeming in so famous an anatomist. For he ought to have told us faithfully what daily changes his own eyes had discovered in the egg before the foetus in it came to perfection. And especially so as he professed to be writing an Historia of the Generation of the Chicken out of the Egg and he illustrated in pictures what happened from day to day. It was, I say,

[20] 4. Cutt up as much as may be in present *ut cum historia peritia innotescat.*

[21] 1v. Anatomia est facultas quae occulari inspectione et sectione partium usus et actiones.

> befitting so much diligence to have informed us on the evidence of his own eyes what is made first in the egg, what later and what things happen simultaneously, *and not by using the example of [building] houses* [domus] *or ships* [navis], *to have put forward some hazy conjecture* [conjecturam umbratilem] *or opinion* [opinionem] *concerning the order and manner of the formation of the parts.*[22]

Fabricius's "petty reasonings" are borrowed from "mechanics" *in the sense of practitioners of the manual arts*, as is made clear by the reference to house and ship building late in the quotation. That is, instead of employing the proper anatomical method, depending on sense and dissection (i.e., on the *anatomia maechanica* we have just discussed), Fabricius turns to the example of the manual arts in order (by analogy) to determine the sequence and manner of the formation of the parts of the chick. Harvey's criticism regards the appropriate method in anatomy for producing *historia* and ultimately (as we will see below) final causal knowledge of the parts. For Harvey, anatomical *historia* is produced by dissection and ocular inspection. Harvey's criticism, here, has a certain irony. Instead of employing manual art himself, Fabricius relies on analogies with the manual art of others.[23]

6.4.2 Machine Analogies in Context

If we turn our attention from the *Exercitationes de generatione animalium* to the *De motu cordis*, we encounter machine analogies that might make us think Harvey is a mechanical thinker in the fifth sense identified above (privileging machine analogies). In Chapter 5, Harvey summarizes the *historia* of the motions of the heart laid out in the previous chapters and identifies the *actio* of the heart. In his summary, he compares the heart to machines.

> Nor is this otherwise done than when, in machines [*machinis*], one wheel moves another and they all seem to move together; or in that mechanical contrivance [*mechanico illo artificio*] which is fitted to firearms where, by compressing the trigger, the flint falls, strikes forcibly upon the steel and brings forth a spark which falls onto the powder which is ignited, enters the touch-hole and explodes, and the bullet flies out and pierces the mark, and all these movements by reason of their swiftness appear to happen simultaneously as in the twinkling of an eye.[24]

Here Harvey employs a machine analogy in his articulation of the motions of the heart. However, the *point* of the analogy is that in both there is a quick, coordinated series of motions producing one action. The point is *not* that in both there is a series of motions produced entirely by the shape, size, and motion of (rigid) parts, and by

[22] Translation is Whitteridge's (Harvey 1981, 18); emphasis added.

[23] Thanks to Cynthia Klestinec for stressing this point to me during the discussion period after my presentation of this material at a conference at the University of Pittsburgh. I suspect that Harvey is also concerned that Fabricius underestimates the possibilities in natural processes and reverses the order of the Aristotelian principle "Art imitates Nature." Fabricius seems implicitly to think rather that nature imitates art.

[24] Harvey 1976, 50–51 (Harvey 1628, 30).

their contact. That is, the machine analogy is *not* being employed to make some kind of "mechanical philosophy" (our fourth sense of "mechanics") more plausible. It is not meant to help the reader appreciate the explanatory power of a restricted mechanical ontology.

Furthermore, it is not Harvey's preferred analogy. He immediately provides a second analogy. This second analogy is not to a machine but to another animal activity:

> So likewise in swallowing, the food or drink is thrown into the gullet by the elevation of the root of the tongue and the compression of the mouth, the larynx is closed by its own muscles and by the epiglottis, the top of the gullet is lifted up and opened by its muscles .... And yet, notwithstanding that all these motions are made by several and contradistinct organs, whilst they are done in harmony and order, they are seen to make but one motion and action which we call swallowing.[25]

It is this analogy that he carries forward into his discussion of the action of the heart: "It clearly happens thus in the motion and action of the heart, which is a kind of swallowing and transfusion of the blood from the veins into the arteries."[26] In fact, careful examination reveals that the comparison between the coordinated and harmonious motions involved in swallowing and those found in the heart is something *more* than an analogy. Harvey says here that the action of the heart "is a kind of swallowing (*deglutitio quaedam est*)."

When we look through Harvey's unpublished notes we find other cases in which he employs analogies for other multi-component activities. In the *Prelectiones*, for example, he draws an analogy between the organs of digestion and chemical apparatus.

> Wherefore Nature has established diverse offices and employs diverse instruments, *just as in boiling in chimistria diverse heats, vessels, furnaces* [are used] to draw away the phlegm, raise the spirit, extract oil, ferment and prepare, circulate and perfect. So Nature makes use of the mouth, stomach, guts, mesenteric vessels, liver and so forth.[27]

However, elsewhere in these notes he develops another (Aristotelian[28]) analogy for the same organs, this time to politics.

> Just as in some rather small state the same man is judge, king and counselor, while in larger states these offices are separate, so is it in animals and their parts; politicians indeed take many analogies from our medical art. And so in the lower belly where are made diverse concoctions needing different heats, different preparation and different nutriment, there are diverse organs besides the heart which provides the heat, and these diverse organs are the tutelary deities and the diverse artificers of the different functions, that is the liver, the spleen, the stomach and so forth—as the alchemists by their diverse furnaces, heats; so diverse organs.[29]

---

[25] Harvey 1976, 51 (Harvey 1628, 30).

[26] Harvey 1976, 51 (Harvey 1628, 30).

[27] 24v (Harvey 1964, 101; translation amended).

[28] See Aristotle, *De motu animalium* 703a28–703b2.

[29] 91 (Harvey 1964, 313; translation amended).

Elsewhere, in his working notes on muscle anatomy, in considering how muscle contraction is brought about, Harvey provides a series of diverse analogies, under the heading *Ratio mechanica*:

> How appetite brings heat, Δ prick ..., ... water.
> How heat brings spirit: Hermes oven and gunpowder.
> How spirit works in fibre, Δ wet rope, barterole of veins.
> How fibre drives tendon, Δ legs of guinea-fowls, peru
> How tendon moves bone, Δ sucking fish, seaweeds, sponges.
> So the motor organs in some animals, and likewise in man, are: spirit, fibre, muscle, nerve and tendon.[30]

Although, in this case, it is not easy to decipher Harvey's hand, let alone his mind, still the diversity of analogies is striking. He compares the components of the process to gunpowder, ropes, various animals, and to other anatomical features (veins).

The analogies we have seen all involve decomposing and localizing processes or activities of the animal and typically stress the way these parts work together to bring them about. In his description of the motion of the heart, Harvey stresses the rapidity and harmony of the multiple motions contributing to one "motion and action." In his description of the organs of digestion, his analogy to the apparatus of the chemists stresses a diversity of subprocesses ordered to the completion of digestion, and the political analogy brings out especially the way these subprocesses are ordered to one end. A similar point seems to occupy Harvey's attention in his decomposition of the motion of the animal. He traces the different steps or components involved in animal locomotion and the parts involved in each.

One feature of such a decomposition, to which Harvey, following Galen, gives special attention, is the action of one part on another, and the resulting distinction between an active and passive motion in a part. In numerous places, Galen emphasizes that some motions in the parts are active and some passive, and that for the study of the parts it is especially important to identify the active motions. For example, in Book XVII of *De usu partium*, Galen writes,

> I have said that action is *active* motion because many motions occur passively and those which happen to bodies when other bodies move them are even called passive. Thus the bones in the limbs have a motion produced by the muscles that are in the limbs and move the bones now outward, now inward at their articulations. With respect to the first principle of motion, which is the authoritative part of the soul, the muscles play the role of instrument, but with respect to the bone moved by them they play both this role and that of the efficient also.[31]

Galen makes the same point, adding an additional example, near the beginning of his *De naturalibus facultatibus*:

> And *activity* [*energeian*] is the name I give to the active change or *motion*, and the *cause* of this I call a *faculty*. Thus, when food turns into blood, the motion of the food is passive, and that of the vein active. Similarly, when the limbs have their position altered, it is the muscle which produces, and the bones which undergo motion. In these cases I call the motion of

[30] 111v (Harvey 1959, 139).

[31] Galen 1968, 724.

> the vein and of the muscle an activity, and that of the food and the bones a symptom or affection, since the first group undergoes alteration and the second group is merely transported.[32]

It is concern with identifying the Galenic action (active motion) of a part that is behind Harvey's comparison of the arteries to bladders or wineskins in the *De motu cordis*.

> [T]he arteries are filled and distended by reason of the inflowing and inthrusting of blood made by the constriction of the ventricles; as likewise, that the arteries are distended because they are filled like waterskins or bladders, and they are not filled because they are distended like a pair of bellows.[33]

Harvey's point here is that the arteries are undergoing a passive motion in their pulsation, like bones in a moving limb, rather than exhibiting an action like the contracting muscles. And of course, the complement of this claim (that the motion of the arteries is *passive*) is that the contraction of the heart is *active*. Indeed, Harvey argues in the *De motu cordis* that the heart is rightly called a muscle, that it produces locomotion in something else as (other) muscles do—though blood rather than bones and limbs. In fact, he suggests that understanding the heart's production of motion in the blood by its contraction reflects a general truth about all locomotion in animals (i.e., that it is produced by contraction) and expresses a hope to publish on the locomotive organs eventually:

> This truth concerning local movement, and that the immediate motive organ in every movement of all animals in which there is from the beginning a motive spirit, as Aristotle says in his book *De spiritu* and elsewhere, is contractile, and in what way *neuron* is derived from *neuo*, that is I nod, I contract, and that Aristotle did recognize muscles and not incorrectly referred every movement in animals to the nerves or to that which is contractile and therefore called those muscular bands in the heart nerves, all this I think will be made clear if at any time I shall have liberty to demonstrate from my own observations these matters concerning the motive organs of animals and the structure of the muscles.[34]

Given, then, that the heart is a muscle that exhibits, in its forceful systole, an *active* motion, it is not surprising that, Harvey will identify in the heart a pulsific "force" or "faculty" in his summary of his views in the second *exercitatio* of the *De circulatione*. As Galen says in the *De naturalibus facultatibus* passage above, to any such *action* of a part, we will find a corresponding *faculty* as its cause. In Harvey the passive diastole of the arteries (so vividly communicated by his analogies to bladders, wineskins, and gloves), is *not* an instance of a systematic effort to eliminate Galenic faculties. Harvey is not a mechanical philosopher. Nor is Harvey a mechanical thinker in the sense of privileging machine analogies.[35] Finally, when he does

---

[32] *De naturalibus factultatibus* I. The translation is A. J. Brock's (Galen 1916).

[33] Harvey 1976, 39 (Harvey 1628, 24).

[34] Harvey 1976, 127 (Harvey 1628, 68).

[35] In fact, any tendency we have to group his glove, bladder, and wineskin analogies with his wheels of a machine and firing mechanism analogies as exemplifying a "mechanization" of the animal is due to our own hope or expectation to find in Harvey evidence of a "mechanical philosophical" rejection of faculties and occult powers. After all, gloves, bladders, and wineskins are hardly *machines*.

employ machine analogies, they are not aimed at establishing the explanatory adequacy of a mechanical philosophy, devoid of Galenic faculties.

6.4.3 Describing Mechanisms and the Goal of Anatomy

Of course, one need not be a mechanical philosopher to be "mechanical" in the final sense distinguished above. Is Harvey, in these various contexts discovering "mechanisms" or, more importantly, providing explanations by describing such mechanisms? One might think that chapters 2 through 5 of the *De motu cordis* are doing just that—describing a mechanism for the transference of blood from the veins to the arteries and thereby providing an explanation of a behavior of the heart. Perhaps Harvey is explaining the action of the heart by identifying and describing the "entities" (the ventricles, valves, etc.) and "activities" (the contractions of the ventricles, the competencies of the valves, etc.) that are (spatiotemporally) organized "such that they are productive of regular changes from start or set-up to finish or termination conditions."

As tempting as this may be, such a mechanistic interpretation does not accurately describe what Harvey is doing in the *De motu cordis*.[36] Of course, anatomy is about the parts. It involves a systematic breakdown of the animal into parts and parts of parts. In this way, it does involve localization of animal activities in the parts and so too the identification of a system of component parts, such that the spatiotemporally structured exercise of the components' capacities constitutes a behavior of the system. However, the goal of anatomy for Harvey is not mechanistic explanation of biological phenomena, but final causal *scientia* of the parts of animals, articulated in terms of their *actiones* and *usus* or *utilitates*.

Recall from above that Harvey defines anatomy as "the faculty that by ocular inspection and dissection [grasps] the *usus* and *actiones* of the parts." We have already had occasion to stress the method or means by which anatomy acquires knowledge of the parts (ocular inspection and dissection). We must now look more carefully at its goal: grasping the actions and uses of the parts. Action (*actio*) and use (*usus*) are Galenic terms[37] important in the anatomical tradition of the sixteenth and early seventeenth centuries. They are particularly prominent in the work of Hieronymus Fabricius ab Aquapendente, professor of anatomy at Padua when Harvey studied there.[38] Fabricius's publications, which exerted a sustained influence

[36] I draw material here from a related discussion in Distelzweig 2014c.

[37] *Actio* translates Galen's *energeia* (and sometimes *ergon*) and *usus* translates *chreia*.

[38] For a general introduction to Fabricius, see Adelmann 1942. For a more detailed discussion of Fabricius' understanding of anatomy see Cunningham 1985; but see also Siraisi 2004 for a critique of Cunningham's view. Distelzweig 2014a and 2014b attempt to provide a more nuanced discussion of the interplay between Galenic and Aristotelian resources in Fabricius's work, building on insights from both Cunningham and Siraisi. For a brief, highly suggestive discussion of Fabricius's influence on Harvey, see Cunningham 2006. See also French 1994. But, for a (brief) more deflationary view of Harvey's relationship to Fabricius, see Klestinec 2011, 146, 164–70.

on Harvey, typically had a three-part structure. First, Fabricius provides a systematic description of the part being studied, including interspecific variation (*historia*); second, he identifies and examines the action of the part (*de actione*); third, he discusses the *uses* of the (varying) features and components of the part (*de utilitatibus*).[39]

As discussed above, the *action* of a part is an *active* motion (a motion for which it is responsible), and prominent examples in the tradition include the hand's action, grasping, and the eye's action, vision. That the action of a part must be an *active* motion precludes (e.g.) the motions of the bones, because they are caused by something else (the muscles), and so are passive. The concept of *use* is closely related to action and its centrality to anatomy derives especially from Galen's *De usu partium*, which takes as its project the study of the *usus* of the parts. Galen distinguishes between the action of a part and its use in Book XVII of the *De usu partium*, immediately before the passage quoted above.

> Now the action [*energeia*] of a part differs from its usefulness [*chreias*], as I have said before, because action is active motion and usefulness is the same as what is commonly called utility [*euchrestia*].[40]

We do not learn much from his gloss of use (*chreias*) as utility (*euchrestia*), but we have other resources with which to flesh out Galen's notion. Galen distinguishes in the same passage between a part's action and the usefulness of that action:

> Hence the usefulness of first importance to animals is that which is derived from actions and the second is that from the parts; for there is no part which we desire for its own sake, and a part deprived of its action would be so superfluous that we should cut it off rather than wish to keep it.[41]

The use of an action is, roughly, the action's contribution to the life of the animal. For example, in his *De usu respirationis*, Galen asks what the use of breathing is. He first notes:

> That it is not a trifling use is clear from our inability to survive for even the shortest time after it has stopped. Hence also it is obvious that its importance is not for any particular and partial activity, but for life itself. For just as our walking is impaired in so far as we are deprived of the means of walking, and our seeing, if we lose the wherewithal for seeing, so, if what is necessary for life is cut off, we die.[42]

Eventually, Galen will argue that the usefulness of respiration is the maintaining of the innate heat (in the heart, especially, but also in the brain) by fanning and cooling and the removal of waste products from the process of combustion of blood in the heart. This is the contribution the action of respiration makes to the life of the animal. However, it is not the action itself.

The understanding of the uses of a *part*, in contrast, involves understanding the way it contributes to some action. In Fabricius's tripartite treatises the final part is

---

[39] Fabricius typically, but not always, prefers *utilitas* to *usus*. I have not found any systematic reason for his occasional use of *usus*.

[40] Galen 1968, 724.

[41] Galen 1984, 724.

[42] Galen 1984, 81.

devoted to discussion of the *utilitates* of *parts*, not those of the *actions*. For Fabricius, this involves systematically examining how the part carries out some action. This will involve identifying the component part that is particularly responsible for the action and how the other component parts aid in that action. He states this in general terms at the beginning of his treatment of action and use in his *De formato foetu*:

> For the *utilitates* of an organ always have reference to its action, and depend upon the action which proceeds from a homogeneous part of it. For this reason, in every organ there is always provided one part from which the action proceeds, while the other parts of the organ are related to the action as useful assistants.[43]

The study of the usefulness of parts is for Fabricius, like Galen, in the first instance a study of the suitability of the part to its action.[44] For Harvey, too, the use of a part is, roughly, the way it contributes to some action. For Harvey, anatomy is the ability to determine these actions and uses of the parts by ocular inspection and dissection.

It is this rich and detailed understanding of "action" and "use" in Fabricius and Galen that Harvey invokes when he says that anatomy grasps the actions and uses of the parts. Harvey opens a later section of the *Prelectiones* devoted to discussing action and use by explaining the place of action and use in anatomy:

> Since the end of Anatomy is to know or grasp the parts and to know [them] through their causes and these [i.e., causes], in all animals, [are the] 'that for the sake of which' and 'that on account of which'; therefore: that on account of which: (1) Action; (2) Use.[45]

Anatomy is the faculty that grasps the actions and uses of the parts *because* the end of anatomy is to have causal knowledge of the parts and the actions and uses *are* these causes. In particular, they are causes *cuius gratia;* i.e., they are the *final* causes of the parts.[46] In addition, Harvey makes it clear here that we are to understand the final causes of a part and its variations not just in humans but *in animalibus*.[47] This is reflected in the titles of Harvey's two main published works: *Exercitatio anatomica*

[43] Etenim utilitates semper ad actionem referuntur, eamque respiciunt, quae a similari parte prodit: propter quam causam in quoquo organo perpetuo datur una pars, quae est praecipuum instrumentum actionis, ut puta a qua action proficiscitur, aliae vero ad ispam, ut ministrae & utilesreferuntur. Translation is adapted from Adelmann 1942, 276.

[44] In this regard I follow May's analysis in her translation of *De usu partium* (Galen 1968, 9). For other treatments, see the discussion in Wilkie and Furley 1984 (58–69) and in Hankinson 1989. I agree with Hankinson that *chreia* is not always best translated "usefulness" in Galen's texts; however, I think that with attention to the distinction between the *chreia* of parts and that of actions, and the possibility of more and less technical uses of the term, much of the diversity of uses in Galen appear coherent. Regardless, in his introductory discussion of the general approach to studying the *chreia* of parts (using the hand as example) opening book 1, it seems clear that what he is seeking to isolate is indeed the fittedness of parts to their actions.

[45] Quoniam finis Anatomae est scire vel cognoscere partes et scire per causas et hae in omnibus animalibus cuius gratia et propter quid, ergo propter quid: 1. actio, 2. usus. (6)

[46] For a discussion that further relates this final causal knowledge to definitional knowledge of the parts, see Goldberg 2012.

[47] This preoccupation with all animals is also reflected in Harvey's criticism, at the beginning of Chapter 6, of anatomists who look only at human anatomy. This approach is characteristically

de motu cordis et sanguinis ***in animalibus***; *Exercitationes de generatione* ***animalium***.[48] For Harvey, anatomy aims at universal, final causal knowledge (i.e., Aristotelian *scientia*) of the parts of animals.

With this account of Harvey's understanding of the goal of anatomy we can appreciate the inadequacy of the interpretation of *De motu cordis* 2–5 as an explanation of the action of the heart by description of a mechanism. For Harvey the *actio* of the heart (transference of the blood from the veins to the arteries and through them back again to the veins and heart) is not *explained* by describing the mechanism for it. Rather this overarching *actio* itself will explain, as final cause, the mechanism in its component parts and activities. For Harvey, the mechanism of the heart described in Chapters 2 through 4 will, ultimately, be the explanandum and the *actio* of the heart the explanans. More immediately, however, chapters 2 through 4, by providing a *historia* of the motions of the heart, aim to *establish*, not *explain*, the *actio* of the heart identified in Chapter 5. For Harvey, it is by means of such systematic ocular inspection and dissection that we come to grasp the causes.

To see this, it is helpful to recognize that the *De motu cordis* reflects the *historia, actio*, *usus* structure so prominent in Fabricius' publications. In chapters 2 through 4, Harvey presents a *historia* of the heart (and arteries), focused particularly on their motion.[49] In chapter 5, after summarizing the motions of the heart (and arteries), Harvey identifies one of its *actions* (he says there may be others).

> The motion of the heart then is after this manner and one of the actions of the heart is the very transmission [from the veins to the arteries] of the blood and its propulsion to the extremities by the intermediacy of the arteries....[50]

In chapters 6 and 7, Harvey identifies the presence of lungs in humans as a source of confusion for past anatomists and defends the universality of the identified action. That is, he argues that in *all* animals (including humans and other lunged animals), an action of the heart is the transference of the blood from the veins to the arteries. In chapter 6 he argues that the identified action is clear in simpler animals and in all animals during fetal development. In chapter 7 he argues that in lunged animals, too, the heart transfers the blood from the veins to the arteries—doing so via the pulmonary transit. Similarly, chapters 8 through 14, in which Harvey presents his central argument for the systemic circulation, are at service of his identification of the *action* of the heart. This is clear from chapter 14, in which Harvey concludes his demonstration of the systemic circulation. Claiming that he has shown by observation and argument the direction and amount of the motion of the blood through the heart, arteries, and veins, he states:

Aristotelian (see Lennox 1987, 1991). On its role in Harvey's *De generatione animalium*, see Lennox 2006; Goldberg 2012.

[48] Emphasis added.

[49] In the *Prelectiones*, Harvey lists motion as one of the features to be studied in the *historia* of a part (5).

[50] Harvey 1976, 51–52 (Harvey 1628, 30).

> It must of necessity be concluded that the blood is driven into a circuit by a circular motion in living creatures, and that it moves perpetually; and that this is an *action or function of the heart*....[51]

For Harvey, these chapters present an argument for a (momentous) refinement of the chapter 5 description of the *action of the heart*.[52]

And for Harvey *action* is a teleological notion. In this he follows Aristotle, Galen, and his teacher Fabricius. Aristotle, for example, says that a part is for the sake of its action, as a saw is for the sake of sawing.[53] Galen, in turn, understands the study of the use of a part to be the study of how all its components and features are designed to contribute especially to the execution, improvement or protection of the part's action. In this way, the action serves as final cause of those components and features. Finally, Fabricius also understands action to be a teleological notion. For example, in his work on the larynx Fabricius introduces his section *de utilitatibus*: "The third part [of the treatise] being the part that pursues the *utilitates*...which always look toward and contemplate the action of the larynx (that is, voice) and *are directed to that action as towards an end*."[54] In Fabricius, this final causal relation between the action and the components and features of a part is articulated in terms of the *utilitates* of these components and features and is treated in the final section of his works.

Harvey also turns to this topic in the final chapter of the *De motu cordis*. Chapters 15 through 17 are framed by Harvey as providing additional arguments in support of the identified action of the heart, the circulation of the blood. Chapter 15 provides general reasons for thinking the circulation of the blood is appropriate and even necessary. Chapter 16 provides a series of arguments *ex consequentiis* in favor of the circulation of the blood. That is, Harvey argues for the proposed action of the heart by showing how it can be invoked as cause in the explanations of a range of

[51] Harvey 1976, 107 (Harvey 1628, 38) Translation adapted and emphasis added.

[52] My suggestion here is best understood as a refinement of Bylebyl's analysis of the structure of the *De motu cordis* (Bylebyl 1973 and especially Bylebyl 1977). Bylebyl sees two structures, one (chapters 8 through 16) inserted into and distorting and obscuring the other (*Prooem*, chapters 1 through 5, chapter 17). Bylebyl, however, seems not to notice that Harvey presents the circulation as the *action* of the heart. This identification determines where in the text the argument for the circulation must appear and provides the overarching unity of the work, a unity centered on articulating *scientia* of the heart. Recognizing this weakens (at least in part) Bylebyl's argument that the *De motu cordis* was written in two stages. Once we see that the circulation is presented as the action *of the heart* it becomes less clear that Harvey's heart-centric descriptions (in the *Prooemium* and Chapter 1) "hardly do justice to the full treatise" (Bylebyl 1973, 446). Of course, more would need to be said to evaluate fully the evidence and arguments Bylebyl employs in his insightful work on this issue. Regardless, even if the *De motu cordis* was composed in two stages, Harvey still chose its final structure and was pleased enough with it to publish.

[53] See, e.g., *Parts of Animals* I.5 645b15-20.

[54] Tertiam partem eam esse, quae utilitates persequitur, tum totius, tum partium organi, jam & vulgo notum, & a me propositum est, quae sane utilitates perpetuo laryngis actionem, hoc est, vocem respiciunt, & contemplantur, in eamque tanquam in finem diriguntur. (Fabricius 1687, 290) Translation and emphasis are my own. See also the opening of Part 2 of *De formato foetu* discussed above.

(mainly medical) phenomena. Near the end of this discussion, Harvey describes chapter 17.

> Therefore in this place, that is to say in the following chapter [Chapter 17], I shall endeavor to refer to their proper uses and true causes, only those things relating to the fabric of the heart and arteries which are visible in the course of an anatomy….[55]

Chapter 17 is thus framed as an extension of the project of chapter 16. In it Harvey shows how the circulation (as an action of the heart) can be invoked in final causal explanations of the heart and arteries and their variation. Thus, Harvey understands the action of the heart (the thrusting of the blood from the veins and arteries and into circulation) to explain the various components and features of the heart—and their variations. The circulation is the final cause of the heart.

It is true that Harvey does not identify, in addition, the final cause of the *circulation* (i.e., the Galenic *usus* of that *action*) but even in the highly polemical context of the 1649 second *exercitatio* of the *De circulatione*, Harvey assumes that this is the ultimate goal. There he insists that one could only have such open questions (*problemata disputanda*) if such facts (as the circulation) could be established before we determine their final cause: "If nothing could be admitted by sense without the evidence of reason, or on occasion against the dictate of reason, there would now be no *problemata disputanda*."[56]

6.4.4 Mathematical Mechanics and De artificio mechanico musculorum

Perhaps the most striking place we encounter the "mechanical" in Harvey's corpus is in his working notes for the never published work on muscles he announced in the passage from *De motu cordis* quoted above. In these notes[57] we find a chapter entitled "On the Mechanical Construction of Muscles" (*De artificio mechanico musculorum*). Harvey introduces this long chapter by interweaving general teleological principles of nature (including from the *De incessu animalium*) and a selective paraphrase of the preface to the *Quaestiones Mechanicae*, along with reflections on their implications for the study of muscles. Here Harvey establishes a parallel between mechanics and muscle anatomy. (See Figs. 6.2 and 6.3.) Mechanics allows us to overcome difficulties and move great weights with small forces in order to accomplish something useful that is *praeter naturam*. Harvey suggests that the same kind

---

[55] Harvey 1976, 117 (Harvey 1628, 63).

[56] Si nihil admitteretur per sensum, sine rationis testimonio, aut contra quandoque rationis dictamen, jam nulla essent problemata disputanda. (Harvey 1649, 97)

[57] Whitteridge has provided a transcription and (free) translation of these notes in Harvey 1959. The transcriptions and translations from these notes here are my own from a microfilm reproduction of the manuscripts. Both my transcriptions and translations have benefited from Whitteridge's.

De Artificio Mechanico
Musculorum

Omnia Dei et Naturae opera perfecta.

Ergo {
nec deficiunt nec redundant
nec quid quam frustra
quod natura et secundum Naturam optimum
quod optimum cuique quod maxime secundum Naturam
si melius hoc modo, secundum Naturam,
si secundum Naturam hoc modo: melius
nil facit per plura : quod potest per pauciora,
nec [?] respicit ad pauciora ubi ad plura

quia In omni actione magis et minus
Natura in fabrica musculorum ad duas
respicit actiones et functiones, seu
perfectionem actionis.
unde in musculo duo animadvertenda sunt
viz {
compositio gratia actionis
arteficium mechanicum gratia roboris
et virium.

Mechanica sicut illud ^quod^ superat ea
a quibus Natura superamur et succurrit
difficultatibus cum quod praeter Naturam
et utilitatem fit Aristoteles ut cum minora
superant maiora et momentum parvum
habentia magna movent pondera.
Sic in musculis Natura nusquam difficultatibus
huiusmodi succurrere deficit.

Unde in musculorum speculationem non solum
temperamentum et quae consecuntur observanda
gratia actionis et contractionis, sed
quomodo gratia virium et functionum factum
et hic tot vere miranda quomodo
musculi vires ossa movent et annexa pondera

Fig. 6.2 Transcription of folio 106 of Harvey's notes on muscle

of difficulty is encountered in the functioning of muscles (i.e., the need to move large weights by small forces and to accomplish varied ends). Nature never fails to assist in these kinds of difficulties. Thus, in the movement of the bones and the weight of the attached limbs, marvelous things are accomplished. By drawing this parallel, Harvey suggests that in the construction of the muscles, nature makes use of the same kind of principles as those invoked in mechanical explanations of machines.

On the Mechanical Construction of the
Muscles

All the works of God and Nature are perfect,
Therefore
nothing lacks and nothing is superfluous
nor [is] anything in vain
what is by nature and according to nature is best
what is best for each is what is most according to nature
if it is better this way, it is according to nature
if it is according to nature this way, it is better
[Nature] accomplishes nothing by many that could be through fewer
nor does it attend to fewer, where many [are better]

Since in every action [there is] more and less,
in the fabric of muscles, nature attends to two things:
action and function, i.e.,
the perfection of action
hence in muscle two things are to be noted:
viz
composition for the sake of the action,
mechanical construction for the sake of strength
and power
Mechanics: just as that ^which overcomes those things
by which we are overcome by nature and aids
in difficulties, when [it] accomplishes something outside of nature
and useful. Aristotle: as when lesser things
overcome greater and things having little power to move
move great weights;
So in the muscles nature never fails to aid in difficulties
of this kind.
Hence in the study of the muscles not only
temperament and what follows it should be observed
for the sake of the action and contraction, but
how they are made for the sake of their powers and functions
and here there are many things about which to marvel how
the powers of muscle move bones and the connected weight

Fig. 6.3 Translation of folio 106 of Harvey's notes on muscle

Harvey articulates the *artificium mechanicum musculorum* by distinguishing between the action of the muscle (which is common to all muscles), and the *functio* of individual muscles, which he calls the perfection or completion (*perfectio*) of the action. This distinction, he says, reflects the fact that in any action there can be "the more and the less." Because "all the works of God and Nature are perfect," parts of the same kind (sharing a basic action) will have variations in that action, depending on their precise role and context in an animal body. In this way, "nothing lacks and nothing is superfluous; nor [is] anything in vain" and "[Nature] accomplishes nothing through many that could be through fewer; nor does it attend to fewer, where many [are better]." Because of the distinction between the action of muscle and its perfection, the anatomist must take note of both the constitution of the muscle, which is for the sake of the simple action (common to all muscles), and the *mechanical construction* of the muscles, for the sake of the perfection of that action.

He further clarifies (106v) the precise character of the mechanical construction by making a distinction within the more and the less in an action. Variation in the more and the less is twofold: according to the intensity in the active motion (*intentionem motu activo*) and according to (something like) its ability to overcome resistance (*effectionem motu resistentia*).[58] In muscle, the former is the power derived from its capacity (*vires a virtute*), and the latter is its strength (*robur*). He suggests that the more and less in the first sense is derived from variations in motive spirit and heat in the muscles. The more and less in the second sense derives from variations in (1) number, (2) shape, (3) size, (4) the location and positioning and interconnections of the muscle parts, and (5) the thickness or fleshiness of the substance of the muscles.[59] These features of muscle anatomy are present and vary for the sake of the appropriate strength (*robur*) of the muscle action, contributing to the perfection of the muscle action. The remainder of the *De artificio mechanico musculorum* is structured around these five categories of variations. He devotes a section to each of these features, sketching how they vary for the sake of the perfection of the action of different muscles.

Harvey's project here is inspired by and responding to Fabricius's efforts to integrate mathematical mechanics into Galeno-Aristotelian teleological explanations of in his publication on the muscles.[60] In his *De musculis* (in the final section, *De utilitatibus musculi*), Fabricius suggests that a "mathematical or better mechanical" reason is needed in order to understand how some features of muscle anatomy are for the sake of the particular action of the muscle. More explicitly than Harvey, Fabricius mentions the *Quaestiones mechanicae*, attributing it to Aristotle, and invokes its account of the grounding of mechanical effects in the nature of circular motion. One notable aspect of Fabricius's integration of mathematical mechanics is its limited and piecemeal character. He only identifies four specific questions for which a mechanical reason can be given. Furthermore, his use of mechanics does not find a natural place in his treatise, but is separated from the main structure of the discussion and placed at the end of the text.

Although Harvey does not work out mathematical details for any of his list of the five categories of variation for the sake of the perfection of the action, his effort is noteworthy for its attempt to develop a conceptual framework with which to articulate and motivate a systematic treatment of the mechanical construction of the muscles and the way he integrates it within his work on muscles. Here, then, we find Harvey employing "mechanics" in the sense of mathematical mechanics (of a

[58] He also calls the first simply *actio* as distinguished from *repassio*, which refers to the second.

[59] He also lists an additional category, but then connects it to one of the others by a line, suggesting he decided it was equivalent. This additional category is difficult to decipher, but Whitteridge reads "compositione, connexione" and Harvey lists under it *tunicis, capite, cauda; ansulis, theca* (tunics, head, tail, retinacula, theca). This he (understandably) identifies with the fourth category, location and positioning of the parts of the muscle.

[60] For more on this, see Distelzweig 2014a. Cf. Baldini 1997, 203–208.

Pseudo-Aristotelian *Quaestiones mechanicae* variety). However, even here Harvey's project is the same Galeno-Aristotelian one articulated in his *Prelectiones* and exhibited in the *De motu cordis*. Harvey is not banishing Galeno-Aristotelian matter theory or conception of the soul (both are abundantly present in the notes). Nor is he attempting to provide explanations of animal motions by describing a mechanism. Finally, while Harvey is, in a sense, expanding the scope of the applicability of mathematical mechanics (from the artifactual to animal realm), it is not best to see in this a contribution to the rise of "physico-mathematics" for at least two reasons. First, for Harvey this expansion is a local and restricted one. He does not see it as a part of a systematic effort to expand the application of mathematics to nature. Second, in the project he sketches in these notes, the application of mathematical tools does not replace, displace or diminish the overarching Galeno-Aristotelian explanatory project. Rather, Harvey aims here to integrate mathematical mechanics into his standard anatomical project: achieving and articulating universal, final causal *scientia* of the parts in animals.

6.5 Conclusion

We are mistaken, then, when considering mechanics and the mechanical in Harvey's texts, to see inconsistency or confusion. We need not think of Harvey as a man divided, with one foot in modernity and one in tradition—at least if we are considering Harvey's own project. If we want to find "Two Harveys" we should look instead at his reception. If we were to distinguish Harvey's self-understanding from how he was received, interpreted, and invoked by his contemporaries, *then* we might find many Harveys—one for each of the ways his contemporaries and near contemporaries criticized or lionized him, resisted or incorporated his discoveries and his insistence on and successful use of *autopsia*, dissection, and vivisection. But Harvey was not Riolan; nor was he Descartes, Hobbes, or Boyle. Regardless of how *they* saw him and his project, Harvey understood himself to be a critical and creative anatomist taking Aristotle as his leader and Fabricius as his guide. He understood the goal of anatomy to be Aristotelian *scientia* of the parts of animals, articulated in terms of the Galenic *actio* and *usus* of the parts. I have argued here that, once adequate attention is given to the semantic range of "mechanics" in the seventeenth century, we can see that this understanding of the goal of anatomy shapes and is reflected in Harvey's attitude towards things "mechanical." Furthermore, and this is an argument for another occasion, in pursuit of this same goal Harvey also developed and self-consciously employed a coherent and highly effective vivisectional and comparative method, one that he sees as a continuation and refinement of the methodological ideas of Aristotle and Galen. Harvey understands his work to be "locally" new, in the specifics of his discoveries, but "globally" continuous with the aspirations and methods of Galen and Aristotle (especially as exhibited in Fabricius's work)—and this because, thinks Harvey, these ancients got so much right.

As Aubrey reports, Harvey thought one could do no better than to turn to the ancients—"the fountain head"—in comparison to which the "neoteriques" are mere "shitt-breeches."[61]

References

Adelmann, Howard, trans. 1942. *The Embryological Treatises of Hieronymus Fabricius of Aquapendente*, 2 vols. Ithaca: Cornell University Press.

Aubrey, John. 1898. *Brief lives,* vol. 1, ed. Andrew Clark. Oxford: Clarendon Press.

Baldini, Ugo. 1997. Animal motion before Borelli, 1600–1680. In *Marcello Malpighi: Anatomist and physician*, ed. Domenico Bertoloni Meli, 193–246. Firenze: Olschki.

Bauhin, Caspar. 1605. *Theatrum anatomicum*. Frankfurt.

Bertoloni Meli, Domenico. 2011. *Mechanism, experiment, disease: Marcello Malpighi and seventeenth-century anatomy*. Baltimore: The Johns Hopkins University Press.

Boyle, Robert. 1772. *The works of the Honourable Robert Boyle*, ed. Thomas Birch. London.

Bylebyl, Jerome. 1973. The growth of Harvey's De motu cordis. *Bulletin of the History of Medicine* 47: 427–470.

Bylebyl, Jerome. 1977. De Motu Cordis: Written in two stages? Response. *Bulletin of the History of Medicine* 51: 140–150.

Cunningham, Andrew. 1985. Fabricius and the 'Aristotle project' in anatomical teaching and research at Padua. In *The Medical Renaissance of the Sixteenth Century*, ed. Andrew Wear, Roger French, and I. M. Lonie, 195–222. Cambridge: Cambridge University Press.

Cunningham, Andrew. 2006. Fabrici and Harvey. In *Harvey e Padova. Atti del convegno celebrativo del quarto centenario della laurea di William Harvey (Padova, 21–22 novembre 2002)*, ed. Giuseppe Ongaro, Maurizio Rippa Bonati and Gaetano Thiene, 129–149. Padova: Universita degli Studi di Padova.

Distelzweig, Peter. 2014a. Fabricius's Galeno-Aristotelian teleomechanics of muscle. In *The life sciences in early modern philosophy*, ed. Ohad Nachtomy, and Justin E.H. Smith, 65–84. Oxford: Oxford University Press.

Distelzweig, Peter. 2014b. *Descartes's teleomechanics in medical context*. PhD thesis, University of Pittsburgh, Pittsburgh.

Distelzweig, Peter. 2014c. *Meam de motu & usu cordis, & circuitu sanguinis sententiam*: Teleology in William Harvey's *De motu corids. Gesnerus: Swiss Journal of the History of Medicine and Sciences* 71.2 (Special Issue: Teleology and Mechanism in Early Modern Medicine): 258–270.

Fabricius ab Aquapendente, Hieronymus. 1625. *Opera Physica Anatomica*. Padua.

Fabricius ab Aquapendente, Hieronymus. 1687. *Opera Omnia anatomica & physiologica*. Leipzig.

Frank, Robert. 1980. *Harvey and the Oxford physiologists*. Berkeley: University of California Press.

French, Roger. 1979. A note on the anatomical *accessus* of the middle ages. *Medical History* 23: 461–468.

French, Roger. 1994. *William Harvey's natural philosophy*. Cambridge: Cambridge University Press.

Gabbey, Alan. 2004. What was mechanical about the mechanical philosophy? In *The reception of the Galilean science of motion in seventeenth-century Europe*, ed. Carla Rita Palmerino and J.M.M.H. Thijssen, 11–23. Dordrecht: Kluwer Academic Publisher.

Galen. 1916. *On the Natural Faculties*. Trans. Arthur Brock. Cambridge, MA: Harvard University Press.

[61] Aubrey 1898, 300.

Galen. 1968. *On the Usefulness of the Parts of the Body*, 2 vols. Trans. Margaret T. May. Ithaca: Cornell University Press.
Galen. 1984. *On respiration and the arteries,* ed. D.F. Furley, and J. S. Wilkie. Princeton: Princeton University Press.
Goldberg, Benjamin. 2012. *William Harvey, soul searcher: Teleology and philosophical anatomy*. PhD thesis, University of Pittsburgh, Pittsburgh.
Hankinson, R.J. 1989. Galen and the best of all possible worlds. *Classical Quarterly* 39: 206–227.
Harvey, William. 1628. *Exercitatio Anatomica de motu cordis et sanguines in animalibus*. Frankfurt.
Harvey, William. 1649. *Exercitatio Anatomica de Circulatione Sanguinis*. Cambridge: Ex officina Rogeri Danielis.
Harvey, William. 1653. *The anatomical excercises of Dr. William Harvey....* London.
Harvey, William. 1886. *Prelectiones anatomiae universalis*. Edited by a committee of the Royal College of Physicians of London with reproduction. London: J. & A. Churchill.
Harvey, William. 1959. *De Motu Locali Animalium, 1627*. Trans. G. Whitteridge. London: Cambridge University Press.
Harvey, William. 1964. *The Anatomical Lectures of William Harvey*. Trans. G. Whitteridge. London: Livingstone.
Harvey, William. 1976. *An Anatomical Disputation Concerning the Movement of the Heart and Blood in Living Creatures*. Trans. Gweneth Whitteridge. Oxford: Blackwell.
Harvey, William. 1981. *Disputations touching the generation of animals*. Trans. Gweneth Whitteridge. Oxford: Blackwell.
Hobbes, Thomas. 1655. *Elementorum Philosophia Sectio Prima De Corpore*. London.
Hunter, Richard, and Ida Macalpine. 1958. William Harvey and Robert Boyle. *Notes and Records of the Royal Society of London* 13: 115–27.
Keynes, Sir Geoffrey. 1966. *The life of William Harvey*. Oxford: Oxford University Press.
Klestinec, Cynthia. 2011. *Theaters of anatomy: Students, teachers and traditions of dissection in renaissance venice*. Baltimore: John Hopkins Press.
Laurentius, Andreas. 1600. *Historia anatomica humani corporis et singularium ejus partium multis controversiis et observationibus novis illustrata*. Paris.
Lennox, James. 1987. Divide and explain: The posterior analytics in practice. In *Philosophical issues in Aristotle's biology*, ed. Allan Gotthelf and James Lennox, 90–119. Cambridge: Cambridge University Press.
Lennox, James. 1991. Between data and demonstration: The *Analytics* and the *Historia Animalium*. In *Science and philosophy in classical Greece*, ed. Alan Bowen, 261–295. New York: Garland Publishing.
Lennox, James. 2006. The comparative study of animal development: William Harvey's Aristotelianism. In *The problem of animal generation in modern philosophy*, ed. Justin Smith, 21–46. Cambridge: Cambridge University Press.
Machamer, Peter, Lindley Darden, and Carl Craver. 2000. Thinking about mechanisms. *Philosophy of Science* 67: 1–25.
Siraisi, Nancy. 2004. Historia, actio, utilitas: Fabrici e le scienze della vita nel Cinquecento. In *Il Teatro dei Corpi: Le Pitture Colorate D'Anatomia di Girolamo Fabrici D'Acquapendente*, ed. Maurizio Rippa Bonati and Jose Pardo-Tomas, 63–73. Milano: Mediamed Edizioni Scientifiche Srl.
Wear, Andrew. 1983. Harvey and the 'Way of Anatomists'. *History of Science* 21: 223–249.

Chapter 7
Descartes on the Theory of Life and Methodology in the Life Sciences

Karen Detlefsen

Abstract As a practicing life scientist, Descartes must have a theory of what it means to be a living being. In this paper, I provide an account of what his theoretical conception of living bodies must be. I then show that this conception might well run afoul of his rejection of final causal explanations in natural philosophy. Nonetheless, I show how Descartes might have made use of such explanations as merely hypothetical, even though he explicitly blocks this move. I conclude by suggesting that there is no reason for him to have blocked the use of hypothetical final causes in this way.

Keywords Descartes • Teleology • Methodology • Hypotheses • Nature of life

Descartes was a practicing natural philosopher. His areas of research included a specific interest in investigating the phenomena of life. He treated human, animal, and plant bodies as distinctive kinds of bodies, and he afforded them separate scientific[1] treatment, both in practice and in his written work. On 18 December 1629, he wrote to Mersenne that he was beginning a study of anatomy (AT I, 102)[2], by which he meant the anatomy of living bodies. The fruits of his anatomical and physiological investigations appeared in various written forms throughout his life, including *Traité de l'homme* (hereafter *Treatise*), the fifth part of *Discours de la méthode* (*Discourse*), a planned but unwritten fifth section of the *Principia Philosophiae* (AT VIIIa, 315/CSM I, 279; *Principles*), the first 16 articles of Part I of *Passions de l'âme* (and various comments scattered throughout the remainder of that text; *Passions*), *La Description du corps humain* (*Description*) which also deals with

[1] I use the term "science" and its cognates for ease of expression, mindful of the fact that our meaning of the term most closely aligns with Descartes' "natural philosophy".

[2] I use the following abbreviations to refer to editions and translations of Descartes' works: AT=Descartes 1964–76; CSM=Descartes 1985a; CSMK=Descartes 1985b; SV=Descartes 1989; SG=Descartes 1998.

K. Detlefsen (✉)
Department of Philosophy, University of Pennsylvania, Philadelphia, PA, USA
e-mail: detlefse@sas.upenn.edu

P. Distelzweig et al. (eds.), *Early Modern Medicine and Natural Philosophy*, History, Philosophy and Theory of the Life Sciences 14,
DOI 10.1007/978-94-017-7353-9_7

animal and plant bodies, *Primae cogitationes circa generationem animalium* (*Generation*), *Excerpta anatomica* (*Excepts*), and assorted letters.

Given this, we can expect that Descartes conceives of living beings as distinct from non-living beings in some way or another. For if this were not true, then Descartes would have no way of isolating a class of bodies taken to be *living* bodies, and he would then not be able to identify any individuals to serve as the subject matter of the life sciences – sciences to which he devoted considerable professional time. And this would render incoherent this aspect of his life as a working natural philosopher. Moreover, he explicitly does acknowledge life as a category. In a letter to Regius of June 1642, for example Descartes talks of many sorts of bodies as machines, but he nonetheless makes distinctions within the broader class of machines,[3] and isolates those that are *living* from the rest (AT III, 566/CSMK 214). He also acknowledges the category of life in other texts. For example, he planned (though never wrote) a fifth section of the *Principles* devoted to "living things, i.e. animals and plants" (AT VIIIa, 315/CSM I, 279), and one can effectively argue that Descartes includes the human body among those that are living given his recognition that human bodies and animals perform many of the same sorts of actions (AT III, 121/CSMK 149), including those detailed in his writings on animals. He also makes a clear distinction between the machines we can build and living machines when he emphasizes that we could never make ourselves a new body because we could never make the matter out of which our bodies are constructed (AT VI, 148).

But there are two difficulties Descartes faces in identifying a separate class of living beings, and both stem from the fact that, for him, metaphysics is ontologically prior to both physics and what we might call the "special sciences".[4] Recall his famous "tree of philosophy" with metaphysics as the roots, giving rise to and placing constraints on physics as the truck, which in turn gives rise to and places constraints on the special sciences, "which may be reduced to three principal ones, namely medicine, mechanics, and morals" (AT IXb, 14/CSM I, 186). There are two aspects of Descartes' metaphysics that cause him potential difficulties in identifying a class of living beings to serve as the subject matter of the life sciences. The first is his austere ontology of the created world, according to which there are just two kinds of substances, material substance (with the essence of extension) and souls (unextended things with the essence of thought). The second is his conception of God's nature and our relationship with him, specifically that fact that we do not have cognitive access to God's ends, or the purposes that guided him in the creation of the material world.

---

[3] On the meaning of "machine", specifically with respect to Descartes' medical philosophy, see Manning 2012.

[4] See Hatfield 1993 and Garber 1992, 13 for this account of the relation between metaphysics and physics. A different way of thinking about the relation between metaphysics and physics is put forth by Stephen Gaukroger who holds that "there was nothing internal to Descartes' project of natural philosophy that required metaphysical foundations, and there was nothing crucial to his natural philosophy that could only be generated from such metaphysical foundations" (Gaukroger 2002, 1–4). I leave aside these two competing visions of the relation between metaphysics and physics, since this debate does not impact my current project.

The first aspect of Descartes' metaphysics noted above leads to the first hurdle in identifying a class of living being – the easier hurdle to overcome. Because he rejects the notion of natural essences beyond material substance as extension and immaterial thinking souls, he loses the ability to ground universals or natural material kinds in the ontology of the world. With every material body having the same essence as every other material body, there appears to be nothing in the nature of bodies themselves that identifies them as distinct kinds of bodies worthy of distinct scientific treatment. Indeed, according to this line of argument, there are exactly two natural kinds in the world – embodied souls, for souls *cannot* non-miraculously exist without human bodies (AT III, 461/CSMK 200) – and all non-ensouled material bodies. So there can be a science of human beings grounded in a distinct ontology, but no other special science grounded in a distinct ontology.[5] Descartes' ontology thus permits special, scientific treatment of only the human being, but not of the living body. But this is a problem for Descartes given that he includes animal and plant bodies with human bodies in his anatomical and physiological writings. In the first section of this paper, I develop what I think should have been Descartes' theoretical conception of life. In doing so, I show that Descartes does not eliminate the class of living bodies from his natural philosophy even while his austere ontology of material substance does result in the ability to explain the phenomena of all living bodies in terms of matter in lawful inertial motion; that is he is a *reductionist* with respect to explanation of life phenomena but not an *eliminativist* with respect to life itself[6] – much as we are today, albeit with a more sophisticated science at our disposal.

Providing a solution to the first problem just noted feeds directly into the second problem. For the theoretical account of living beings that I think Descartes must be – and implicitly is – committed to relies upon making claims to God's ends or purposes *vis a viz* the created material world. But this flies in the face of the second aspect of his metaphysics noted above, specifically that we cannot know any of God's ends with respect to his creation of the material world, and so we cannot rely upon knowledge claims regarding those ends in natural philosophy.[7] I think this problem is surmountable given resources Descartes has within his natural philosophy, and I show (in Parts 2–4 of this essay) how Descartes could have overcome this difficulty had he called upon these resources. I am particularly interested in showing: (a) that there is a way of attributing weak sorts of internal[8] ends to material bodies

[5] Stephen Menn (2000, 139–41) and Dennis Des Chene (2001, 30, 62 and 64) both suggest that this may well follow from Descartes' ontology.

[6] On this point, see Gaukroger 2000 and 2010. T.S. Hall (1970, 55–56) also points to the fact that Descartes provides reductionist explanations, and while Hall does not explicitly mention that Descartes does not thereby eliminate the category of life altogether, it is strongly implicit in his discussion of Descartes' account of living bodies.

[7] For a few of the many articles on Descartes' ideas on final cause in natural philosophy, see Brown 2013; De Rosa 2007; Detlefsen 2013; Distelzweig 2015; Hatfield 2008; La Porte 1928; Schmaltz (manuscript); and Simmons 2001.

[8] I avoid the use of "intrinsic" and "extrinsic", using "internal" and "external" instead to avoid the technical meaning of the former pair in Descartes' philosophy. See Manning 2012 and Manning forthcoming. I engage with Manning's discuss of intrinsic and extrinsic denominations in Sect. 7.3 below when I expand on what I mean by "internal ends" in Descartes.

considered *not* in terms of their metaphysical essence but rather in terms of their built structures; and (b) that Descartes' own friendliness to hypotheses in natural philosophy could have allowed him to appeal to such internal ends (even though he explicitly blocks this move).

In the process of completing this work, I aim to underscore Descartes' role in two historical trends that are especially interesting in the history of the life sciences. First (only implicit in Descartes), once Aristotelian substantial forms are ousted from an account of living bodies by mechanists such as Descartes, there appears to be no easy way to ground ends within the nature of wholly material bodies. And yet, pre-theoretically, and in accordance with common sense, built machines must have some sort of end internal to them; Aristotle implicitly acknowledges this, even with respect to artifacts, and it is implicit in Descartes' writings too. The crucial difference is that Aristotelianism has the ontology to account easily for this while it is less clear how this teleology can be accommodated on a Cartesian ontology. Second, in scientific epistemology, there is the emergence of a respectable category of the probable according to which the probable is not automatically associated with the merely speculative. This category is associated with the use and testing of hypotheses, and Descartes himself embraced the use of hypotheses, and thus embraced (however uneasily) the category within scientific epistemology of the respectably probable. He just didn't capitalize on his embrace of this trend as fully as he might have in his life sciences.

Before starting the main work of this paper, I make the following two preliminary points. First, there are two distinct theories of the *origins* of living bodies to be found in Descartes' corpus. One is the idea that living forms emerged from an initial chaos through non-purposeful motion of that material chaos (e.g. VI: 42/CSM I, 132; XI: 34-5/ CSM I, 91; and VIIIa: 102-3/ CSM I, 257).[9] The other is the idea that God formed those beings. In this paper, I proceed on the assumption of the latter idea, even while I think there is much promise in Descartes' chaos idea. Dealing with the chaos theory is work for elsewhere.

The second preliminary point is that I propose we think about Descartes' general approach to the life sciences in the follow way. Dennis Des Chenes' insight expressed thus is helpful:

> No doubt some sort of distinction between living and nonliving things comes to us early in life. In every human culture the classification of things into living and nonliving is among the most basic. Though some judgments have changed, Aristotle's division between living and nonliving, those of Aristotelian authors, Descartes', and our own, overlap a great deal. But broad agreement on the domain of life coexists easily… with grossly dissimilar concepts of life. The list of things that Hobbes, Descartes, and Regius would call plants and animals differs little from the lists that Toletus, Suárez, or Eustachius would give. The concept of the living in the new philosophers, on the other hand, differs as greatly from the Aristotelians' as do their concepts of body and natural change.[10]

[9] For a discussion of some of the material I cover herein with the chaos theory in mind, see Hatfield 2008.

[10] Des Chene 2000a, 20.

Descartes' own way of proceeding as a natural philosopher seems to follow the general approach captured by Des Chene. First, Descartes pre-theoretically identifies the domain of the living. Second, he then subjects the individuals within this domain to scientific investigation. The investigations may well problematize pre-theoretical intuitions about what does and does not fall into the domain of the living – as it does for working scientists today. But the first two steps do seem to capture Descartes' actual approach as a working life scientist. Further, it is clear what Descartes takes to be the items that serve as the subject matter of the life sciences: plants, animals, and human bodies considered (counterfactually) in isolation of their souls.[11] These are the bodies that he implicitly identifies as living when he studies these and only these in his active scientific practice and in his theoretical biological[12] writings. He also explicitly identifies animals and plants as living, and he does so within the context of his treatment of human bodies indicating that the latter are living too. In the *Description*, for example, he is explicit that human bodies, animals, and plants should be categorized together *as living* when, for example, he extends his discussion of nutrition beyond the human: "…we must bear in mind that the parts of those *living bodies* that are maintained through nourishment, that is, animals and plants, undergo continual change…" (XI: 247/ CSM I, 319; emphasis added). The domain of life, then, includes all and only plant, animal and human bodies. My task now is to reconstruct a theoretical account of life that is consistently capable of picking out all and only members of this domain, and that is consistent with Descartes' texts and own conceptual commitments, including the metaphysics that is at the foundations.

7.1 Descartes' Conceptions of Life

Ann Wilbur MacKenzie is right when she proposes that "Descartes did not provide any systematic and general analysis of 'x is alive'",[13] because he did not abstract sufficiently enough from his specific claims about individual living beings to derive a general theory. Still, as she and others have shown, it is possible to infer a number of different possible conceptions of life, which Descartes may have embraced. In this section, I draw upon the insights of MacKenzie and others who bring some elements of Descartes' conception of life to our attention.[14] I consider three possible

[11] Given my focus on the human *body*, along with other non-ensouled living bodies, my project departs somewhat from a project that focuses exclusively on *medical* philosophy to the extent that the latter is a field concerned with the health and illness of human beings.

[12] As with my use of "science", I use the term "biology" mindful of the fact that this term and the cluster of sciences we now recognize by this term did not emerge until the late eighteenth century. I use this for ease of expression to capture Descartes' writings about living bodies.

[13] MacKenzie 1975, 2–3.

[14] Ablondi 1998; Bitbol-Hespériès 1990; Canguihelm 1965; Distelzweig 2015; Des Chene 2000b; and Shapiro 2003.

theories of life for which there is textual evidence in Descartes' corpus. I show that all three capture crucial elements of the theoretical account of living bodies to which I believe Descartes must have been committed.

7.1.1 Living Bodies as Those with Heat as Their Corporeal Principle of Action[15]

In a letter to Henry More of 5 February 1649 Descartes writes: "I do not deny life to animals, since I regard it as consisting simply in the heat of the heart…" (AT V, 278/CSMK 366; c.f. AT IV, 686; AT XI, 226/CSM I, 316; AT XI: 333/SV 23; AT XI 407/SV 76–7). Since this is the most explicit statement regarding the principle of life to be found in Descartes, it is tempting to simply take Descartes at his word and accept this as the defining criterion of life.

But this criterion will not serve the purpose for it cannot unfailingly pick out all and only living bodies. Some of the apparent difficulties with this criterion are surmountable with a large dose of charity in interpretation, but not all the difficulties can be overcome. First, while Descartes locates this heat in the heart of the living organism, it is not clear that all living organisms have hearts; plants are the clearest case.[16] Still, one may salvage the heat criterion by acknowledging that Descartes also allows for heat generally conceived (and not located in any specific organ), to act as the principle of life since he also says that it is the principle common to animals, plants, and human bodies (letter to Mersenne: AT III, 122/CSMK 149), even before any organs, including the heart, have begun to form at all (AT XI, 534). But, and second, one may object to the claim that all organisms are in fact hot, and again plants are an obvious example as are cold-blooded animals. Descartes explicitly faces this objection. In response to Plempius' claim that fish do not have hot hearts (AT I, 498), Descartes responds that "although we do not feel much heat in fish, their hearts feel hotter than all other organs in their body" (AT I, 529/CSMK 83; c.f. AT II, 66/CSMK 94–5). Likewise, he takes the heat found in animal hearts to be analogous to the heat in hay before it dries (AT XI, 121/SG 100 and 254/CSM I, 322; AT VI, 46/CSM I, 134), and charitably read, this can be taken as a case of plants so newly cut as to retain some vestige of life (namely, heat). More explicitly, Descartes claims that tree bark and fruit (presumably both examples of plant life) can exude vapors due to their internal heat (AT II, 67/CSMK 95–6). Whatever the empirical validity of these observations, it is clear that Descartes wishes to extend heat to all human, animal, and plant bodies seemingly in order designate them all as living machines.

The heat criterion, however, is not an adequate principle of life because it allows too many individuals into that category. Fred Ablondi draws our attention to this

[15] Bitbol-Hespériès 1990, *passim* takes heat as Descartes' theory of life.

[16] This is MacKenzie's (1975, 3–5) objection to the conception of life as heat in the heart. Ablondi (1998, 181) makes this objection too.

difficulty, noting the problematic case of the steam engine.[17] Similarly damaging are Descartes' own examples, such as when he likens the heat found in living bodies to that which occurs during the fermentation of wine (AT XI, 254/CSM I. 322), indicating that this heat is also found in the nonliving. Heat from a fire without light, then, is consequently not up to the task of identifying all *and especially only* members of the class of living machines since it is also found in some non-living bodies and processes as well.

One may wish to take this as evidence that there is, in the final analysis, no clearly delineated category of living bodies for Descartes given his explicit association of heat with life.[18] I resist this conclusion, for we must pay heed to Descartes' own words and practice, acknowledge that he is committed to a science of life, and therefore acknowledge the need for the category of life. Consequently, we must dismiss the "heat without light" candidate as a viable one for Descartes' theory of life.

7.1.2 Living Bodies as God-Made Machines with a Complexity Specific to Them

Less explicit than the heat criterion is the suggestion that living bodies are machines made by God and thus have a kind of complexity that distinguishes them from non-living machines. Here are two texts suggesting this conception:

> Those who know how many kinds of automata, or moving machines, the skill of man can construct with the use of very few parts, in comparison with the great multitude of bones, muscles, nerves, arteries, veins and all the other parts that are in the body of any animal.... will regard this [animal] body as a machine which, having been made by the hands of God, is incomparably better ordered than any machine that can be devised by man, and contains in itself movements more wonderful than those in any machine made by man (AT VI, 55-6/ CSM I, 139).

And:

> We see clocks, artificial fountains, mills, and other similar machines, which, even though they are only made by men, have the power to move of their own accord in various ways. And, as I am supposing that this machine [made with the explicit intention of being as much like us as possible] is made by God, I think you will agree that it is capable of a greater variety of movements than I could possibly imagine in it, and that it exhibits a greater ingenuity than I could possibly ascribe to it.
>
> I shall not pause to describe to you the bones, nerves, muscles, veins, arteries, stomach, liver, spleen, heart, brain, not all the other different parts from which this machine must be composed, for I am assuming that they are just like those parts of our own bodies having the same names.... [S]o that it remains only for me to explain these movements [that depend upon the parts] to you here in the proper order and by these means to tell you which of our functions these represent. (AT XI, 120-1/CSM I, 99)

---

[17] Ablondi 1998, 183.

[18] See Bitbol-Hespériès 1990, 71.

There are a number of ways of interpreting this criterion. Certainly, there seem to be two central features of it: living machines are "incomparably better ordered" and so exhibit "a greater ingenuity" than is to be found in non-living machines such as those made by humans; and living machines are "made by the hands of God". The first feature just noted might be interpreted in one of two ways: living bodies might have a *degree* of complexity that far surpasses that of non-living bodies; or they might have a *kind* of complexity far superior to any that a human could achieve when building a machine.

Locating the source of the uniqueness of living bodies in a difference in degree is suggested in the first passage where Descartes refers to human-made machines as having "very few parts" in comparison with God's machines. This is a promising route to take, especially for a theologically minded philosopher of the seventeenth century. For one might claim that the difference in kind between living and non-living derives from a difference in degree between infinitely complex living bodies (that only an infinitely capable builder, i.e. God, could make) and merely finitely complex non-living bodies (that humans may well be capable of making). This will be one way through which both Malebranche and Leibniz secure the distinction between living and non-living. But it is not Descartes' way for he is reluctant to associate the infinite with anything other than God himself (e.g. AT VIIIa, 14–15/CSM I, 201–202). According to Descartes, God's machines are only "incomparably" better ordered. Perhaps, then, Descartes believes that living bodies are complex *enough* (but not infinitely so) to demarcate living bodies. While this accords with Descartes' own position on the infinite, it fails to secure a conception of life. For without the difference in degree being a difference between the finite and the infinite, there can be no decisive difference in kind. Somewhere along the continuum of increasingly complex machines, a line is supposedly crossed that demarcates the living from the non-living, but it is not clear where this line lies such that a principled distinction can be drawn.[19] Maybe Descartes could shore up this second approach by saying that what makes living bodies unique is not *simply* that they have an incomparable (though not infinite) degree of complexity, but that they have this due to their having been made by God. But this will not suffice, for God made many other machines besides living bodies, and so we must still be able to distinguish between his living and his non-living machines. But then the burden for this distinction must fall somewhere, and, once again, an incomparable yet not infinite degree of complexity is not up to the task of doing the work necessary to make the distinction.

So perhaps Descartes' intention is to locate the source of the uniqueness of living bodies in a special *kind* of complexity found in God-made living machines that is

[19] Thomas Fuchs makes this point (2001, 125). Genevieve Rodis-Lewis (1978) approaches this point too when considering AT II: 525 which allows that crystals may have a middle nature between living and non-living. It may be possible for Descartes to tolerate these grey areas in the same way that we tolerate difficult cases that seem to straddle the life-nonlife divide (such as viruses), but there is no need for this since there is a better theory of life forthcoming which does not require Descartes to accommodate the sort of grey area identified here.

not shared by any other machines, God-made or other. This does seem to capture better Descartes' intention as articulated in the texts above. Then the obvious question arises: what *is* that special kind of complexity in structure that God has made that can demarcate the living from the non-living? The texts cited above offer two answers. According to one answer, living bodies have a "great multitude" of certain *kinds of parts* in common (also, probably, disposed to one another in certain ways): hearts, arteries, livers and so forth (as Descartes lists in the passages above). The second passage cited is dealing with the supposed replica of a *human* body that Descartes is asking the reader to imagine. As human, the list of very specific body parts offered as unique to such a body makes sense. The first passage is more troublesome, however, for that passage is meant to apply to the body of "any animal", and it is not immediately clear that all animals (monkeys, turtles and oysters alike) possess the same collection of body parts. Moreover, if we were to take this conception to be the theoretical conception of life Descartes is committed to, then it would have to apply equally to plants as well as to animal and human bodies. But on the face of it, plants do not have hearts or livers or spleens or bones.[20] So *prima facie*, a special kind of complexity that identifies specific body parts as necessary to that complexity, is not adequate as a conception of life since it cannot reliably pick out all members of the domain of life.[21]

According to the second answer to the question "what *is* that special kind of complexity in structure that God has made that can demarcate the living from the non-living?", living bodies have the sort of structure – including the sorts of body parts – that can permit "movements more wonderful than those in any machine made by man". This answer certainly makes reference to the structure, but the structure remains entirely abstract[22] – a living body's structure is *whatever structure is necessary* to give rise to specific, wonderful movements, and many, diverse structures might fit that bill. Additionally, in this answer, the structure is subordinate to and in service of the life-specific functions or behaviors of the body. And it is these functions or behaviors, which do the real conceptual work in distinguishing the living from the non-living; the abstract structure is only a means to the definitive functions. So this second answer is really a third and distinct conception of life: living bodies are those that behave or function in specific ways. I turn to this third, extremely promising, conception shortly.[23]

So, as with the heat theory, the present theory of life fails to identify all and only living bodies in a reliable and principled fashion. Taken as a theory about the *degree* of complexity of structure, this theory fails for there is no way to establish a difference

[20] There were attempts in the early modern period to find structural equivalents of major organs across all living beings, including plants. The fact of these attempts might blunt the current criticism somewhat. See Delaporte, François [1979] 1982.

[21] See Des Chene 2001, 54ff for difficulties in identifying *parts* in Descartes.

[22] This is MacKenzie's point. She holds that one causal component in Descartes' definition of life must be this fully abstract structural complexity, which permits the behaviors definitive of living bodies (MacKenzie 1975, 9).

[23] See Ablondi 1998 for an enlightening discussion of the structural complexity criterion.

in *kind* between living and non-living without recourse to an infinitely complex body. And taken as a theory about the *kind* of complexity, where reference to specific body parts is essential to that theory, it once again fails because it cannot pick out all and only members of the domain of life given the immense diversity in the parts of different living bodies. And so this second theory by itself, cannot be Descartes' considered theoretical conception of life.

7.1.3 Living Bodies as Machines that Function in Ways Unique to Plants, Animals, and Human Bodies

As Descartes' experiments and writings on living bodies suggest, the behaviors or activities of life are more or less those that Aristotle associates with the vegetative soul and some of those Aristotle associates with the sensitive soul. The most general functions associated with all living bodies (e.g. AT XI, 202/CSM I, 108; AT I, 263/ CSMK 40) are foetal formation (or generation), growth (which includes transformation as opposed to mere accretion of matter [XI: 596–87]), nutrition and self-maintenance, reproduction, and response to the surrounding environment; in animals, this ability to respond to the environment includes the abilities to sense, remember, and learn in so far as these psychological abilities are conceived of solely in corporeal terms (e.g. AT VII, 436/CSM II, 294; AT X, 416/CSM I, 43; AT III, 433-34/CSMK 196; and *Passions passim* when Descartes discusses habituation).

MacKenzie includes life functions as one among a few that together make up Descartes' complex theory of life in her view, which includes both causes and effect. "A creature is alive if and only if it has some principle of motion (or other) which, together with some arrangement of parts (or other), enables that creature to engage in some set of activities (or other) which in turn enable that creature to carry out a set of life functions".[24] The life functions she recognizes are nutrition, growth and generation, and all living bodies display these functions. She also recognizes more determinate activities that only specific *kinds* of living beings exhibit as contributing to the more general life functions. Examples of these more determinate activities (e.g. in animals with hearts) might include digestion, the heartbeat, and respiration.[25] According to MacKenzie's approach, then, an adequate account of life must make reference to two causes – a principle of motion (such as heat), and a suitable disposition of organic parts – and a complex of effects – specific behaviors unique to a sub-class of living machines (e.g. animals with hearts) that give rise to general life functions, exhibited by all and only living machines. Heat, then, is better seen as the principle of motion within living bodies, and not the principle of life itself, an option Descartes explicitly offers in the *Passions*: "While we are alive there is a continual heat in our hearts, which is a kind of fire that the blood of

[24] MacKenzie 1975, 10.

[25] *Ibid.* 8–9.

the veins maintains there. The fire is the corporeal principle underlying all the *movements* of our limbs" (AT XI, 333/SV 23).

Recently, Distelzweig has provided another distinction that can help fill out Descartes' conception of life, a distinction derived from the historical medical context in which Descartes was writing. Specifically, Distelzweig notes that the

> medical tradition employs *functio*… to refer to and categorize a familiar, long established set of characteristic activities of living things. *Usus*, in contrast, refers to the contribution a part of activity makes to the exercise of some *functio*. Both parts and *functiones* have *usus*. The *usus* of a part is the contribution it makes to the exercise of some *functio*. The *usus* of a *functio*, in turn, is the contribution that *functio* makes to some larger or more fundamental *functio*, terminating ultimately in the list of main natural, vital and animal *functiones*.[26]

The distinction that Distelzweig draws our attention to focuses on the hierarchical nature of *usus* and *functio*, while MacKenzie's distinction between life behaviors and life functions focuses on the differences between activities that a sub-class of living beings exhibit and activities exhibited by all living bodies. But they can be related to one another precisely because more localized parts and activities often tend to be unique to sub-classes of living beings, as Mackenzie's specific examples underscore.

These basic distinctions seem right to me, though I differ from MacKenzie on a few points. First, I specify that growth is of a specific form, namely growth with bodily transformation – most notably the constant turnover of constitutive matter – and not mere growth by aggregation. In the *Description*, for example, Descartes writes: "we should bear in mind that the parts of all living bodies which require nutrition to sustain them (that is, animals and plants) are continually undergoing change" (AT XI, 247/CSM I, 319). Importantly, once foetal formation is complete, the visible organic structure is maintained despite the constant change in the subvisible constitutive matter of organisms. Today, of course, we call this process metabolism, and it is crucial to the enduring health and survival of living bodies. No other bodies grow in this fashion; it is a form of growth unique to plants, animals and human bodies.

Further, I include two more elements in the list of life functions beyond the three identified by MacKenzie (i.e. nutrition, growth, and generation). These are, first, the ability to react to the surrounding environment (including animals' abilities to sense, remember and learn considered as material, and not mental, processes) and, second and related, the ability to maintain the unified structure of the body despite the wear and tear that follows from interaction with the surrounding environment. Lisa Shapiro identifies these elements as providing a promising non-teleological criterion of health for both human bodies and animals – specifically, she claims that human bodies and animals have integrated structures that are stable and able to preserve themselves. Moreover, she connects staying healthy with the fact of a body's being and staying *alive*. So I take it that she would endorse this criterion as a necessary component of Descartes' conception of life. Distelzweig (2015), too, accepts this account of life, emphasizing the *self*-stabilizing aspect of all and

[26] Distelzweig 2015.

only living beings, which is presumably captured by Shapiro's mention of *self*-preservation.[27]

These additions are significant for they indicate a crucial aspect of Descartes' theory of life: living bodies perform their activities (e.g. digestion) to contribute to life functions (e.g. growth with transformation) *which helps them achieve the further goal of self-maintenance of a unified structure of inter-related parts.* This self-maintenance, in turn, permits the continuation of the life-specific behaviors and functions. So in addition to the sub-processes of localized parts within a specific subset of kinds of living bodies, which contribute to the most general, whole-body functions of all living bodies, I propose that Descartes' conception of living bodies includes, as Shapiro notes, the further element of self-maintenance of a unified structure of inter-related parts – or, more familiarly, self-preservation. Indeed, the other behaviors of living beings all contribute to this ultimate, most general behavior.

There is evidence that Descartes takes the self-maintenance of a unified structure adequate to permit continuing self-maintaining activities as a defining feature of living bodies. In *Passions*, for example, Descartes writes: "For the [human] body is a unity which is in a sense indivisible because of the arrangement of its organs, these being so related to one another that the removal of any one of them renders the whole body defective" (AT XI, 351/SV 35). Once this removal of an essential organ happens, death occurs (AT XI, 330/SV 21). Similarly, in *Treatise*, Descartes suggests that the living human body forms an integrated whole which, because of its "good condition" of parts into a whole, is able to maintain that whole from disintegrating (AT XI, 143-44/CSM I, 102–3; c.f. AT VIIIa, 318/CSM I, 282; and AT VI, 153). Such passages indicate that the proper dispositions of parts to one another form a structurally integrated whole – what Des Chene calls "dispositional unity".[28] This whole of parts properly disposed to one another permits the machine to function in specific ways, which further allow it to maintain a stable structure, which is tantamount to engaging in self-preservation.[29] Notice that Descartes' emphasis in the *Passions* quotation is on the human *body*, and so nothing turns on the presence of a soul. As a result, claims he makes here are equally relevant to other living bodies in so far as they exhibit a similar unified arrangement of parts. These passages suggest that living machines could be those that are able to maintain a unified structure of essential organic parts, and that they are able to do so through an internal principle of motion. Crucial to this account of life is the fact that living bodies are able to maintain their unified structure through their *own* functions, and do not require the interference of an external builder to maintain that structure.

Living machines, therefore, are distinguished from non-living machines as follows. First (as with MacKenzie), I believe Descartes must appeal to both causes and effects in his account of what makes a body a living body. There are two causes one can find in Descartes' texts (these are the two criteria Ablondi takes as necessary and sufficient for demarcating the living in Descartes). The first cause is that living

[27] Shapiro 2003, 433–434, including footnote 34.

[28] Des Chene 2001, 125ff.

[29] Shapiro 2003, *passim*.

bodies have their own internal source of motion, and given a charitable interpretation of Descartes' own texts, this is the heat produced (even in plants) by rapidly moving particles. The second cause is that living bodies have a unique kind of God-made complexity. As with MacKenzie, I believe this complexity must be conceived of abstractly, and it is simply any kind of complexity that permits a specific collection of effects. And so, the effects are as follows. As with MacKenzie and Distelzweig (and bringing their two insights together), specific subclasses of living bodies engage in specific activities, which are often confined to local parts and processes. These are necessary preconditions for, and contribute to the more general, often whole-body, life functions that all plants, animals, and human bodies engage in. These life functions are nutrition, growth, and generation (as with MacKenzie – though growth is of a unique kind whereby the body transforms as it grows), as well as the ability to respond to the environment, and the ability to maintain the complex structure of the body in the face of some wear and tear. Taken together, these abilities contribute to the ultimate living function of self-preservation or self-maintenance of a stable structure (Shapiro), which in turn permits the continuation of the activities and life functions identified above.

I ought to underscore one final point about these living machines. Descartes expects – and even goes to considerable lengths in order to try to realize this expectation – that all these elements of living bodies can be fully explained in terms of bits of matter-as-extension within living bodies moving according to simple laws. That is, he fully expects us to give reductionist explanations of living phenomena, but this does not amount to the elimination of the category of living beings. Descartes' austere ontology of the material world allows these powerful mechanical explanations within the life sciences, but does not thereby threaten the life sciences by stripping them of a subject of study. The first problem mentioned at the outset of this paper is thus resolved.

There are significant difficulties for a Cartesian metaphysics with this conception of life. One, which I shall not address here, concerns issues in the metaphysics of individuation. In brief, Descartes' own strict criterion of individuation of physical bodies as found in the context of his discussion of motion at *Principles* II, 25 (AT VIIIa, 53-4/CSM I, 233) does not permit the constant flux of constitutive matter in a body considered to be the *same* body through time. Thus, the (non-ensouled) living body cannot be an enduring individual for Descartes, according to this conception of a material individual. I bracket this problem as one to be dealt with elsewhere, and I turn instead to a second difficulty.

This is the problem of the role of teleology in Descartes' theory of life. For there is at least one juncture – and quite possibly more – at which teleology seems to enter in the conception of living bodies I have just developed as the conception to which I believe Descartes must be committed so as to vindicate his practice as a working scientist. But, to reiterate a well-known feature of Descartes' natural philosophy, he cannot make claims to teleology (taken specifically as a reflection of God's purposes) in natural philosophy, for God's purposes are opaque to us. And so, Descartes' theory of life may well rely upon illegitimate appeals to teleology. This difficulty will occupy the remainder of this paper.

7.2 Descartes' Theory of Life and Teleology

In this section, my general approach is as follows. If any aspect of Descartes' theory of life requires an appeal to teleological purposes, then the second problem just identified arises. While there may be more than one way in which Descartes' theory of living bodies relies upon such purposes, all I need in order to proceed with my investigation of Descartes' theory of life and the related topic of method in his study of living bodies is one case where his theory relies upon appeals to teleological purposes. So I proceed by identifying just one such case and progressing to my proposed solution to the problem that arises for him as a result of this teleology.

Now it may seem that there is no difficulty since, despite appearances, Descartes does not rely upon teleology in his account of living bodies. He may rely heavily on *functionality*, but this is quite distinct from teleology. Shapiro (2003), as indicated above, provides a non-teleological account of the apparently normative concept of health, and to the extent that good health indicates continuing *life*, her account can extend to life as well. More pointedly, Deborah Brown (2013) explicitly offers a powerfully argued, non-teleological account of functions in Descartes' discussion of living bodies. I return to aspects of Brown's paper below.

However, Distelzweig argues that some of the uses of organic parts that appear in Descartes' medical writings rely upon final causal explanations of those parts. For at times, Descartes discusses parts and processes in terms of their uses in contributing to a function – that is, the parts are present *because they serve the purpose of* fulfilling certain functions. These are examples of illegitimate reliance upon teleology. Distelzweig discusses two such cases, namely Descartes' discussion of the number of membranes in the mitral valve of the heart in the fifth part of the *Discourse* (where his concern is with the human *body* and not the human composite) and his discussion of the senses in the sixth part of the *Meditations*. According to Distelzweig, in these cases Descartes holds that the body has specific parts or processes *so as to be able to* achieve at least some of the functions, which are definitive of them as living bodies. Ignoring the case of the senses (for this introduces the troublesome case of the human composite, which I will not address in this paper), the fact that Descartes employs teleological explanation in the case of the heart is problematic. For this example shows that in the case of the human body's heart and its mitral valves, a part and the processes that part undergoes, exist *so as to realize a specific end or purpose*. Thus a specific living activity of a subclass of living beings relies, in Descartes' analysis, upon a teleological explanation. If this is so, then at least some members of the domain of life (human bodies) are identified by at least one part and related process that are depicted teleologically. There may be other such examples, but as mentioned above, one is all I need for my purposes. Such teleologically-based explanations cannot be permitted on a Cartesian natural philosophy. So one of the effects, which go into the theoretical account I have provided of living bodies in Descartes' corpus, runs into difficulties.

Distelzweig further points out that teleological explanations might be grounded in one of a couple of different ways, neither of which is open to Descartes. The way

I portray the nest of issues in what follow departs somewhat from Distelzweig's own way of laying out the conceptual terrain, but my portrayal is meant to bring out certain features of the terrain that I will need for what follows. I do not think that what I write here distorts Distelzweig's own understanding. The first way one might ground a teleological explanation relies upon the ontological priority of the whole to its parts such that the parts, systems and living functions and behaviors are there *because of* and *to serve the purpose of* preserving the whole animal. Distelzweig further argues that Descartes' theory of generation precludes this option, because according to his theory of generation, the parts come into being one after another and only *after* they have come into being does the whole begin to function. There are two ways in which this *temporal* priority of parts to whole might co-exist with an *ontological* priority of whole to parts. Both routes rely upon saying that there was always a plan of the whole, and that the plan included the fact that the whole would function so as to be self-preserving. The plan is what determines that the parts come into existence, one by one, and take on their finished, whole form. One way in which this general strategy could play out is to rely upon the Aristotelian substantial form, passed from male to female in sexual reproduction; this form carries with it the plan of the whole such that the parts form precisely in order to generate the whole and to serve the purpose of the self-preservation of the whole. Descartes' austere ontology precludes this approach; there can be no such form. The other way in which this general strategy could play out is to suppose that the plan is in the mind of a conscious builder of the whole such that the parts again are there because they serve the plan of creating a whole that is able to preserve itself through its life functions. This is the second option Distelzweig claims is closed to Descartes, for the conscious mind in the case of living bodies is God's – God intended for the parts, systems and their functions to be so-and-so in order to contribute to God's further purposes which may include the ability of a living body to preserve itself. And yet, Descartes unequivocally precludes making reference to God's intentions. It is this second option that I will interrogate in the remains of this paper.

Up to, and perhaps throughout, the early modern period, there were two general forms of teleology, even while there may also have been more forms that blended features of these two basic forms together. We may think of these as Platonic and Aristotelian forms of teleology.[30] In brief, according to Aristotelian teleology, some natural beings embody an immanent drive to fulfill purposes or achieve an end or goal that is their own end or goal, and they usually do so non-consciously or non-intentionally. Moreover, according to Aristotelian teleology, the intrinsic *drive towards an end* means that the efficient cause is end-directed; it is not the uniform, non-directed inertial motion we find in, for example, Descartes' conception of efficient cause.[31] The Aristotelian model thus includes the belief that some natural beings have an intrinsic teleological *nature* such that explanations of their purposes

[30] For some helpful texts on thinking about different conceptual and historical issues in teleology/final causation, see for example: Lennox 1985; Lennox 1992; Johnson 2005; Mayr 1992; and Detlefsen 2013.

[31] See Carriero 2005.

can be grounded in the nature of the being itself, and not in something external to the being. According to Platonic, unnatural teleology, created beings have been designed by an external, conscious and intentional agent to fulfill the goals or ends of the agent; the craftsman model is paradigmatic. The Platonic model thus includes the belief that beings created by a craftsman may have no internal teleological *nature* such that explanations of their purposes must be grounded in claims about the intentions of its maker, and not in something internal to the being itself.[32] As noted, there are blended forms of these two basic types of teleology. Aquinas, for example, believes that God creates natural beings with purposes in mind (Platonic teleology), but that he conveys these purposes in non-conscious form upon the natural beings such that they can share in God's purposes – albeit non-consciously – thus having an intrinsic teleological nature (Aristotelian teleology).[33]

Descartes' living bodies are ontologically *only* matter (taking on various sizes, shapes, speeds of motion and so forth) in lawful, inertial motion. For this reason, with respect to living bodies (the case of the ensouled human being may well be very different, of course), Descartes cannot rely upon Aristotelian teleology as the appropriate form of teleology to explain his reliance on *teleological* functions in his conception of life – wherever his functional accounts are, indeed, teleological, as in the case of the mitral valves in the heart.

Rather, if he is going to rely upon either of the forms of teleology under consideration, it would seem to have to be Platonic teleology: God built living bodies, he had purposes in mind with respect to those bodies and their parts when he built them (i.e. that they would function in specific ways), and those purposes are in the mind of God and in no way (unconsciously) held in the body. Bodies have no internal teleological nature, and so explanations about their purposes must make reference to the mind of God as the source and sole location of those purposes. In one sense, this is promising because Platonic teleology is wholly compatible with the ontology of living bodies as matter in lawful motion. But in another sense, this approach may seem doomed – and this is the source of Distelzweig's dismissal of this approach as a viable option for Descartes. That is, Descartes cannot seem to go this route because, according to Descartes, we do not know the purposes that God had in mind when he constructed the bodies of the world, and those purposes are to be found nowhere else but in the mind of God. Consequently, Descartes famously argues, we cannot rely upon those hidden purposes when investigating natural bodies:

> When dealing with natural things, we will, then, never derive any explanations from the purposes, which God or nature may have had in view when creating them and we shall entirely banish from our philosophy the search for final causes. *For we should not be so arrogant as to suppose that we can share in God's plans*. We should, instead, consider him as the efficient cause of all things… (AT VIIIa, 15-6/CSM I, 202; emphasis added).

So Descartes seems to have no theory of teleology to rely upon in order to explain the functions of the living body – the functions which *define* living bodies – should

[32] For a development of these points and their impact on Descartes' conception of the mind-body human composite, see Detlefsen 2013.

[33] See, for example, Aquinas [1265–72] 1952–4.

these functions be teleological. But there *is* at least one such case of a teleological function, i.e. the case of the mitral valves in the human body's heart. So Descartes relies upon teleology in this case, but seems to have no viable theory of teleology at hand to support this reliance. There's the problem.

But there is a part of my elaboration of the Aristotelian versus Platonic scheme given above that I think is too stark and understanding how it is so opens up a new possibility for thinking about teleology in Descartes. The overly stark characterization is in the claim that the Cartesian/Platonic model includes the belief that material bodies (ultimately consisting of only extended matter) that are created by God have no internal teleological *nature* and that thus, explanations of their purposes must be grounded *entirely* in claims about the God's intentions, and not in something internal to the bodies themselves.[34] I don't think this is true, and I don't think Descartes could have held it to be true. Rather, I think Descartes is implicitly – and correctly – committed to the belief that wholly material bodies (where matter is extension) can, and in some way do, *embody their builder's purposes*. Specifically, for my current purposes, I think Descartes is implicitly – and correctly – committed to the belief that living bodies can, and in some way do, embody God's purposes such that we can make claims to those purposes without relying upon especially robust knowledge claims about purposes in God's mind. In the next section, I will provide textual and conceptual evidence for this claim as well as situating my claims about Descartes in historical developments about bodies and teleology.

7.3 Natures and Teleology

On the face of it, the claim that wholly material bodies on a Cartesian ontology can, in a sense, have internal ends communicated to them by God would seem to be dead in the water. It would seem to be indisputable that matter conceived of as only extension *cannot*, by its very ontological nature, *embody* purposes of a mind – whether that be the mind of a human who builds a clock, for example, or the mind of God who builds a living body. Moreover, the claim that such matter can embody the purposes of a conscious mind may seem to fly in the face of Descartes' own enunciation of that non-purposive ontology of matter as found in Meditation VI, especially when we focus on the italicized portions of this passage (and breeze over the underlined portions, which I will discuss below):

> *[A] clock constructed with wheels and weights observes all the laws of its nature just as closely* when it is badly made and tells the wrong time as when it completely fulfills the wishes of the clockmaker. In the same way, I might consider the body of a man as a kind of machine.... I can easily see that if such a body suffers from dropsy, for example, and is affected by the dryness of the throat which normally produces in the mind the sensation of

[34] Manning (2012, 252) notes that it is a "serious misreading" to interpret Descartes' extrinsic denominations, such as the health or illness of a human being, as entirely mind-dependent and in no way in the human being itself. I agree, though I do not focus on extrinsic denomination.

> thirst, the resulting condition of the nerves and other parts will dispose the body to take a drink, with the result that the disease will be aggravated. *Yet this is just as natural as the body's being stimulated by a similar dryness of the throat to take a drink* when there is no such illness and the drink is beneficial. Admittedly, when I consider the purpose of the clock, I may say that it is departing from its nature when it does not tell the right time; and similarly when I consider the mechanisms of the human body, I may think that, in relation to the movements which normally occur in it, it too is deviating from its nature if the throat is dry at a time when drinking is not beneficial to its continued health. But I am well aware that 'nature' as I have just used it has a very different significance from 'nature' in the other sense [as applied to the human composite]. *As I have just used it, 'nature' is simply a label, which depends on my thought; it is quite extraneous to the things to which it is applied.... But by 'nature' in the other sense I understand something, which is really to be found in the things themselves; in this sense, therefore, the term contains something of the truth.*
>
> *When we say, then, with respect to the body suffering from dropsy, that it has a disordered nature because it has a dry throat and yet does not need drink, the term 'nature' here is used merely as an extraneous label. However, with respect to the composite, that is, the mind united with the body, what is involved is not a mere label, but a true error of nature, namely that the body is thirsty at a time when drink is going to cause the body harm.* (AT VII, 82-5/CSM II, 57–9; emphases added; trans alt.)

Focusing especially on the passages emphasized in italics, Descartes says explicitly that the supposed goal-directed 'nature' of clocks and human bodies considered solely in terms of their matter is a mere label, reflecting only purpose in my mind and is "is quite extraneous to the things to which it is applied". Material bodies are contrasted with mind-body composites, or human beings, in this passage, and human beings, unlike mere bodies, *do* have goal-directed natures.[35] So the obvious question is: how possibly can I suggest that non-ensouled bodies can embody *in their natures* the purposes given to them by a conscious mind – how possibly can I suggest that wholly material bodies can have internal ends – when Descartes appears to deny precisely that?

To answer that question, let me distinguish among the following three topics: the natures of things and teleology; epistemology and teleology; and methodology and teleology. With respect to the natures of things and teleology, I make the further distinction between the ontologically basic nature of matter (ground floor metaphysics, if you will, or matter-as-extension in the case of Descartes) and the derivative nature of matter (the nature of visible physical bodies made up out of matter-as-extension). As Gary Hatfield has pointed out, Descartes himself acknowledges these two different kinds of natures, including that living bodies have natures *qua* visible living wholes (e.g. VIIIa: 53; IXb: 14).[36]

With respect to the natures of things and teleology, it is fruitful to ask whether or not a material being can embody, in its very nature, the purposes of a conscious mind (its maker, for example); can material beings possess internal ends? But that question can be further specified to ask two sub-questions: can a body considered

[35] Manning (2012) deals with this section of Meditation VI by focusing on the historical meaning of "extrinsic denomination" and "intrinsic denomination". My project, as will come clear, is a different one, and I believe it is, for the most part, compatible with Manning's approach. There is one point of departure from Manning's reading, which I address below.

[36] See Hatfield 2008, 416–17.

solely as matter-as-extension have, in its very nature, internal ends?; and can a body of a clock or a dog, for example, made up out of matter-as-extension, have internal ends?

With respect to epistemology and teleology, it is fruitful to ask whether or not we can know the purposes of a thing, whether those purposes be embodied in a material being (internal ends) or whether those purposes be in a conscious mind and thus wholly external to the material being.

With respect to methodology and teleology, it is fruitful to ask whether there is a methodologically respectable way of relying upon appeals to purposes when explaining features of material beings (that is, in natural philosophy), whether those purposes belong to the natures of material beings or not, and whether we can definitively know what those purposes are or not. In this section, I deal with natures and teleology. In the next (final) section, I deal with the latter two issues of epistemology and methodology.

The italicized portions of the above-cited passage suggest that there is one kind of internal ends only, i.e. that which is found in the mind-body composite. Bodies without souls (or considered counter-factually in isolation from a soul, as in the case of the human body) do not possess such ends. Material bodies are, in their ontological nature, only extension of various sizes, shapes, moving in various directions and at various speeds, always in accordance with three basic laws of inertial motion. They have no goal-directed nature within themselves.

The underlined portions of the above-cited passage, however, *rely upon* bodies without souls possessing internal ends. If there were truly no difference in the nature of different clocks or different living human bodies, then the distinction between a clock that is "badly made and tells the wrong time" and a clock that "completely fulfills the wishes of the clockmaker" would be nonsensical; no such distinction could meaningfully be made. The same can be said for the distinction between a living human body and a dead human body. Descartes makes this strict parallel when he writes:

> And let us judge that the body of a living man differs as much from that of a dead man as a watch or other automaton when it is in good working order and has in itself the corporeal principle of the movements for which it is instituted with all that is required for its action, [differs from] the same watch or other machine when it is broken and the principle of its action has ceased to act. (AT XI, 331/SV 21)[37]

Making the distinctions between a clock or a human body that works well/is healthy and alive and a clock or a human body that works poorly/is ill or dead relies upon those bodies possessing some kind of internal finality, or embodying the purposes of their makers, such that when those purposes are realized by the body,

---

[37] I have chosen to focus on living and dead humans, and their symmetry with working and broken watches, rather than to focus on the dropsy case because of the special, theological, context of the Sixth Meditation, where Descartes is trying to make sense of God's goodness in the face of apparent biological mistakes. While important (Brown 2013, 90ff), and I shall address this passage briefly below, I wish to keep the focus on the nature of living bodies and the ways in which understanding clocks can help us understand certain features of living bodies.

the body works well/is healthy and alive, and such that when those purposes are not realized by the body, the body works poorly/is sick or dead. Or, to quote Descartes himself, a clock itself can "fulfill the wishes of the clockmaker", or fail to fulfill those wishes, which include the purposes the clock maker had in mind when building the clock, *and the success or failure is a feature of the clock itself.* To maintain the strict parallel at work in this passage, a living body can fulfill the wishes of God who made that body, or fail to fulfill those wishes, which include the purposes God had in mind when building the body, *and the success or failure is a feature of the living body itself.*

To approach the point from a different – and I think highly instructive – direction, consider the following example. Suppose I wish to make something that can convey to you, with a fair degree of precision, where the sun is in the sky relative to your location on earth. How do I do that? One way I can do it is by building a machine with two long sticks of slightly differing lengths that sweep around a circular surface such that when the sun is directly overhead of the spot on the equator where you find yourself, for example, the two sticks point straight up, and such that when the sun is either dipping down over the horizon or popping up over the horizon, the big hand points up and the little hand points down, and so forth. Similar descriptions can be given for a sun dial and other mechanisms built with the intention of telling time. I *cannot ever* convey to you where the sun is in the sky relative to your position on earth – that is, I can never *tell* you the time – by spilling one small drop of coffee on the floor in a room that has no access to natural light, no matter how insistently I say that producing a time-telling device was my purpose in spilling that single drop of coffee. The former machine can embody my purposes *vis a vis* time-telling, and it can convey those purposes to you in a way that the drop of coffee can never do. These facts remain true regardless of what I claim my intentions are. So: I have an intention (e.g. build something that tells the time) that requires I use material of specific sorts, organized in specific ways, and that once I build that thing such that it can successfully convey my intention to another conscious mind, then the object I have built embodies those intentions in a way that a drop of coffee, on this example, cannot do.

Where is the difference between these two material bodies, given that on a Cartesian ontology, the built machine and the drop of coffee both have the same, ontologically basic material nature (extension), and both inviolably obey the same laws of motion? Here is where the further distinction in the discussion of natures and teleology is helpful. For the temptation to say that material bodies are simply not the sorts of things that can embody internal finality, especially in light of the Sixth Meditation passage cited, can surely be said about material bodies considered in terms of their ontologically basic nature but this need not apply to bodies considered in terms of their derivative, built natures. For if there were no internal ends embodied in built machines (clocks built by humans or living bodies built by God), then no sense could be made of the idea of a clock *being broken* (i.e. failing to convey my purposes to you) when the hands don't move, or the idea of the body being defective when the mitral valves in the heart, for example, fail to open. Yet Descartes takes these ideas of deviation from well-working/health in the case of wholly material bodies as givens and as completely sensible (and he is right to do so). The underlined portions of the passage cite above establish this.[38]

Descartes recognizes these facts, and he does so specifically with respect to living bodies. For example, he claims that we humans could never build a bird, because we could never make matter that is appropriate for building a living bird (AT III, 163). Similarly, he claims we could never make ourselves a new body, for we cannot make such matter (AT VI, 148). In these claims, he recognizes that, despite the ground-floor ontological sameness of bird bodies, human bodies, clocks and so on, matter in its derivative forms can allow or not allow certain machines, presumably with certain purposes, to be built.

So the Sixth Meditation passage makes distinctions among three – not two – different ways of thinking about bodies. One kind of body is the ensouled body of the composite, and these bodies have a goal-directed nature internal to them; they have internal finality of a unique sort grounded in their unique ontological nature as ensouled bodies. I will say no more about this special, and theoretically complicated, being in Descartes' ontology. There is a second kind of body that can be thought of in two different ways. This is the wholly material body, such as the human-built clock or the God-built living body. Thought of in terms of its ground-floor metaphysical nature, i.e. its constitutive matter-as-extension, such a body is in no way goal-directed; it is only matter in lawful inertial motion. But thought of in terms of its derivative physical nature, i.e. matter of a derivative kind structured in very particular ways, such a body can have internal ends, sharing in the purposes that a human or God had in mind when building it, even while this sort of internal finality may be very different from that found in the human composite.[39]

Let me be explicit about what I am and am not claiming about this third way of thinking about bodies, according to which wholly material bodies can have natures that include internal ends. Some conceptual-historical background will help here. Aristotelian ontology makes a difference in kind between living bodies and artifacts because the former have all four causes internal to them. Indeed, he even goes so far as to say that the formal cause (or substantial form), the efficient cause (or internal principle of change), and the final cause (the drive to a telos or end point) are one and the same cause within living bodies. In artifacts, by contrast, bodies themselves have only the formal and the material causes within themselves. Efficient and final causes are external to bodies, namely in the craftsman who builds the artifacts. According to one crude depiction, Cartesian mechanisms makes all bodies, including living bodies, into Aristotelian artifacts. But Aristotle is moved by a pre-theoretical, and entirely common sense, understanding of artifacts, and that is that a craftsman can't build just anything out of any old matter. I cannot build a statue of a deer out of warm water; warm water does not have a suitable *nature* to be fashioned into a statue of a deer. Warm water is not the sort of matter that can bear

[38] For historical context that helps to bolster this idea, see Manning on extrinsic denominations (2012).

[39] On this point, I depart from a number of commentators. See Hoffman 1986 and 1999; Ariew 1983; Grene 1986 and 1991; Gueroult 1952; and Rodis-Lewis 1950.

my purposes in this case. Aristotle has a theoretical way to account for this pre-theoretical and common sense intuition: no matter is completely un-informed in his view. Matter is always *informed* with some form or another. That is what makes warm water different, in kind, from marble. Ridding his ontology of this robust, Aristotelian conception of form may rob Descartes of this way of accounting for the pre-theoretical, and entirely common sense, intuition. But it does not dispense with the intuition in the first place. And it is an intuition that Descartes shares with Aristotle. This is shown by his acknowledgment that we could never build a living bird or make ourselves a new body, for we cannot make matter of the appropriate nature. No less than Aristotle (or any common sense, pre-theoretical view), Descartes believes that craftspeople must use material of a particular nature if they wish to build an artifact that can bear their ends.[40]

What I am claiming is the following. Descartes maintains the Aristotelian intuition about material natures that can embody a builder's ends – that can have internal ends, that is. This nature is presupposed by the distinctions he makes between a clock that works well and a clock that doesn't (indeed, to use my example, between a clock that is able, in the first place, to tell time and a drop of coffee that is not so able), and between a human body that is alive and one that is dead. Maintaining this intuition is perfectly sensible, and it would a be non-starter were Descartes to deny this pre-theoretical, common sense, understanding of bodies. For Descartes, this material nature capable of bearing internal ends is *not* to be found in the ontological essence of matter as extension but is rather to be found in the derivative nature of medium-sized matter shaped in various ways. However, I am not able, here, to further spell out the precise ontology of this derivative nature that has internal ends; indeed, given Descartes' rejection of Aristotelian ontology of informed matter, I am not sure his new austere ontology can allow for a derivative nature of material bodies with internal ends.[41] Specifically, I am not sure his own conception of matter of three kinds, depending upon the relative size and speed of motion of their constitutive parts, is up to the task of accounting for the kind of matter needed to build living bodies. But he assumes bodies with such natures, and it is a sensible assumption.[42]

One final, crucial comment is in order. The question of God's making a body to fulfill certain purposes is distinct from, albeit intimately connected with, the question of the *value* or *normative goodness* of how well a body fulfills those purposes.

---

[40] For helpful material on Aristotle on many of these points, see Kosman 1987.

[41] Michael Della Rocca has suggested (in correspondence) that in creating the eternal truths, God has imposed natures on things, thereby endowing them with an intrinsic character. Indeed, in the case of God's creations, it might be more plausible to make the claim that his products can embody internal purposes. This would bolster my interpretation here, though my argument proceeds by analogy from the familiar case of human-made machines to the case of God-made machines.

[42] Tad Schmaltz has recently developed a convincing argument in favor of an unconscious, Aristotelian-type internal finality in human composites. See "Descartes's Critique of Scholastic Teleology" (manuscript). The current conception of intrinsic ends relies more upon a conscious agent's ability to *signal* her purposes, through very specific uses of matter, to another conscious agent. The current form thus leans more toward a Platonic form, albeit with the Aristotelian element of the purposes also being embodied in a non-conscious being.

Brown draws our attention to the issue of normative value in her non-teleological account of bodily functions, underscoring that a non-teleological account of living functions has the benefit of accounting for life activities without God being culpable for mistakes in the body such as dropsy and death.[43] My account here cannot similarly avoid this difficulty so easily, though this is not to say that there is therefore no solution to this problem. But the fact remains that Descartes does rely upon teleology in the case of the mitral valves. One way to explain that reliance is to couple his own strict parallel between clocks (for example) and living bodies with his embrace of the pre-theoretical acknowledgment that clocks can embody or fail to embody our purposes, and then conclude that for Descartes, living material bodies can similarly embody or fail to embody God's purposes. This may saddle Descartes with the problem of God's culpability, but the alternative would be to leave his teleological claims unexplained.

7.4 Epistemology, Method, and Internal Ends

Still, to complicate the current account and to set the stage for thinking about epistemology and teleology, as well as methodology and teleology, imagine someone who has never seen a clock before and has no previous knowledge of human-made time telling devices.[44] Such a person may come across my large and heavy clock and wonder what it is and what it does. She may carefully observe it working over the course of several days and stumble across my true purpose in building it by noticing that the two sticks both point upwards when the sun is directly overhead or in the depths of night, and that the big stick points up and the little stick points down just when the sun rises or sets on the horizon and so forth. That person might come to understand that my clock will be very handy in conveying to her certain information about the position of the sun when she is in a basement room without access to natural light. Alternatively, this person may notice that my large and heavy clock is very handy in holding open the door to her basement room, and she might then conclude that it was made for this purpose. Indeed, the physical clock is made of materials that can serve this purpose too, though that was not *my* purpose when I built it.

Crucially, the clock is different from the living body for Descartes in the sense that the decoder of the clock can always ask me, or another human with knowledge of the true purpose that clockmakers have in mind, what internal ends the clock is supposed to embody according to its builder. Then the knowledgeable person can directly convey the purposes of the clockmaker. God's purposes are, according to Descartes, inscrutable and buried in the abyss of his wisdom. There is no asking God what he intended, and we cannot, without hubris, pretend to know his purposes,

---

[43] Brown 2013, 89–90.

[44] In *Dialogues on Natural Religion*, David Hume, of course, considers this question and provides a response that is especially interesting for the chaos theory, which I note is beyond the scope of this current project.

even though they may seem to be on full display in the parts, processes and behaviors of the living bodies he has made (such as Gassendi claims: AT VII, 309; CSM II, 215). This brings us to the issues of teleology and epistemology, and teleology and methodology.

We have arrived at this point: conscious minds can convey purposes to bodies themselves, not in so far as we consider body as matter-as-extension, but in so far as we consider body as structured, visible machines made up out of matter-as-extension.[45] Still, in the case of living bodies for Descartes, we cannot *know* what purposes God might have conveyed to them when he created them, and so we cannot make use of teleology in our natural philosophy at all. This epistemological block seems to thwart my attempt to vindicate Descartes' teleologically laced conception of living bodies.

There is no doubt that Descartes' insistence that we cannot *know* God's purposes is meant to translate into the requirement that we never use teleological explanations in natural philosophy, including in our theories about living beings. But I think he goes too far in his precluding teleological explanations in natural philosophy, and I think there is a way he *could have*, and perhaps even *should have* made use of such explanations for a richer, more powerful natural philosophy. Specifically, he could have included teleological explanations on the basis of their being hypothetical explanations grounded in natural investigations of the internal ends found within living bodies themselves. It is true that Descartes explicitly rejects this methodological tool, but he may have been wrong to do so, and he loses so much more than he gains as a result of his rejection.

To set the stage for my suggestion, imagine again the person with no previous knowledge of human-made time-telling devices. She has decided that the purpose of my large and heavy clock is to prop open the door to her basement room. Suppose then, she comes across a small analogue pocket watch that resembles my clock in many ways though not in its size and weight, and that keeps time in perfect tandem with my clock. The similarity in most aspects of these two machines' structures, and in the behaviors they exhibit, are not lost on the imaginary observer. She then concludes that, while my clock does indeed do a wonderful job of holding open her door, there may well be a different purpose in the mind of the clock's builder – and embodied in the clock itself – than the one she originally attributed to the builder and clock. The more she comes across similar devices, the more she may investigate what may be the true purpose of my clock, at least as I intended it, and of similar machines; she may even latch upon my true purpose should she conclude that my clock is meant to tell people where in the sky the sun is. Of course, if she does not ask me my true purpose that I have embodied in the clock, she cannot claim that she

[45] Manning's (2012, 262) approach to the issue of health in the human and extension of this teleological notion to non-human living bodies, is to employ the historical conception of extrinsic denominations to attribute teleological notions of health and illness to human bodies themselves, and then extending these conclusions to animals due to their likeness to the living human body. My approach is to focus on the process of making machines, and the intentional imparting of purposes in that process, and to find a way we can depend upon that without depending upon *knowledge* claims about God's purposes.

indubitably *knows* that she has latched upon my true purpose embodied in the clock. But the more evidence she gathers in the form of different examples of a wide variety of devices, which behave in a uniform way (i.e. successfully conveying the position of the sun relative to the person standing on the equator) despite all their material variety, the more she will be justified in thinking that her belief about my clock is probably a true belief; she may come to have what Descartes calls "moral certainty" that her belief is true.

This approach captures the way of hypotheses to which Descartes is very friendly starting at least from the time of the *Discourse* and texts attached to the *Discourse*.[46] Descartes believes that first principles of philosophy set the confines for all of natural philosophy. But those principles radically underdetermine what could be true of bodies in the natural world. Most crucially, matter-as-extension and the three simple laws of motion, could have given rise to many different phenomena, most of which do not obtain in our actual world (e.g. AT VI, 64; CSM I, 144). And so the natural philosopher observes what is true of our world and proceeds to hypothesize (or suppose or guess) about the exact mechanisms, which might have given rise to the world we have – all the while respecting the first principles. This general approach to and reliance upon hypotheses carries through to Descartes' later works where he develops details of his approach to hypotheses more fully.

Historically, there have been two key directions in which thinking about hypotheses developed, indeed from Ancient times, and certainly throughout the seventeenth and eighteenth centuries as well. According to one approach – typified by Ptolemy in pre-modern thought and sometimes associated with 'save the phenomena' type explanations – hypotheses are posited merely because they are useful instruments, mere mathematical calculating devices especially useful for prediction and scientific practice. The aim with hypotheses, according to this approach, is not to propose a *true* account of the nature of things, since reaching true conclusions about the world is not necessarily relevant when formulating hypotheses according to this tradition, which focuses more pointedly on prediction. According to the second approach – typified by Aristotle in pre-modern thought and sometimes associated with causal explanations – hypotheses are posited in order to provide an explanation of how experienced effects might have come about. The aim is to give a *true* account of the nature of things, especially the causal nature of things.[47] In the *Principles*, Descartes comes down much more firmly on the side of hypotheses aiming for a true account of causes rather than on the side of hypotheses aiming simply to save the phenomena.[48] His reasoning in the later work captures something

[46] This aspect of Descartes' method is far more complex – and interesting – than I make out here. For some work on Descartes and hypotheses, see Clarke 1989 and 2011; Lauden 1981; McMullin 2000 and 2008; Sakellariadis 1982; and Detlefsen forthcoming.

[47] For more on these two approaches to hypothesis, including the understanding of those such as Kepler and Galileo who believed these methods to be compatible, see McMullin 2000 and Friedman 2008, 71.

[48] There is a moment in the *Principles* when he seems to allow for the latter use of hypotheses, but a careful reading of this passage leaves open the distinct possibility that what is going on in the passage is Descartes' recognition of their lack of certainty, not their mere instrumentality.

implicit, yet crucial, found in his letter to Morin of 13 July 1638 where he suggests that any hypothesis which accounts for multiple effects, including those not originally under investigation, is likely 'the true cause from which they [effects] result' (AT II, 199/CSMK 107). That is, should hypothesized causes explain a plethora of effects, including others not initially under investigation, then this simplicity and systematicity indicates that the hypotheses are probably true. He repeats this idea in the *Principles* (PP III, §43-4; AT VIIIa, 98-9/CSM I, 255).

This point connects with a significant feature of Descartes' account of hypotheses in the *Principles*, and this captures a development in scientific epistemology, which Desmond Clarke and Ernan McMullin have recently detailed. They note, that is, that some natural philosophers were moving away from treating less than certain knowledge in the form of hypotheses as merely speculative and thus unhelpful in scientific investigations. Rather, these natural philosophers believed that hypothetical claims carry important, even if not indubitably true, scientific information. That is, such philosophers were moving toward treating such knowledge as more or less probable, and therefore, more or less respectable. The degree of probability enjoyed by such hypotheses depends upon a number of factors, including, as suggested by Descartes, how simple and systematic they are.[49] Clarke thus points out that, throughout the 1600s, a new scientific epistemology emerged which allowed for a respectable, because not wholly speculative, category of the probable. Shortly after Descartes' time, this more palatable notion of probability is clearly articulated by Edme Mariotte in his *Essai de logique* (1678): 'An hypothesis of one system is more probable than that of another if, by assuming it, one explains all the phenomena or a greater number of phenomena more exactly, more clearly and with a stronger link with other known things…'.[50] Three quarters of a century later, Émilie Du Châtelet would provide a theoretical account of hypotheses and their role in science, which fully articulated this powerful new scientific eistemology.[51] In his later work, Descartes seems to embrace such a conception of probability, retreating from an all-out claim to the certain truth of hypothesized causes (PP IV, §204; AT VIIIa, 327/CSM I, 289), even while claiming 'moral certainty' of their truth (PP IV, §205; AT VIIIa, 327-28/CSM I, 289–90). That is, while not metaphysically certain, Descartes' own posited hypotheses and conclusions derived from them are, in his view, not thereby mere arbitrary speculation. They are scientifically useful despite not being indubitably true.[52]

(See PP III, §44; AT VIIIa, 99/CSM I, 255). The preponderance of Descartes' claims indicates that he takes the role of the natural philosopher to be the pursuit of true causes of phenomena.

[49] For accounts of Descartes' maturation on the relation between hypotheses and scientific epistemology, see Clarke 1989, chapter 7, and 2011 and McMullin 1990, 2000 and 2008. For a much earlier account of many of these themes recently developed by Clarke and McMullin, including a discussion of hypotheses, see Garber 1978.

[50] Mariotte 1678, 624.

[51] Du Châtelet 1740, chapter 4.

[52] For discussions on why Descartes' hypotheses are not merely speculative, see for example, McMullin 2008, 89 and Clarke 1989, 141–4. The latter makes a distinction between arbitrary and reasonable hypotheses, with reasonable hypotheses being assumptions, which can be systematized and unified into a system, ideally bound by laws.

In the clock example above, I suggest a crudely parallel approach. And presumably the life scientist could employ the method of hypothesis posing and testing with respect to the purposes – the internal ends – that she finds in at least some living bodies. To put this in the context of the three teleological issues: God could have conveyed purposes upon living bodies, not with respect to their ground-floor metaphysical natures as matter-as-extension but with respect to their derivative natures *vis-a-vis* the sort and structure of their matter (natures and teleology), and although we can never *know* what those purposes are, we can develop beliefs about those purposes that are fairly likely true (epistemology and teleology), and we can do so through a method to which Descartes is very friendly, namely the method of positing hypotheses and seeing how they hold up to additional empirical data (method and teleology). That is, perhaps she could amass other example of hearts similar in structure to the human's heart to see if the mitral valve operates similarly therein. Or perhaps she could posit an hypothesis to test for other ends served by the heart's structure so as to determine whether or not that specific structure serves a number of other purposes beneficial to keeping the whole organism healthy and alive.

The merely probable, and not certain, nature of the purposes of teleological features in living bodies would not undermine Descartes' separation of living beings within a broader class of self-moving machines. Recall the general approach I suggested that Descartes takes to the study of living bodies. First, pre-theoretically, he identifies the kinds of bodies that belong to the class of living bodies, and these are plants, animals and human bodies. Then, he theorizes (or one can theorize on his behalf) about members of this domain. This theorizing can isolate (as argued in section I) a set of causes (heat and structure of a suitable nature) and a set of effects (a hierarchy of life behaviors), which are able to demarcate all and only these individuals. The job of the life scientist is to investigate these causes and effects, and in some cases (e.g. the mitral valves of the heart), features of the living being turn out to be explicable only by making appeal to a plan held by a conscious craftsman of those bodies, a plan that guided the construction of those bodies. The theoretical account of living beings turns out to be irreducibly teleological. Given Descartes' strictures against claiming knowledge of God's purposes with respect to the natural world, at best the natural philosopher could only hypothesize about God's purposes with respect to his plans for living bodies. She may be wrong about those purposes, but being so does not mean that Descartes loses the category of life *tout court*. It means simply that the natural philosopher will sometimes be mistaken in her explanations of some parts or processes that embody, in their very nature, internal ends – an embodiment that contributes to that which distinguishes the living from the nonliving. Moreover, one can use Descartes' own account of probability with respect to hypotheses to argue that the larger number of effects that can be explained by the supposed purpose, the greater the probability that the natural philosopher has hit upon the true teleological explanation of the behavior under investigation. One can never reach certainty, but a hypothesis can be thought to be more probably true with more and more effects accounted for by the hypothesis. This is in keeping with Descartes' participation in the historical emergence of a new scientific epistemology (recognizable to us today), according to which the probable is a respectable category and not to be discarded as merely speculative. So, far from posing a problem for Descartes by, for example, undermining the distinction between the living and the non-living, the role played by hypotheses regarding purposes is a scientifically powerful tool, which has the promise of spurring on empirical investigations of the behaviors of living things in order to grant greater and greater degrees of probability to the hypotheses posed.

Alas, Descartes rejects this approach – while we can hypothesize about God's purposes when engaged in moral philosophy, we cannot do so in natural philosophy (AT VII, 375; CSM II, 258). This is strongly implied by the nature of the hypothetical causes that Descartes specifies in the *Principles* – specific sizes, shapes and so forth of subvisible parts of matter (AT VIIIa, 325–6; CSM I, 288), and not God's plan with respect to the construction of the built machine. So, from Descartes' point of view, what I suggest above is illegitimate; we cannot use teleological explanations in so far as they are grounded in claims about God's purposes *even as merely likely true beliefs* in our explanations about the natural world, and so we cannot explain the teleological nature of (at least some) life activities by relying upon hypothetical claims to God's purposes as embodied in (at least some) living bodies. This is the core of Descartes' difficulty as I see it in his conception of living bodies: he does not exploit his scientifically powerful tool of the method of hypotheses in the realm of teleology and the life sciences. For without extending his friendliness to hypotheses regarding micro-mechanisms in natural philosophy to hypotheses about the internal ends of living bodies – for this would ultimately require making claims about God's likely purposes – it is impossible to make teleological claims about living bodies. But as I have shown in section II above, the theoretically robust conception of living bodies that one can develop on Descartes' behalf, and for which there is textual evidence in Descartes' corpus, *depends upon* making at least one teleological claim about living bodies. So some crucial aspects of what I take to be Descartes' theoretically robust explanation of living bodies that can reliably pick out all and only plants, animals, and human bodies, run afoul of Descartes' metaphysics of God's mind and what we can know of it.

Descartes does not reject hypothesizing about God's purposes *tout court*; he explicitly allows that we can engage in such an endeavor in the field of ethics by hypothesizes about God's purposes for us as moral beings. He does not, as we have seen, extend this use of hypotheses to the purposes of natural beings. To understand why, we should note that he offers two objections to using final causes in natural philosophy. His first objection is that it is hubris to suppose we know God's purposes; I have provided a way that Descartes could have side-stepped this worry given his friendliness to hypotheses. For in hypothesizing about God's purposes, we do not claim to *know* them. His second objection is that final causal explanations are the wrong kinds of explanations to offer in natural philosophy (AT V, 158; CSMK 341).[53] But these sorts of causal explanations can co-exist with Descartes' favored form of explanation grounded in efficient causes, and they can co-exist within a

[53] For discussion of Descartes' reluctance to include final causes in natural philosophy, see Simmons 2001 and Hatfield 2008.

Cartesian ontology of matter. One can hypothesize about God's likely purposes as he embodied them in living bodies, and these purposes can be compatible with a mechanical ontology of the material world, where all bodies are ultimately made up out of matter (as extension) in inertial lawful motion. One can further provide efficient causal explanations for how God's purposes are carried out in the living machines that he has built, as we witness over and over again in Descartes' reductionist explanations of life phenomena. It is true that precluding final causal explanations *forces* the natural philosopher to give efficient causal explanations if she is to engage at all in explanation – that is, if she is to engage in that crucial aspect of natural philosophy. But allowing final causal explanations does not thereby automatically preclude her giving efficient causal explanations as well. Descartes' overly cautious approach to the hypothesizing about God's purposes with respect to the natural world might have been sensible given the intellectual climate that forged him as a thinker, but it was unnecessary, and in going this route, he gave up on one crucially powerful tool for use within the sciences of life.

References

Ablondi, Fred. 1998. Automata, living and non-living: Descartes' mechanical biology and his criteria for life. *Biology and Philosophy* 13: 179–186.

Aquinas, Thomas. [1265–72] 1952–4. *Truth*. Trans. Robert W. Mulligan, S.J., James V. McGlynn, S.J., and Robert W. Schmidt, S.J. Chicago: Henry Regnery Co.

Ariew, Roger. 1983. Mind body interaction in Cartesian philosophy: A reply to Garber. *Southern Journal of Philosophy* 21(supplement): 33–37.

Bitbol-Hespériès, Annie. 1990. *Le Principe de Vie Chez Descartes*. Paris: J. Vrin.

Brown, Deborah. 2013. Cartesian functional analysis. *Australasian Journal of Philosophy* 90(1): 75–92.

Canguihelm, George. 1965. *La connaissance de la vie*. Paris: J. Vrin.

Carriero, John. 2005. Spinoza on final causality. *Oxford Studies in Early Modern Philosophy* 2: 105–147.

Clarke, Desmond. 1989. *Occult powers and hypotheses: Cartesian natural philosophy under Louis XIV*. Oxford: Clarendon.

Clarke, Desmond. 2011. Hypotheses. In *The Oxford handbook of philosophy in early modern Europe*, ed. Catherine Wilson and Desmond Clarke, 249–271. Oxford: Oxford University Press.

De Rosa, Raffaella. 2007. A teleological account of Cartesian sensations? *Synthese* 156: 311–336.

Delaporte, François. [1979] 1982. *Nature's second kingdom: Explorations of vegetality in the eighteenth century*. Trans. Arthur Goldhammer. Cambridge: MIT Press.

Des Chene, Dennis. 2000a. *Life's form: Late Aristotelian conceptions of the soul*. Ithaca: Cornell University Press.

Des Chene, Dennis. 2000b. Life and health in Cartesian natural philosophy. In *Descartes' natural philosophy*, ed. Stephen Gaukroger, John Schuster, and John Sutton, 723–735. New York: Routledge.

Des Chene, Dennis. 2001. *Spirits and clocks: Machine and organism in Descartes*. Ithaca: Cornell University Press.

Descartes, René. 1964–76. *Oeuvres de Descartes*, 11 vols, eds. C. Adam and P. Tannery. Paris: J. Vrin. Cited with abbreviation AT, followed by volume and page number.

Descartes, René. 1985a. *The philosophical writings of Descartes*, 2 vols. Trans. John Cottingham, Robert Stoothoff, and Dugald Murdoch. Cambridge: Cambridge University Press. Cited with abbreviation CSM, followed by volume and page number.
Descartes, René. 1985b. *The philosophical writings of Descartes*, vol. 3: The correspondence. Trans. John Cottingham, Robert Stoothoff, Dugald Murdoch, and Anthony Kenny. Cambridge: Cambridge University Press. Cited with abbreviation CSMK, followed by page number.
Descartes, René. 1989. *The passions of the soul*. Trans. Stephen H. Voss. Indianapolis: Hackett Publishing Company. Cited by abbreviation SV followed by page number.
Descartes, René. 1998. *The world and other writings*. Ed. Stephen Gaukroger. Cambridge: Cambridge University Press. Cited by abbreviation SG followed by page number.
Detlefsen, Karen. 2013. Teleology and natures in Descartes' sixth meditation. In *Descartes' meditations: A critical guide*, ed. Karen Detlefsen, 153–176. Cambridge: Cambridge University Press.
Detlefsen, Karen. forthcoming. Du Châtelet and Descartes on the roles of hypothesis and metaphysics in natural philosophy. In *Feminism and the history of philosophy*, eds. Eileen O'Neill and Marcy Lascano. Kluwer Academic Press.
Distelzweig, Peter. 2015. The use of *Usus* and the function of *Functio*: Teleology and its limits in Descartes's physiology. *Journal of the History of Philosophy* 53: 377–399.
Du Châtelet, Émilie. 1740. *Institutions de physqiue*. Paris: Prault Fils.
Friedman, Michael. 2008. Descartes and Galileo: Copernicanism and the metaphysical foundations of physics. In *A companion to Descartes*, ed. Janet Broughton and John Carriero. Malden: Blackwell.
Fuchs, Thomas. 2001. *The mechanization of the heart: Harvey and Descartes*. Trans. Marjorie Grene. Rochester: Rochester University Press.
Garber, Daniel. 1978. Science and certainty in Descartes. In *Descartes: Critical and interpretive essays*, ed. Michael Hooker, 114–151. Baltimore: The Johns Hopkins University Press.
Garber, Daniel. 1992. *Descartes' metaphysical physics*. Chicago: University of Chicago Press.
Gaukroger, Stephen. 2000. The resources of Descartes' mechanist physiology and the problem of goal-directed processes. In *Descartes' natural philosophy*, ed. Stephen Gaukroger, John Andrew Schuster, and John Sutton, 383–400. London: Routledge.
Gaukroger, Stephen. 2002. *Descartes' system of natural philosophy*. Cambridge: Cambridge University Press.
Gaukroger, Stephen. 2010. Descartes' theory of perceptual cognition and the question of moral sensibility. In *Mind, method and morality: Essays in honour of Anthony Kenny*, ed. John Cottingham and Peter Hacker, 230–251. Oxford: Oxford University Press.
Grene, Marjorie. 1986. Die Einheit des Menschen: Descartes unter den Scholastikern. *Dialectica* 40: 309–322.
Grene, Marjorie. 1991. *Descartes among the scholastics*. Milwaukee: Marquette University Press.
Gueroult, Martial. 1952. *Descartes selon l'ordre des raisons*, vol. 2. Paris: Editions Montaigne.
Hall, T.S. 1970. Descartes' physiological method: Position, principles, examples. *Journal of the History of Biology* 3: 52–79.
Hatfield, Gary. 1993. Reason, nature, and God in Descartes. In *Essays on the philosophy and science of René Descartes*, ed. Stephen Voss, 259–287. Oxford: Oxford University Press.
Hatfield, Gary. 2008. Animals. In *Companion to Descartes*, ed. Janet Broughton and John Carriero, 404–425. Malden: Blackwell.
Hoffman, Paul. 1986. The unity of Descartes's man. *The Philosophical Review* XCV: 339–370.
Hoffman, Paul. 1999. Cartesian composites. *Journal of the History of Philosophy* 37: 251–270.
Johnson, Monte Ransome. 2005. *Aristotle on teleology*. Oxford: Oxford University Press.
Kosman, L.A. 1987. Animals and other beings in Aristotle. In *Philosophical issues in Aristotle's biology*, ed. Allan Gotthelf and James Lennox, 360–391. Cambridge: Cambridge University Press.
La Porte, Jean. 1928. La finalité chez Descartes. *Revue d'Histoire de la Philosophie* 2(4): 366–396.

Lauden, Larry. 1981. *Science and hypothesis: Historical essays on scientific methodology*. Dordrecht: D. Reidel Publishing Company.
Lennox, James. 1985. Plato's unnatural teleology. In *Platonic investigations*, Studies in philosophy and the history of philosophy 13, ed. Dominic J. O'Meara, 195–218. Washington, DC: The Catholic University of America Press.
Lennox, James. 1992. Teleology. In *Keywords in evolutionary biology*, ed. Evelyn Fox Keller and Elisabeth A. Lloyd. Cambridge: Harvard University Press.
MacKenzie, Ann Wilbur. 1975. A word about Descartes' mechanistic conception of life. *Journal of the History of Biology* 8(1): 1–13.
Manning, Gideon. 2012. Descartes' healthy machines and the human exception. In *The mechanization of natural philosophy*, ed. Sophie Roux and Dan Garber, 237–262. Kluwer: New York.
Manning, Gideon. forthcoming. Extrinsic denomination. In *Descartes Lexicon*, ed. L. Nolan. Cambridge: Cambridge University Press.
Mariotte, Edme. 1678. *Essai de logique*. In *Oeuvres*, volume ii.
Mayr, Ernst. 1992. The idea of teleology. *Journal of the History of Ideas* 53(1): 117–135.
McMullin, Ernan. 1990. Conceptions of science in the scientific revolution. In *Reappraisals of the scientific revolution*, ed. David C. Lindberg and Robert S. Westman, 32–44. Cambridge: Cambridge University Press.
McMullin, Ernan. 2000. Hypothesis. In *Encyclopedia of the scientific revolution: From Copernicus to Newton*, ed. Wilbur Applebaum. New York: Garland Publishing Inc.
McMullin, Ernan. 2008. Explanation as confirmation in Descartes's natural philosophy. In *A companion to Descartes*, ed. Janet Broughton and John Carriero, 84–102. Malden: Blackwell.
Menn, Stephen. 2000. On Dennis Des Chene's physiologia. *Perspectives on Science* 8(2): 119–143.
Rodis-Lewis, Geneviève. 1950. *L'individualité selon Descartes*. Paris: J. Vrin.
Rodis-Lewis, Geneviève. 1978. Limitations of the mechanical model in the Cartesian conception of the organism. In *Descartes: Critical and interpretative essays*, ed. Michael Hooker, 152–170. Baltimore: Johns Hopkins University Press.
Sakellariadis, Spyros. 1982. Descartes's use of empirical data to test hypotheses. *Isis* 73(1): 68–76.
Schmaltz, Tad. Manuscript. Descartes's critique of scholastic teleology.
Shapiro, Lisa. 2003. The health of the body machine? Or seventeenth century mechanism and the concept of health. *Perspectives on Science* 11(4): 421–442.
Simmons, Alison. 2001. Sensible ends: Latent teleology in Descartes' account of sensation. *Journal of the History of Philosophy* 39(1): 49–75.

[illegible]

Chapter 8
Mechanism, the Senses, and Reason: Franciscus Sylvius and Leiden Debates Over Anatomical Knowledge After Harvey and Descartes

Evan R. Ragland

Abstract By the mid-seventeenth century, philosophy, anatomy, and chymistry were inextricably bound together, and concentrated in lively debates over the action of the heart. In the wake of Harvey's anatomical demonstration of the circulation of the blood, and Descartes's provocative but error-prone anatomical speculations, Dutch physicians adopted varied positions on the sources and status of anatomical knowledge. This article attends to Leiden professor Franciscus Sylvius's central place in this history, beginning with his early demonstrations of the circulation and his dissections and disputes with Descartes. His collaboration with Johannes Walaeus produced innovative experimental work on the circulation and the origins of the blood in digestion. Sylvius and his colleagues were generally comfortable with mechanical explanations, which they had already met in Galen's depictions of the mechanical anatomy of Erasistratus, but only as far as they squared with sensory experience. Even prominent mechanistic anatomists such as Sylvius's student Nicolaus Steno would accept ideals and methods of mechanistic explanation, while rejecting extant mechanisms for their sensory and experimental inadequacy. Our own, anachronistic sense of early moderns' errors is of little use to our historical understanding, but their perceptions of error, especially in the combination of philosophical systems and the autoptic anatomical tradition, were essential to their history.

Keywords Mechanism • Senses • Reason • Descartes • Harvey • Sylvius

---

E.R. Ragland (✉)
Department of History, University of Notre Dame, Notre Dame, IN, USA
e-mail: eragland@nd.edu

P. Distelzweig et al. (eds.), *Early Modern Medicine and Natural Philosophy*, History, Philosophy and Theory of the Life Sciences 14,
DOI 10.1007/978-94-017-7353-9_8

8.1 Introduction

In 1639, the Dutch anatomist and physician Franciscus Dele Boë Sylvius (1614–1672) thrust experimental anatomy before the public eye when he demonstrated the Harveian circulation of the blood in the Leiden public gardens. These anatomical investigations unfolded within and partially constituted the growing controversy over the motion of the heart and blood. The two principal players here were, of course, William Harvey and René Descartes. A great deal of scholarly effort has gone into analyzing the dispute over the motion of the heart between Harvey and Descartes.[1] I will not rehearse the arguments here, but I will follow Geoffrey Gorham in pointing out the poverty of characterizing the debate in simplistic terms as an 'empiricist' Harvey vs. a purely 'rationalist' Descartes, as a purely empirical dispute, or a metaphysical contest between 'vitalist' thought and 'mechanist' philosophy. None of these strict dichotomies will do.[2] But, following the publication of Descartes's anatomical thinking in his 1637 *Discours*, anatomists and physicians criticized the accuracy of Descartes's anatomical claims, and debated the usefulness of his approach to anatomy, a discipline shaken and shaped by Harvey's recent autoptic discoveries.[3] At the same time, many of these physicians also embraced Cartesian mechanism as an ideal of explanation and ontology. As we will see, we can understand the controversy over the motion of the heart only by taking into account the confluence of medical and philosophical traditions that shaped it.

8.2 Harvey and Descartes on the Heart

First, sketches of Descartes's and Harvey's accounts of the heart's action. Throughout his works, Descartes held that the blood rushed out of the heart into the arteries as rarefying blood expanded the ventricles, so that the ventricles and the arteries dilated at the same time.[4] The series of movements, of the influx of the blood, the closing of the valves at different stages as the blood pushes back on them, and especially the dilation of the ventricles as the blood expands and the outrushing of

---

[1] Toellner 1972; Grene 1992; Clarke 1982, 148–154; Gorham 1994; Fuchs 2001.

[2] Comparing to the account of the motion of the heart found in the 1637 *Discourse* to his letters reveals Descartes changing the details of his mechanism in response to empirical and experimental criticism. Descartes *Discours* AT VI 46–55; CSM I 134–39.

[3] McGahagan 1976; French 1994, ch. 8. This article greatly expands our understanding of Sylvius's work and his place in these debates, and also adds new material on Walaeus and corrections to previous accounts. Unfortunately, for example, French, pp. 186 and 207, mistook Sylvius's later disputations of 1659–1663 for his earlier *Dictata* of 1640–1641.

[4] Descartes, *Passions*, AT XI 333–334; CSM I 331. There, Descartes writes only that "there is a continual heat in our hearts, which is a kind of fire that the blood of the veins maintains there. This fire is the corporeal principle underlying all the movements of our limbs. … Its first effect is to make the blood which fills the cavities of the heart expand."

the blood during expansion, all proceed mechanically.[5] The heart itself remains passive.

This mechanistic account of the heart's motion was not merely an interesting diversion for Descartes. In his view, "it is so important to know the true cause of the heart's movement that without such knowledge it is impossible to know anything which relates to the theory of medicine."[6] In an earlier letter to Mersenne, Descartes put even greater weight on his account of the heart's motion:

> Those who take a merely superficial view of things hold that what I wrote is the same as Harvey's view, simply because I believe in the circulation of the blood; but my explanation of the movement of the heart is radically different from his ... I am prepared to admit that if what I have written on this topic or on refraction—or on any other subject to which I have devoted more than three lines in my published writings—turns out to be false, then the rest of my philosophy is entirely worthless.[7]

In the debate over the heart's action, Descartes held to a comprehensive standard of falsification. To show that his account of the heart's motions was false, an opponent needed contrary empirical evidence, as well as a rival *general* theory of the world that fit the evidence.[8] This allowed Descartes to discount some objections, and also claim a vital connection between his metaphysics and his medicine.[9] In contrast to Descartes's remarks here, we will see that Dutch anatomists could sever this supposedly necessary interconnection, and rejected his model of the heart while often accepting an ideal Cartesian ontology. He was quite right that his model was very different than Harvey's, but physicians and philosophers could reject his account of the heart without rejecting the ontology or mechanism at the heart of his project.

In contrast, Harvey's account of the heart's motion depended on a vital faculty in the heart which initiated the muscular contraction of the ventricles to expel the blood into the arteries.[10] This was a very important innovation, which Realdo Colombo had also discussed. Descartes, in contrast, had the dilation of the heart and arteries occurring at the same time, which was much closer to the older Galenic model.[11] For Harvey, the heart was a powerful muscle, contracting to force the blood into the arteries. In his "Second Letter to Riolan," Harvey made a "pulsific faculty" (*facultas pulsifica*) responsible for the contraction of the right auricle.[12] The contraction of the heart does not come from an external source, such as the 'spirits' or

---

[5] Descartes explicitly compared the motion of the heart to that of a clock, its "movement follows just as necessarily as the movement of a clock follows from the force, position, and shape of its counterweights and wheels." AT VI 50; CSM I 136.

[6] *Description* AT XI 245; CSM I 319.

[7] Letter to Mersenne, 9 February 1639, AT II 501; CSMK 134.

[8] Sakellariadis 1982.

[9] Gorham 1994; Aucante 2006.

[10] Pagel 1967.

[11] Pagel 1967, ch. 9; Galen 1968, 316.

[12] Harvey 1649, 115; Harvey 1993, 131.

'vapors' of J. C. Scaliger and Jean Fernel, but rather from an "internal principle."[13] Whereas for Descartes all motion of bodies must have an extrinsic cause (either from the collision of another body or the human will), for Harvey the motion of the heart and other muscles is intrinsic.

Descartes certainly appealed to experiments, especially in his 1638 letters with Vopiscus Fortunatus Plemp.[14] Comparing these letters to the account of the motion of the heart found in the 1637 *Discourse* reveals Descartes temporarily changing the details of his mechanism in response to empirical and experimental criticism.[15] As Dan Garber has shown, experiments had a necessary place in Descartes's system since they allowed the Cartesian philosopher to choose which of a set of possible mechanisms constructed from certain first principles was correct.[16] More recently, Jed Buchwald argues that Descartes's investigation of the rainbow integrated observations, experiments to isolate dependencies, and possible micro-level mechanical models to generate new explanations for the existence and order of colors.[17] In his anatomy, Descartes modified his account of the mechanism of the heart from one of the balloon-like expansion of the whole heart, caused by the rarefaction of large masses of blood by the heat of the heart, to accepting that even smaller pieces of the heart can be moved by a rarefaction of tiny drops of blood, dilated by heat produced from a 'fermentation' in the heart. It is not clear whether Descartes continued to hold to this view, since any mention of a ferment or fermentation is absent in his final work, *The Passions of the Soul*.[18]

Throughout his works, however, Descartes held that the blood rushed out of the heart into the arteries as rarefying blood expanded the ventricles, so that the ventricles and the arteries dilated at the same time. This coincidence of ventricular systole and arterial dilation would encounter fierce opposition from the Dutch experimental anatomists, as we shall see. Descartes's mechanisms for the heart's action, even with concessions to 'ferments,' remained necessarily mechanistic. The fermentation sustained a 'fire without light' hidden in the crevices of the heart's walls, a fire that was no different than that found in wet hay.

---

[13] Bono 1995, 85.

[14] AT I 521–34; CSMK 79–85 and AT II 62–69; CSMK 92–96. Some of these experiments appear in the *Description of the Human Body* AT XI 242–243; CSM I 317–318. A recent analysis of this controversy in Leuven stresses the importance of theological concerns in shaping the acceptance or rejection of the Cartesian explanation of the heart's action: Petrescu 2013. In Leiden, I think things proceeded differently, since even those who rejected the Cartesian explanation on empirical grounds accepted something very like Cartesian ontology and mechanical explanation as an ideal.

[15] *Discourse*, AT VI 46–55; CSM I 134–39.

[16] Garber 2001, ch. 5.

[17] Buchwald 2008.

[18] *Passions*, AT XI 333–334; CSM I 331. Though, as an anonymous reviewer kindly points out, there are mentions of a leavening agent or *levain* in the *Description of the Human Body*. Descartes insists that the source of the heat in the heart is perceptible, and that it is no special sort of heat, but only that which is generally caused by a mixture of some liquor, or by some leaven. AT XI 228.

Although both Harvey and Descartes employed experiments to buttress their arguments, it seems fair to characterize Harvey's methodology as more experimental.[19] Harvey's works discuss dozens of experiments in detail, from Galen's anatomical experiments with arteries and the pulse to his own tests of blood flow, the nature of the blood, and generation.[20] Descartes may have performed relatively fewer experiments, but he used his anatomical experiments on the heart to help decide between rival systems and developed his embryology along paths picked out by observation.[21] Still, Harvey performed more and far more precise anatomical experiments.[22] Harvey also objected to Descartes's account precisely on the grounds that he had not observed rightly.[23] A decade earlier, Descartes wrote that he had an "experiment [*experimentum*] by which the opinion of Harvey about the motion of the blood is killed by cutting its throat [*iugulatur*]."[24] Yet Descartes was often dismissive of experimental arguments, especially those aimed at his own explanations and hypotheses.[25] This pattern also appears in the story presented here, and from his first encounters with Dutch anatomists, Descartes objected that their adherence to the "animal motion" and muscular contraction of the heart betrayed an insufficient understanding of Mechanics.[26] In contrast, Harvey began with the

[19] Gorham 1994; Aucante 2006, 71, 95–96, 121, and 428; but also see 146–148 and 314–329 for an analysis of Descartes's experiments on generation, which provided him with an account of the generative phases that departed from those of his sources.

[20] Goldberg 2012, 244–245.

[21] Aucante 2006, 200–206, 314–329.

[22] Aucante 2006, 200.

[23] Harvey 1649, 135. Descartes and the others with him "hardly observe rightly" ("haud rectè mecum observant"). For Harvey, when the heart is rigid, raised up, and envigorated, then it is contracting in systole. In Harvey's view, Descartes's relative inexperience in anatomy allowed him to mix up systole and diastole. When Descartes insists on the same cause for both systole and diastole, rather than contrary causes for contrary effects, Harvey concludes he is not following proper anatomical method. After all, "all anatomists know sufficiently that opposite muscles are antagonists. Thus for contrary, and diverse, motions, contrary and diverse active organs have been fabricated necessarily by nature" (Harvey 1649, p. 136). To Harvey, Descartes follows Aristotle ("secundum Arist.") in holding that the efficient cause of the pulse is the same for systole as diastole, namely the effervescence of the blood, brought about as if by boiling.

[24] Descartes AT I 527; CSMK 82. Descartes' experiment here involved vivisecting a young rabbit, cutting off the tip of the heart, and then supposedly observing that the chambers of the ventricles grow larger at 'diastole' and smaller at 'systole.' Descartes claims that Harvey has it the other way around, but Harvey was quite clear that the ventricles of contracting heart grew smaller in systole, forcing out the blood into the arteries.

[25] Sakellariadis 1982. Descartes to Mersenne, 18 December 1629, AT I 100; Descartes to Mersenne, 16 April 1634, AT I 287; CSMK 43; Descartes to Huygens 1643, AT III 617. Descartes's legacy in regard to experimentation is complex. For a very recent review, see Ragland 2014.

[26] Descartes to Regius [before mid-October 1641], AT III 440–441; A recent edition makes several significant corrections to the dating and contents of the Descartes- Regius correspondence: Bos 2002, 83–4: "When your letter was sent, I was not here, and now that I have first returned home I am taking it up. Sylvius's objections do not seem to be of any great moment, in my opinion, and they bear witness to nothing other than that he has an insufficient understanding of Mechanics; nevertheless I wish you to respond to him more gently. I noted in the margin by transverse lines the

forceful systole of the heart, and built up his *De motu* with detailed and careful experiments, attempting to isolate and characterize the action, and perhaps reason to the *utilitas*, of the heart.[27] Robert Boyle (1627–1691), an icon of experimental philosophy himself, considered Harvey's discovery and the subsequent elaboration and confirmation of the circulation to be exemplary of the experimental method.[28]

Harvey himself presented his work as grounded in sensation and experience rather than speculation, a rhetorical distinction taken up by many later anatomists. In his "Second Letter to Riolan," Harvey wrote of the circulation:

> Finally, this is that which I was endeavoring to recount and lay open by observations and experiments, not to demonstrate by causes and probable principles, but I wanted to render it confirmed by sense and experience [*per sensum et experientiam*], as by the greater authority, according to the way of the Anatomists [*anatomico more*].[29]

Leading up to this passage, Harvey pointed to the ancient origins of his emphasis on testing by the senses, after long experience, citing Aristotle and Plato in the same letter:

> Aristotle advises us much better when, in discussing the generation of bees (*De generatione animalium*, Book 3, Chap. 10), he says: "Faith is to be given to reason if the things which are being demonstrated agree with those which are perceived by sense: when they have become more adequately known, then sense should be trusted more than reason." Hence we ought to approve or disapprove or reject everything only after a very finely made examination. But to test and examine if things are rightly or wrongly spoken, ought to lead to sense, and to confirmation and establishment by the judgment of sense where nothing false will remain hidden. Whence Plato, in his *Critias*, states that it is [not] difficult to explain the things of which we shall be in a position to claim experience. And listeners who are devoid of experience are not fitted for science.[30]

Harvey certainly reasoned from established principles, constructing demonstrative conclusions for the circulation from the expulsion of the blood, the distension of the arteries, and the frequency of the expulsion—if the blood did not move in a

---

passages that seemed somewhat harsh." Descartes to Regius [November 1641], AT III 390–392; Bos 87–89: "Those who say that the motion of the heart is an Animal motion, say nothing more than that they should confess that they do not know the cause of the motion of the heart, since they do not know what an Animal motion is. When, moreover, the dissected parts of eels are moved, in truth the cause is nothing other than when the dissected point of the heart pulses, nor different than when the sinews/nerves [*nervi*] of a tortoise are dissected into particles, and existing in a hot and humid place, contract in the likeness of worms, although this motion is said to be Artificial, and the former Animal; in all of these things the cause is the disposition of the solid parts and the motion of the spirits, or more fluid parts, permeating the solid parts." These and all other translations, unless otherwise noted, are my own.

[27] For a similar but contrasting account, see French 1994, 100–104. Cf. Goldberg 2012, 191–252. Goldberg's treatment gives a more complete account of Harvey's Aristotelian-Galenic methodology. See also Wear 1983.

[28] Rose-Mary Sargent argues that Harvey's example "typified the experimental program and became the paradigm that Boyle would follow in all of his investigations." Sargent 1995, 83.

[29] Harvey 1649, 118–119.

[30] Harvey 1993, 130. Cf. Harvey 1628, 110–111.

circle, the force of systole would burst the arteries, or swell and stop the ventricles.[31] But the senses grounded and checked the premises, and seeing for oneself was far more certain than the phantasms of the mind: "This collective demonstration of mine is true and necessary, if the premises are true: moreover, that these are true and not false, the senses ought to make us more certain, and not the things admitted by reason; *autopsia*, not the agitation of the mind."[32]

In contrast, Descartes famously questioned the reliability of the senses, especially in *The World*, composed around 1632, but also in the *Discourse* and the *Meditations*.[33] In his anatomy, Descartes's model of the heart's action faced persistent experimental criticism from his contemporaries and immediate successors, but he did not simply ignore empirical evidence. Most recently, Vincent Aucante argues persuasively that Descartes's medical views on generation developed in concert with empirical observations, and that he changed his mind in the face of evidence that contradicted his early theorizing.[34] Still, Descartes stayed well within the bounds dictated by his metaphysics, and was consistently reluctant to describe the heart as a muscle working like other muscles, generating an animal motion in response to a stimulus. Heat or a material ferment drove the rarefaction of the blood, which drove the heart.

8.3 Walaeus and Van Hogelande: Early Investigations of Anatomy, Chymistry, and Mechanism

Let us now turn to the Low Countries. From the 1640s to the 1650s, experimentalist Dutch physicians continued to draw from leading anatomists to investigate the circulation of the blood and the origins of the blood from food. From Johannes Walaeus's letters from the end of 1640, we learn that his collaborations with Sylvius and other anatomists involved significant research into the process of digestion.[35]

[31] Harvey 1993, 132–133.

[32] Harvey 1649, 117; Harvey 1993, 132. I have modified Franklin's translation. For further such passages from Harvey contrasting the "phantoms of the mind" versus the senses, see Wear 1983, 239.

[33] Clarke 1992, 260–261. CSM I 126–131; AT VI 31–40. Although in the Sixth Meditation Descartes restores some of the senses' reliability for everyday life in general, they still do not render up "the way things really are," and the inspection of clear and distinct ideas remains far more reliable. See Garber 2001, 280.

[34] Aucante 2006, 314–322.

[35] Educated in medicine at the University of Leiden, Johannes Walaeus (Jan de Wale) had recently received his M.D. (1631) when Sylvius arrived at the university in 1633. Since Walaeus began teaching as an extra-ordinary professor in 1633, it is likely that Walaeus instructed Sylvius at some point. We know little of Walaeus's work until 1639, when Sylvius's demonstrations of the circulation of the blood turned Walaeus from a harsh critic to a zealous supporter. His 1640 writings supported Harvey's account of the heart and the circulation. Schouten 1972, 14–19, 80, 108.

Sylvius had in fact met and perhaps dissected with Descartes in 1639, but they soon parted ways over the action of the heart. Descartes urged Regius to ignore Sylvius, since Sylvius trusted too much to anatomical demonstration and had "an insufficient understanding of Mechanics."[36] Against Descartes's repeated rebukes, Sylvius continued to maintain that the heart contracted as a muscle, with an "animal motion." Here, he continued to follow Harvey against the Cartesians. In fact, it was Sylvius' public demonstration of the circulation near the Leiden botanical gardens that convinced Walaeus to stop his criticism of Harvey, and become instead one of his most vigorous supporters.[37]

> ...the circular motion of the blood, then first introduced by us into this Academy (let it be said without ill-will), and also shown to those present, whom we name for the sake of honor, *Adolph Vorstius, Professor of Medicine Primarius and presently Magnificent Rector, once our Preceptor, now our honored colleague*; as also *Johannes Walaeus, our Most Skilled Predecessor, and a Man of great brilliance*, so that a little after he had publicly fought quite harshly against this *Harveian* motion of the blood, this very same man was conquered and captured by the evidence of experiments we performed, so that he then fought for it with equal zeal and fervor. To these, I say, and other contemporary Men who are lovers of Honor and Truth, coming together with a large troop of Students, we have frequently demonstrated to the eye that circular motion of the Blood in the Academic Garden and elsewhere...[38]

The presence of an investigative community of anatomists in Leiden around 1640—composed of at least Sylvius, Walaeus, Thomas Bartholin, and Johannes Van Horne—is indicated in this passage. Public gardens, private rooms, and the anatomical theater were all sites for regular work similar to that of scientific societies elsewhere in Europe.[39] All of these young anatomists collaborated over the experimental demonstration of the Harveian circulation of the blood. They also dissected the brain and digestive system, and in Bartholin's *Institutiones* (1641, 1645, etc.) and his *Anatomia reformata* (1651, 1655, etc.) Sylvius and Walaeus frequently appear as authorities on the anatomy of the viscera and brain, and on the motion of the heart.[40]

---

[36] Descartes to Regius [before mid-October 1641], AT III 440–441; Bos 2002, 83–4.

[37] Thomas Bartholin to Anton Deusing, 20 November 1663, in Bartholin 1740, 416. See also Schouten 1974, 259–279 and Pagel 1978, 113–135.

[38] Sylvius 1679, 22. Cf. Walaeus 1645, 477. Walaeus 1641, 408, has slightly different text, singling out Sylvius as "most accurate in dissections," and listing the names of three figures left out of the witnesses in the 1645 and later editions: Philippus de Glarges, Roger Drake, and Henricus à Schaeck. The 1645 and subsequent editions add the names of Johannes van Horne and Ahasuerus Schmitnerus. It is interesting that the printed versions of the letters—which retain identical dates throughout—show these and other changes from the 1641 to the 1645 editions (see, e.g., Walaeus 1645, 445 for an addition, as well as the illustrations added to the 1645 edition). The 1645 versions have a title page indicating they are the "Fourth Edition" (*Editio Quarta*) of the letters.

[39] Rupp 1990, 263–282.

[40] Bartholin and Bartholin 1641, 290 and 395. Also see the end of the preface: *"In novis Cerebri iconibus calator accuratissimi* Francisci Sylvii *ductum sequutus est & cultrum, cui hac in parte debemus quidquid cerebrum vel augmenti habet, vel ornatus: sicut ad* Cl. Walaeum *grati referimus universi operis & nitorem & renatae vitae causas*." Bartholin 1655, 331–337 and *passim*.

Walaeus's and Sylvius's experimental approaches elaborated on Harvey's methods. Primarily using vascular ligatures, Walaeus sought to give firm experimental support for the connection between the arteries and the veins, a potential weak point in Harvey's original case for a circular flow. In Walaeus's third experiment, for example, both the vein and the artery in the leg were lifted out from the surrounding muscle and ligated. When he made small incisions on either side of the venous ligature, blood poured out from the hole on the distal side of the ligature, but only came out in drops from the side toward the heart. When another ligature was bound on the vein distal to the first, the flow of blood from the incisions ceased immediately—all showing the one-way flow of venous blood back toward the heart.

Walaeus and his collaborators also performed experiments to illuminate the origins of the blood. They continued to believe that the liver generated blood from chyle, but Gaspare Aselli's 1622 discovery of the lacteals had opened up the field to new investigations of the generation of chyle and its path in the body. Walaeus ligated the lacteals and showed that they swelled on the side of the ligature toward the intestines—showing the directionality of the flow of the chyle. Chymistry provided Walaeus and Sylvius with incisive tools for thinking about digestion. Nearly-verbatim lines from the experimental chymistry of Van Helmont appear in Sylvius's *Dictata* from 1640 to 1641.[41] At the end of his first letter supporting the circulation, Walaeus even compared the circular motion of the blood to chymical distillation, talking in chymical terms, and discussed digestion *per minima*.[42]

---

[41] Sylvius 1679, 882: "Hoc Chyliticans Fermentum in recens natorum Vitulorum Ventriculis reperitur crassiusculum, diciturque Coagulum. Sensim autem minuitur, ac in Adultis Glutinis instar offenditur liquidiusculum, inter Ventriculi rugas haerens." Compare to Joanne [Jan] Baptista Van Helmont, *Febrium Doctrina Inaudita* (Cologne 1644), 181: "Vitulus namque, dumtaxat lac maternum bibens, ostendit mox à nece, quod lac statim grumescat in coagulum acescens, & liquorem aqueum acidum: utrumque caseis parandis expetitum." I have checked for similar passages in Van Helmont's earlier writings, and find none. This is strong evidence for the circulation of some of his writings in manuscript several years before they appeared in print.

[42] On the use of the term 'chymistry,' see Newman and Principe 1998. Walaeus 1641, 406 and Walaeus 1645, 445–448 [*sic*, 446]. "A little later both the more tenuous and thicker food is cut into *minima* as if plucked off into little torn bits; in dogs, even the very shells of eggs. Without a doubt this happens due to a certain acid humor, which has the power of dissolving. Thus we tested [*experiri*] a ventricle heavy with a mass or thickness of food, which felt alleviated by taking in vinegar, lemon juice, and oil of sulphur or vitriol. And this should not be referred to anything in the saliva or bile regurgitated into the ventricle, since bread soaked in hot saliva or cow bile, seemed in a few hours to be softened, but by these, moreover, it is not broken into little pieces. In a hundred dogs and more, which, for this reason we have dissected while still living, we have found only in two that some bile flowed into the ventricle, one of which spent three days fasting, and in his ventricle, marvelous to see, the bilious froth was so dense, as if boiling, of such kind as we see to float in lye, in the washing of laundry-women. ... Thus the food is thoroughly mixed by the liquor *per minima*, arriving at the consistency of thin barley gruel in a length of time. When it has arrived at this point, then the food is pushed out into the intestines." Talk of foods thoroughly digested *per minima* was characteristically Helmontian language. Like Van Helmont, Walaeus thought that the acid humor arose from the spleen, a conjecture he confirmed by perceiving a sharp humor there and noting that a bit of boiled spleen aids the digestion of meat. See Pagel 1978, 130.

But Walaeus and his colleagues also took their investigations further than these singular chymical experiments or the artificial 'digestion' of bread by saliva and bile outside of the animal. Walaeus reported systematic observations of the times it took to digest various foods. Lighter fare, such as milk and broth, receive their own perfection through digestion in the space of an hour.[43] This can be shown without any dissection by noting that the animal voids urine quite soon after eating. Vegetables are digested more slowly, and bread seems to be a middling substance.[44]

Catching, feeding, then vivisecting in order to observe and time digestion was surely a very poor and violent way to treat the hungry dogs of Leiden. It was also, tragically, probably the only way that Walaeus and his colleagues could observe the lacteals and the process of digestion *in situ*. This type of careful, timed observation, which must have been carried out in scores of animals to get general times for different substances, is exemplary of the early Dutch experimentation on living animals.

Also characteristic was a rejection of the ingenious yet speculative explanations of Cartesian physiology. In the 1641 edition of Walaeus's letters, printed in Thomas Bartholin's revised edition of his father's *Institutiones anatomicae*, Walaeus spends several pages defending Harvey's account, in which a muscular heart expels the blood into the arteries, which dilate when the ventricles of the heart contract in systole.[45] He contrasts his repeated observation of the muscular contracting and pushing out of the blood into the arteries with another, incorrect story: "And certain men assert that they have seen the blood exit the dilation [of the heart] in the dissection of living animals, and in this they have judged wrongly, because they consider it to be dilation, which in truth is constriction."[46]

---

[43] This passage does not appear in the 1641 edition and I have not found it in the 1645 edition. It is likely that Walaeus added this particular passage to the 1647 edition, but I have not been able to examine a copy to confirm this. I have used Walaeus 1655, 534.

[44] Walaeus 1655, 534. "By reason of its being cooked, bread seems to have a middle sort of substance, and after an hour and a half is seen to be changed very little, and in the following hour becomes entirely rare like a wet sponge. When that second hour has passed, then it is divided into the very smallest little morsels [*in minima dividi frustilla*], and is thoroughly mixed with the draught so that it appears wholly liquid, and is soon greatly concocted. At last, between the fourth and fifth hour after it was eaten, what has been concocted from the bread is propelled by the pylorus from the stomach into the intestines. Some relic of the bread remains, and this gradually receives its proper perfection, just as happens if any other food was ingested with the bread, which makes its concoction more difficult. We have observed these foods to be concocted in this order: First beans, then fish, then soon flesh, which is perfected and excreted within the sixth or seventh hour; beef within the seventh or eighth hour; and indeed the membranous parts of animals and eggshells more slowly. We saw that bones remained in the stomach into the third day, in which time they were made like cartilage. ... We readily observed these things in dogs which we cut up alive at various times after they had eaten food."

[45] Walaeus 1641, 400–402.

[46] Walaeus 1641, 401.

In the 1645 edition these attacks are amplified and made much more explicit:

> Certain men famous for their intelligence [*ingenium*] judge that the blood is thrust out because it is immeasurably rarefied by the heat of the heart, and so demands a bigger place, and then dilates and lifts up the heart. Since it cannot be contained in the dilated heart, it is poured out into the venous artery and the aortal artery with such force that it distends all the arteries and makes them pulse. ... [T]here is indeed an entirely light rarefaction from a certain tepid warmth in the heart, but no ebullition or sudden diffusion. And in truth the blood does not leap out from the heart on account of the rarefaction, as we have often seen in strong dogs with the tip of their hearts cut off. When, on account of the outflow of the blood the heart was not half filled, it being erect, it was not filled by rarefaction. But in the following constriction the portion of the blood that was in the heart was ejected more than four feet, so that we and our neighbors in the large crowd were befouled. Whence it is evident, that the blood is propelled by the part.[47]

This passage nicely illustrates Walaeus's blood-spattered experimentalism and his depiction of his Cartesian opponents as men too dependent on their brilliant minds. Even though they may have observed some dissections, these men "judged wrongly" due to intellectual bias. On the contrary, wrote Walaeus, direct observation of the action of the heart—in the presence of many witnesses, whom he names—showed that the heart contracted to expel blood when it was only half full. Even a relatively small amount of inflowing blood irritated the walls of the ventricles, causing the constriction.[48]

The 1645 edition of the first letter also included a very interesting passage on mechanical explanation in anatomy in general, which was also a specific rejection of another Cartesian mechanism. (Walaeus may have some of the details of his opponents' explanations wrong here, but his remarks are worth inspecting.) Walaeus opposes this explanation with the testimony of his senses, and also gives the idea an Erasistratean pedigree:

> There are also those who judge that the blood once carried out from the heart goes back, and returns again by the arteries to the heart. It seems that they assert this so that a mechanical cause can be given, according to which the valves of the heart in the mouth of the arteries fall down and close. We, indeed, have always estimated this to be a brilliant custom of Erasistratus, to explain all the things which happen in our body Mechanically, but we judge that it is rash for him to measure divine wisdom by his own. Rather, those things are to be

[47] Walaeus 1645, 465–466: "Protrudi sanguinem viri quidam ingenio praeclari arbitrantur, quod calore cordis immensum rarescens, majorem locum exposcat, ideoque eum cor dilatare & attolere; cumque nec in dilatato corde contineri queat, in venam arteriosam arteriamque aortam tali effundi impetu, ut omnes distendat arterias & faciat pulsare. Suae autem opinionis hoc argumentum adferunt, quod cor anguillae alteriusve animalis, ubi pulsare desinit, si à substrato calefiat igne denuo pulsum edere conspiciatur. Sed an is pulsus fieri non posset, quod spiritus à calore vegetior factus, melius ei causae possit inserviere quae in corde pulsum facit? non aliter ac calefactis in vivorum sectione intestinis, musculisque, in quibus tamen nulla ebullitio est, restitui motus videtur. Omnino enim levis tantum quaedam rarefactio à tepore quodam in corde est, nulla ebullitio, aut diffusio subita. Et revera ob rarefactionem sanguinem è corde non exilire, in validis saepe canibus conspeximus, quorum cor discisso mucrone; cum ob effluxum sanguinis dimidia parte non repleretur, id erectum, à rarefactione repletum non fuit: sed accedente constrictione, portio illa sanguinis quae in corde reliqua erat, ultra quatuor pedes fuit ejecta, ut in magna frequentia nos & vicini conspurcaremur. Vnde evidens est, sanguinem à parte propelli."

[48] Walaeus 1645, 465.

> established as machines which manifest reason and above all the senses show to be such. Here the senses observe the contrary, that the blood is moved through the arteries from the heart, not toward the heart. … Indeed, this contraction of the fibers of the heart frequently stands out as obvious in inspection.[49]

First, we should note that Walaeus presents Erasistratus as the founder and exemplar of mechanical anatomical thinking. Indeed, Galen portrayed the purely material-efficient operation of the heart's valves as Erasistratus' innovation.[50] Walaeus seems to censure Erasistratus and other rash mechanists for jumping to quick judgments about mechanical processes that have not been verified by sensory experience. Mechanisms that correspond to sensory phenomena, however, are perfectly acceptable. Sensory confirmation is the only proper ground for proposed anatomical features or processes, and neither Galenic pulsific faculties nor Erasistratean material mechanisms are ruled out *a priori*.

We find a similar demand for rigorous observation in the writings of Thomas Bartholin. Not long after Walaeus's second edition of the letters in 1645, Bartholin published a revised anatomical compendium as his *Anatomia reformata* in 1651. This work incorporated many of the findings of Harvey, Sylvius, and Walaeus, and other modern anatomists. In 1649, Descartes's *The Passions of the Soul* appeared, which reiterated his earlier account of the heart's action from the *Discourse* and included a summary of his teachings on the pineal gland as the impressionable, mobile seat of the soul. For Descartes this gland was a highly mobile theater for the soul, the place where it received material impressions of spirits and directed the flow of spirits into the pores of the nerves.[51] From their first writings in 1640, Sylvius and his colleagues were not impressed with Descartes's speculative anatomy, and by 1651 Bartholin included a long list of reasons to reject this "new and ingenious opinion." First on the list were Sylvius's observations that the gland was firmly fixed by a little nervous thread and that he often found it full of detritus, such as sandy grains or even pea-sized calculi.[52] Later, Bartholin introduced a series of objections with the rebuke that the gland was "too slender and obscure [*obscurus*] a body to

[49] Walaeus 1645, 475–76: "Sunt quoque qui arbitrantur sanguinem è corde delatum retrorsum cedere & per arterias denuo ad cor redire. Quod illis ideo videtur statuendum, ut causa dari mechanica possit, qua cordis valvulae in orificio arteriarum, decidant & occludantur. Nos equidem praeclarum semper Erasistrati institutum aestimavimus, omnia quae in copore nostro contingunt Mechanice explicare, sed divinam sapientiam sua metiri temerarium judicamus. Eas vero machinas esse statuendas quas evidens ratio & potissimum sensus ostendant. Hic contra sensus observant à corde non ad cor per aterias sanguinem moveri … Ea quippe fibrarum in corde contractio passim obvia in conspectum prodit."

[50] Lonie 1964, 431 n. 18.

[51] Descartes AT XI 354; CSM I 341.

[52] Bartholin 1651, 336: "According to the Observation of *Sylvius* a *little nervous string* fastens this gland firm between the testes [structures of the brain]. Who also observed more than once some grains of sand in this pineal gland, and sometimes also a little stone as big as the fourth part of a pea, and somewhat round."

represent clearly the species of all things."[53] Five other objections rounded out his arguments against taking the pineal gland as anything other than a small, probably unimportant gland, in appearance much like other glands.

It should be clear by now that the continued insistence of Bartholin, Sylvius, and other leading anatomists on a strict adherence to observationally-established anatomical knowledge contrasted with the approach of some leading Cartesians, and especially their view of Descartes's own anatomy.[54] The differences between the Cartesians and the experimental anatomists were displayed quite clearly in anthologies of the period, such as the *Recentiorum disceptationes de motu cordis sanguinis, et chyli, in animalibus* (Leiden 1647), which put Harvey, Primerose, Drake, Regius, and Walaeus into print together.[55]

By the mid-1640s, the dispute between the Cartesians and the experimental anatomists following Harvey had reached such a pitch that one leading Cartesian, Cornelis van Hogelande, could occupy more than one position in the debate.[56] Van Hogelande (1590–1676) was deep within the circle around Descartes, and apparently remained a trusted friend (when Descartes departed for Sweden he left Van Hogelande with a trunk of his letters).[57] As part of his medical studies, Descartes and Van Hogelande may have consulted together on Van Hogelande's rounds, and Descartes certainly lodged with him in Leiden.[58]

---

[53] Bartholin 1651, 336–37: "Sed multa sunt, quae ab opinione hac nova & ingeniosa me dimovent. Nam

1. Nimis exile est corpus, & obscurum, quàm ut omnium species clarè repraesentet.
2. Species omnium sensuum huc non appellunt, quia nervi non tangunt glandulam.
3. Posita est excrementorum loco, qui per tertium & anteriores duos ventriculos expurgantur, ubi species rerum inquinarentur.
4. Species sentiuntur potius, ubi deferuntur. At ad principium spinalis medullae quilibet nervus sensorius defert species suo quovis loco, unde singuli suo loco in principio medullari ab anima dijudicantur & reperiuntur. Est praeterea haec medulla magna satis globosa, durior, & illustrior colore.
5. Fieret in exili hoc corpusculo idearum diversarum confusio. Oculos quidem etiam minimus sine confusione species recipit, sed visibiles tantum, quum hic sensuum diversorum diversae species debeant recipi.
6. Nullus hinc ductus apertus ad nervos, aut cognitus, sicut à principio medullari, nec ulla communio cum quibusdam nervis sensuum externorum."

[54] We might also add the anatomy of Henricus Regius, which articulated a strong doctrine of empiricism, yet entertained many speculations about subvisible mechanisms. Gariepy 1990, 211: "Also, the sheer number of anatomical errors in the Physiologia was astounding." For Regius' doctrine of empiricism, see Bellis 2013.

[55] Harvey et al.1647.

[56] Cf. French 1994, 214–220.

[57] Bos 2002, xxi.

[58] Clarke 2006, 212–3. As Clarke reports, Descartes and Van Hogelande consulted together on the case of a girl with rickets, one Johanna de Wilhem, on 6 June 1640.

The two shared interests in medicine and Cartesian philosophy, as well as a Catholic faith.[59] Van Hogelande's approach to Cartesian physiology and medicine in his 1646 work is summed up well by the title: *Cogitationes quibus Dei Existentia item Animae Spiritalitas, et possibilis cum corpore unio, demonstratur: nec non, brevis Historia Oeconomiae Corporis Animalis, proponitur, atque Mechanice explicatur* [Thoughts by which the Existence of God and the Spirituality of the Soul, and its possible union with the body, is demonstrated: and a brief History of the Oeconomy of the Animal Body is set forth, and is explained Mechanically].[60]

In the first section of his treatise, Van Hogelande argued for the existence of God by pointing to signs of the rational creation and governance of the world, then to our own rationality, and finally to an omnipotent, rational God. But an extended discussion of the 'animal economy' comprised the great majority of the work. In his explanations of bodily processes, Van Hogelande favored the law-bound, corpuscular Cartesian pictures of phenomena celebrated by other Cartesians in the Netherlands, such as Regius:

> … we hold that all bodies acting in any way whatsoever, must be considered as machines, and their actions and effects must be explained or made explicable as if mechanical and corporeal, and consequently only mechanically, that is, according to mechanical laws.[61]

This approach included especially the rejection of anything like a final cause or internal teleology.[62] God's providential control guaranteed a divine teleology, in that the cosmos worked according to God's laws and ends.[63]

The confidence of Van Hogelande and others such as Johannes de Raey in their Cartesian corpuscular speculations was no doubt buttressed by Descartes's own somewhat inconsistent assertions in the *Discourse* and the *Principia philosophiae* (1644).[64] In the *Discourse*, Descartes claimed certainty only for the first principles of his physics of extended matter in motion, with observations becoming increasingly

[59] The details remain to be worked out. Descartes remarked to Elizabeth that Van Hogelande "does just the opposite of Regius, in that everything Regius writes is borrowed from me and yet manages to contradict my views, whereas everything Van Hogelande writes is quite alien to my own views (indeed I think that he has never even read my books properly) and yet he is always on my side, in that he has followed the same principles." Descartes to Elizabeth, March 1647, AT IV 627; CSMK 315.

[60] Van Hogelande 1646.

[61] Van Hogelande 1646, 276: "… omnia corpora quôcunque modô agentia, tanquam machinas consideranda, eorundémque actiones atque effectus, tanquam mechanicos & corporeos, & per consequens non nisi mechanicé, id est, secundùm leges mechanicas, explicandos aut explicabiles existimamus."

[62] For internal and external teleology, see Lennox 1992. For the Dutch context, see Jorink 2010.

[63] Van Hogelande 1646, 14, 83, 94. French, *Harvey's natural philosophy*, p. 215.

[64] Clarke 1992, 258–85. See also Garber 2001, ch. 5, who argues for a shift from intuition over experiment to hypothetical argument as Descartes moved from the *Discourse* to the *Principles*. Garber also recognizes that some key passages in the *Principles* continue Descartes's early privileging of intuition over experimentation.

necessary in order to select the correct corpuscular explanation.[65] Thus, at this stage, corpuscular explanations for phenomena are only morally certain—the sort of certainty we experience with sure conclusions in everyday life. By the penultimate proposition of the *Principles* he seemed to be claiming more than moral certainty for even the complex and speculative corpuscular explanations illustrating phenomena throughout the text:

> ...there are some matters, even in relation to the things in nature, which we regard as absolutely, and more than just morally, certain. This certainty is based on a metaphysical foundation, namely that God is supremely good and in no way a deceiver, and hence that the faculty which he gave us for distinguishing truth from falsehood cannot lead us into error, so long as we are using it properly and are thereby perceiving something distinctly. Mathematical demonstrations have this kind of certainty, as does the knowledge that material things exist; and the same goes for all evident reasoning about material things. And perhaps even these results of mine will be allowed into the class of absolute certainties, if people consider how they have been deduced in an unbroken chain from the first and simplest principles of human knowledge.[66]

The French version of this text is even clearer about the demonstrative, mathematical status of these Cartesian explanations:

> I think that one should also recognize that I proved, by a mathematical demonstration, all those things which I wrote, at least the more general things concerning the structure of the heavens and the earth, and in the way in which I wrote them. For I took care to propose as doubtful all those things which I thought were such.[67]

At least in the beginning of his text, Van Hogelande adopted similar rhetoric in praise of the certainty of reasoning from first principles. A person reached true causes via ratiocination, which the immortal soul performs without any images or ideas proceeding from the body or impressed on the brain.[68] The structure of the heart and its valves provided Van Hogelande with his premier example for how to explain all natural things mechanically [*mechanicé*]. First, from the expulsion of the blood into the aorta "by the rarefaction of the blood, or by the constriction of the heart, or through both ways at once," all the arteries of the body dilate and distend

[65] Descartes AT VI 63; CSM I 143: "I also noticed, regarding observations [*expériences*], that the further we advance in our knowledge, the more necessary they become. At the beginning, rather than seeking those which are more unusual and highly contrived, it is better to resort only to those which, presenting themselves spontaneously to our senses, cannot be unknown to us if we reflect even a little. The reason for this is that the more unusual observations are apt to mislead us when we do not yet know the causes of the more common ones, and the factors on which they depend are almost always so special and so minute that it is very difficult to discern them."

[66] Descartes AT VIIIA 328–29; CSM I 290–91.

[67] Descartes AT IXB 325, trans. in Clarke 1992, 278–79.

[68] Van Hogelande 1646, 24–25. "Quae ratiocinationis actio, cum nullô modô à corpore, qualicunque ratione agitatô vel motô; neque ab ullis imaginibus aut ideis, incertô casu per sensus illatis vel oblatis cerebróque impressis, prodire possit (licet ipsemet homo, imagines sive ideas, etiam corporeas, intentioni suae inservientes, sibi ipsi liberè & pro arbitrio suo proponat; atq; ut ad optatum ratiocinationis finem, quaesitam scilicet veritatem perveniat, liberè & pro arbitrio sup sibi proponere debeat, non autem incerto casu oblatas vel illatas accipere."

in a moment.[69] Second, when the blood ceased its dilation or the heart its constriction, a noticeable portion of the blood would return to the heart, unless it was hindered. Third, since the valves of the heart are so constituted that, when a liquor flows against them, they close, they must block the return of the blood. This is all the result of material necessity, and Van Hogelande's mechanistic rejection of internal teleology applied especially to the heart's action. Why, then, would a seemingly *a priorist* Cartesian mechanist allow for a constricting motion of the heart?

Of course, philosophical sympathies did not keep early modern physicians or philosophers in intellectual straitjackets, and the reception and appropriation of Descartes's writings enjoyed a complex history.[70] Along with what he considered to be Descartes's deductive method from first principles of motion, Van Hogelande also kept his basic model of the action of the dilating heart. But he made interesting additions, and only arrived at a tepid acceptance of constriction after a journey from *a priori* deduction through anatomical experiments.[71]

The action of the heart is the subject of a "digression" which comprises 41 pages in the midst of Van Hogelande's chain of reasoning from God's nature to the fermentation by subtle matter in the heart.[72] In his earlier *digressio*, the heart had three states: the natural state, when the blood is about to enter the ventricles; dilation, when the blood has rarefied in the heart by fermentation and it distends and dilates the ventricles; and constriction, which was always "accidental or fortuitous" since it occurred mostly to restrain the dilation of the passive heart, stretched by the action of the fermenting blood.[73] The fermentation of the blood, in turn, resulted from "the figures and magnitudes of the pores and particles" of the blood, chyle mixed in the blood, and the internal parts of the heart.[74]

Some observations offered useful analogies. He appropriated some of his examples from chymistry, and favored chymical analogies for the corpuscular processes of the heart's action and for digestion.[75] Chymistry supplied similitudes, but all action could only result from passive matter in motion, governed by the laws of motion.[76] For Van Hogelande, our clear and distinct conceptions pointed to the truth,

---

[69] Van Hogelande 1646, 273.

[70] See, most recently, Dobre and Nyden 2013. Ragland 2014.

[71] Van Hogelande 1646, a5–a6: "Great man, what in this kind of argument I can put forward given the thinness of my powers, is yours to judge. For it is yours, or ought to be joined to you, the plain and easy rule [*ratio*] of reasoning, which I followed in order to track down the truth. And indeed, I gave this one work so that I would assert nothing without doubting, which was not sufficiently perceived by me. By your example, I myself proposed the opinions of no one, nor obscure dubious reasonings, but sought the perspicuous and easy principles, according to the motion of matter, and its form and magnitude, in the oeconomy of the animal body."

[72] Van Hogelande 1646, 272–276.

[73] Van Hogelande 1646, 54 and 124. Contraction could also come about due to disease, external cold, or old age, but was usually due to the fermentation of the blood.

[74] Van Hogelande 1646, 53.

[75] Van Hogelande 1646, 75–6, 81, 118.

[76] Van Hogelande 1646, 81: "Thus whenever blood has flowed into the heart, in a way similar or analogous to that which spirit of niter excites when infused into butter of antimony, rarefaction and effervescence follow; (as will be shown more clearly in the following) All philosophers teach that

and since it was impossible even to *conceive* of self-moving matter, the faculties of the Galenists, the substantial forms of Aristotelians, and the *archei* of the Paracelsians were all rubbish.[77]

Every constrictive motion of every piece of the heart resulted from the motion of particles. Parts of hearts could be cut, re-cut, and then set in motion not only by a sharp point or stick, but even by a finger's touch or the heat of a nearby hand. Heat, after all, was only the motion of the smallest particles.[78] The vaporous, particulate exhalations of the heat of the hand brought near stirred the blood particles into motion again, causing the fibers of the dissected pieces of the heart to contract. There was nothing unique about the heart—little pieces of the skin, fibers, and dry nerves contracted, too, when heated.[79] The heart remained passive, and the blood entered the arteries when the heart distended with the fermenting blood.[80] Walaeus and Sylvius would not approve.

Jumping into the local controversy over the action of heart, Van Hogelande hastily added a formal experiment:

> Let anyone well enough pleased with the things said above clearly enough understand that the mentioned particles of the heart receive their own motion from the fermentation of the blood, intermixed with its parenchyma, through the coronary artery; as a favor, moreover, for those distrustful of every ratiocination, however, and believing experience alone, I will not be amiss in adding on this following and easy-enough experiment, in confirmation of the truth of the aforesaid [conclusion]. With a moderate pressure, I separated all the blood from the bisected heart of a large eel, which had been retained for some time in water, so that its parts, placed for long enough in a round wooden dish (along the surfaces by which they [the parts] were wounded or bisected), would not produce any further motion; and then by the same method, blood having been poured upon the dish where the pieces are collected together, so that little by little, i.e. at first for some time there is no motion at all, and then very slowly, and afterward indeed they are moved with a very speedy motion: and I even repeated this experiment with its particularities with similar success; while the printer hurries. Which I add for this reason, lest anyone assume that I collected those preceding reasons *a posteriori* (that is, with the experiment having first been done).[81]

---

entities should not be multiplied without necessity: Here, the continuous heat of the heart and of the blood (which is united with the blood, mediated by the arteries, and communicated to the whole body, and commonly said to be innate or native) is analogous or similar to the heat of the butter of antimony and the spirit of nitre, and it seems that it should be attributed to their manifest action and motion (that is, to the rarefaction and effervescence, born from *the diversity of the motions of the internal parts, and the certain quantity and quality of the motive particles*, otherwise accidentally called *antipathy*), rather than to any other sort of incomprehensible or occult quality."

[77] Van Hogelande 1646, 98–99.

[78] Van Hogelande 1646, 144–146, and 154, where he lists experiences to support the claim that heat consists in motion.

[79] Van Hogelande 1646, 144–146.

[80] Van Hogelande 1646, 143. Van Hogelande added an experiment from one "Doctor Honinga" that the heart of a hunting dog expanded both laterally and longitudinally without an influx of blood, pp. 160–162.

[81] Van Hogelande 1646, 147–149: "Licet quilibet ex supra dictis clarè satis intelligat, citatas cordis particulas à fermentatione sanguinis, per arteriam coronariam ejusdem parenchymati immissi, motum suum recipere; in gratiam tamen omni ratiocinationi diffidentium, solique experientiae

Van Hogelande's experiment was performed only "for those distrustful of ratiocination" and was repeated when the book was already in press ("while the printer hurries"). He could very well have had Walaeus, Sylvius, and even Harvey in mind. In the final sentence, Van Hogelande expressed his anxiety that someone might misunderstand his method. His conclusions were not at all *a posteriori* or derived from experiment. At this point, it seems that pure ratiocination from first principles—with a few observational analogies—was enough for the enlightened Cartesian.

This does not mean that Van Hogelande could not appreciate the force of experiments. For him, Harvey "infallibly teaches" the movement of the blood into the arteries and through the body, and back to the heart in the veins.[82] He demonstrated most of the circulation with "firm enough reasons and sensible experiments," though proof of the connection between the arteries and veins was lacking.[83] But Van Hogelande still retained his distinction between those who could reason properly with the laws of mechanics, and those who lacked trust in ratiocination, and so needed sensible experiments.

> The Neoterics, in truth, with the originator the most brilliant Harvey (to whose benign generosity we owe this knowledge), not only according to the manifest laws of mechanics, solidly and firmly reasoning from the width of the aforesaid vena cava, the capacity of the auricles, and the injection of the blood into the [ventricles of the] heart from the packed and excited auricles, but, instead, for the sake of those not trusting enough in reasoning [*ratiocinatio*], they turned themselves to girding and adding on to these [laws] scrutinizing investigations or real and sensual disquisitions, and to dissecting living animals of every kind: thus they noticed that such a quantity of blood is poured into the heart of the various animals, so that, a comparison having been made between the magnitude and constitution of those [hearts] with the human [heart], they have estimated by probable conjecture that easily half an ounce of blood is poured into the human heart with each opening of the auricle.[84]

A physician and close associate of Descartes, Van Hogelande nicely illustrates a tension in the development of Cartesian anatomy and medicine. The recent autoptic

---

credentium, sequens hoc & satis quidem facile experimentum, in praedictae veritatis confirmationem hisce adjungere non gravabor. A corde itaque anguillae majoris bisecto, atque in aqua aliquamdiu detento, omnem sanguinem mediocri pressione ita separavi, ut partes ejus orbi ligneo (secundum eam superficiem quâ laesae sive bisectae erant) satis diu impositae, nullum amplius ederent motum; donec eôdem à sanguine ei infuso madentem collocatae, paulatim, i.e. prius aliquamdiu null, deinde lentissimô, posteà [149] verò velocissimô moverentur motu: atque hoc experimentum eisdem particulis simili etiam successu iteravi; dum typographus festinat. quod ideo addo, ne quis existimet, me praecedentes rationes à posteriori (scilicet praedictô experimentô prius factô) collegisse."

[82] Van Hogelande 1646, 198.

[83] Van Hogelande 1646, 200.

[84] Van Hogelande 1646, 195–6: "Noeterici verò, autore Clariss. Hervejô [*sic*] (cujus benignae liberalitati hanc cognitionem debemus) non solum secundam manifestas mechanicae leges, ex dictae venae cavae amplitudine, auriculaeque capacitate, ac crebra sanguinis in cor ex citata auricula injectione, solidè firmiterque ratiocinantes; verum etiam in gratiam ratiocinationi non satis fidentium, ad investigationes vel disquisitiones reales & sensuales sese accingentes, easdémque aggredientes, omnis generis animalia viva dissecando: Tantam sanguinis copiam in variorum animalium corda infundi notarunt, ut comparatione secundum magnitudinem & constitutionem eorum cum homine factâ, facilè dimidiam unciam sanguinis singulis auriculae aperturis in cor humanum infundi, probabili conjecturâ aestimaverint."

traditions in anatomy, and, to some extent, chymistry, prized sensory experience and experiment.[85] Descartes offered rhetorical and conceptual support for subordinating experience and experiment to intuition and reasoning. In his own work and that of his supporters, the action of the heart was a persistent problem. The demands of Cartesian metaphysics conflicted with common observations of a muscular heart which contracted in systole as the arteries filled. How could mechanists satisfy the demands of philosophical ideals and the disciplinary practices of autoptic anatomy?

8.4 Sylvius and His Students: Ideals of Mechanism and the Witness of the Senses in Anatomy and Chymistry

Sylvius' later work, in part, was an attempt to resolve this tension. In his treatise on medical method, Sylvius set down which qualities were legitimate in true explanations.[86] Here, as elsewhere in his metaphysical writings, Sylvius endorsed strict mechanism as an ideal of ontology and explanation.[87] All things in the world were endowed with only the common sensibles—shape, size, motion, etc. Sylvius gave one prime example of this method, which aimed to deduce the real changes of natural things from their varied shapes, in which he reasoned from the observed effects of fire, which pierces and rarefies, to the geometry of its tetrahedral particles.[88] This method of this example, which no doubt seemed cribbed from Plato's *Timaeus*, was not much followed in Sylvius's medicine.

All knowledge had to begin in the senses, and the most certain knowledge was not the sort reached by Cartesian meditation or speculation, but whatever was immediately derived from the senses: "Whatever is in the external senses, from which the beginning of all our cognition naturally comes about, is so certain to me, as what is as certain as possible; whence even by the ancient Philosophers this was solemnly decreed, most truely, Nothing in the intellect, which was not firstly in the sense."[89] Even if, as Garber argues, Descartes moved from emphasizing the certainty of knowledge reached via intuition in the *Discourse* to relying on

[85] See, e.g., French 1994; Wear 1983; Ragland 2012.

[86] Sylvius 1679, 128: "Whatever we might consider in regard to the Proper sensible Qualities, and then the Common, and should propose for their Remedies, I would not wish anyone to be carried away with madness, so that he would think that we should be so insane, from these things said, that we would consider that from the Proper sensible *Qualities*, as such, *Changes are produced* in Natural things, and other Functions in Men other than those External Senses; with us, as with the judgment of others, all those Qualities are not *Real*, as the Philosophers speak, but rather *Intentional*. That is, natural Things do not act on other natural Things so that they are changed by the power of the Proper Sensible Qualities; but merely in the External senses, as the Soul notices them, and according to their representation to it."

[87] Sylvius 1679, 896 and 128. For mechanism as ontology and explanation, see Des Chene 2005.

[88] Sylvius 1679, 128–129.

[89] Sylvius 1679, 896: "Quicquid in sensus externos, à quibus omnis cognitionis nostrae initium naturaliter fieri tam mihi certum est, quàm quòd certissimum; unde etiam ab antiquis Philosophis sancitum verissimè, Nihil in Intellectu, quòd non priùs fuerit in sensu."

hypothetical arguments, checked by experience, in the *Principles*, Sylvius's emphasis on experience as the sole and sure *foundation* of anatomical and medical knowledge is clearly different, and followed more closely the anatomical and Aristotelian traditions. Emphasizing the certainty of the senses, properly trained by long experience in anatomy, was a hallmark of Harvey's approach.[90] Sylvius also borrowed his basic scheme and some of his terminology on cognition and the senses from the Paduan Aristotelian Jacopo Zabarella.[91]

The only way to come to know the quantitative mechanisms of the world was through the senses, and especially though the witnesses of sight, touch, and taste in anatomical and chymical experiments.[92] The experience of working with the sensible changes in bodies—animate or inanimate—moved Sylvius to endorse the approach of the chymists. He repeatedly rejected the pores-and-particles explanations used by Cartesians to explain glandular secretion, and taught rather that glands worked by alchemical tincture—just as a pinch of the Philosophers' stone could create more gold, so a little leftover bile creates more bile in the gallbladder. The approach of the chymists, with their attention to the proper sensibles—flavors, smells, colors—was a better fit with bodily phenomena.

> Thus, in things obscure and far from the external senses, those who explain the changes in our body according to chymical operations evident to the senses seem to me to philosophize more aptly and in greater accommodation to natural things, and thus I will follow them. Though they do not remove all difficulties, yet they offer more light to our darkness than the others, who, although they use sensible similitudes, do not use such that sufficiently correspond [*quadrare*] with the things that happen in our body.[93]

Against the reductive emphasis on the commons sensibles of many clever men [*Ingeniosi*], Sylvius *in practice* preferred the principles and methods of the chymists, with all their attention to the proper sensibles.[94]

The method of Harvey and the "Sensible Philosophers" was perhaps also important to Sylvius, and it resembled the approach of the chymists. For Sylvius, philosophers and physicians needed to combine reason and experience—*Ratio et*

---

[90] Wear 1983 and Goldberg 2012, 235–239.

[91] Wear 1983. Sylvius takes his distinction between *primae notiones* and *secundae notions* from Jacopo Zabarella. Zabarella 1608, c. 6 B; cf. Sylvius 1679, 896, 395, 415, and 647. As these references from the 1658 *Oratio* through his posthumous works show, Sylvius expressed a broadly Aristotelian view of the generation of universals from sense experience throughout his career.

[92] For more on Sylvius's use of the senses, see Ragland 2012.

[93] Sylvius 1679, 311: "XXVI. Quapropter videntur mihi, quos idcirco & Ego secutus sum, aptius & naturalibus rebus accommodatius philosophari, qui in rebus obscuris & ab externis Sensibus abstrusis secundum Chymicas mutationes Sensibus patentes Operationes in corpore nostro factas explicant, licet ne sic quidem omnem tollant difficultatem: plus tamen luminis afferunt nostris tenebris, quam caeteri, qui, utut similitudinibus sensilibus utantur, non utuntur tamen talibus, quae satis quadrant cum iis, quae fiunt in corpore nostro."

[94] This qualitative approach to principles and substances is especially clear throughout the works of Sylvius's immediate chymical predecessors and sources, Daniel Sennert and Van Helmont. Ragland 2012.

Experientia—to generate true natural knowledge.[95] In his own investigations, Sylvius presented accounts of anatomical or chymical action by reasoning from first principles, but they always had to be checked by the senses and experiment.[96] Anatomists and chymists both, in Sylvius' view, took proper consideration of the testimony of the senses, while Descartes's account of the heart's motion contradicted the witness of the senses. Harvey taught "according to the custom of the Physicians, as much as the Sensible Philosophers, and according to the testimony of the external Senses."[97] Descartes, on the other hand, "trusting more in the laws of his own Mechanics, rather than in his external Senses, suspected and believed that the Ventricles of the Heart and the Arteries were Dilated and Contracted simultaneously."[98]

Sylvius also objected to Descartes's account of respiration, and his colleague Johannes de Raey's insistence on teaching it to the Leiden medical students. This explanation of respiration in terms of a circular movement of the air—pushed by the chest or abdomen, moving a circle from the chest through the air to the nose and into the lungs—was known as the 'Cartesian circle.'[99] In fact, Jan Swammerdam's tardy medical thesis, finally given in 1667, *De respiratione*, would use experiments to argue *for* the Cartesian circle.[100]

Sylvius, on the other hand, used very similar experiments to argue against the Cartesian circle, adding in a bit of ridicule about the discriminating pathway of the air in the final line. His target here was no doubt his colleague de Raey, who continued to teach a course on medicine. The other medical professors had objected in 1659, arguing that "medicine must be founded on experience, not on philosophical speculations."[101] When Vander Linden's course did not attract enough students, the university Curators allowed de Raey to continue teaching.[102]

Faced with the continued insult of overly-speculative mechanistic medicine, Sylvius went even further to reject a central tenet of the extreme Cartesians: that all motion was by impulse or impact alone. This did not mean he returned to older

---

[95] Sylvius 1679, 896 and 900.

[96] E.g., Sylvius 1679, 30–31.

[97] Sylvius 1679, 43: "Docuit autem *Harvejus*, Medicorum, utpote Sensilium Philosophorum, sequutus morem ac Sensuum externorum testimonium, *Arteriaq Dilatari, quando Cordis contrahuntur Ventriculi!* & vice verâ, *Contrahi* easdem *Arterias, quando Dilatantur Cordis Ventriculi*."

[98] Sylvius 1679, 43: "*Cartesius* verò, Mechanicae suae legibus, quàm Sensibus suis externis magis fidens, suspicatus est & opinatus, *Dilatari simul, & Contrahi simul Cordis Ventriculos & Arterias*."

[99] Regius had defended just this account in a June 1640 disputation, and Descartes approved of the account. Van Hogelande also took up this account. See Bos 2002, 47.

[100] Swammerdam 1667. Swammerdam's treatise also added other experiments, and observations on muscular action and hermaphroditic snails.

[101] Quote and trans. in McGahagan 1976, 307.

[102] He taught Cartesian physiology into the 1660s, as evidenced by Ole Borch's lecture notes, in which de Raey even argued that "no one before Descartes had demonstrated that there was something corporeal in those natural actions of the body, or how it could be distinctly conceived." March 1661, Borch 1983, vol 1, 47: "... adesse sed neminem ante Cartesium, quid esset praeterea, demonstrasse, aut quomodo distinctè id posset concipi."

models of attraction: "And I, certainly, do not follow in my mind the possibility of *Attraction* in natural things (as used in the language of the Schoolmen), but, in addition, I do not accede to those who teach what is primary, and I consider that it is observed in many examples or experiments, that *every Motion is to be accounted to Impulse alone*."[103] Ultimately, Sylvius professed to follow his senses, and even his common-sense reasoning. His opponent could neither "clearly nor distinctly demonstrate that a thing is *Pushed*, which the external Senses acting soundly and rightly—not deceived—testify is *Pulled*."[104] The Cartesian account was simply too fantastical; did a bulb made from a stomach, attached to a tube inserted into the trachea, really shrink in size due to the impulse of the external air, pushed by a rising chest or abdomen? Or did it collapse as the action of inspiration pulled the air out?

> What if a little tube is equipped such that there can be no transmission of the Aer due to the great density of its outer covering, and it has been connected to a similar Stomach, not, I say through the equal breadth or depth, but through the side wall and then again the Stomach is very tightly closed around the tube passing through, while the trachea has been drawn tight around the other end of the tube—should we really say then that the Aer, pressed by the expansion of the chest and abdomen seeks and finds that Stomach, through I know not what paths, by journeys very long and imaginary, which it then presses so powerfully, that it forces part of the aer contained in it to enter the Lung? Let the Jew Apella believe it, not I![105] I, who abhor and am wary of prejudices that are evidently contrary to Experience.[106]

Sylvius certainly harbored no love for Descartes's specific explanations of the body. His experimental and jocular public attack on de Raey's teaching points to continued tension in his intellectual allegiances. While he accepted something like a Cartesian ontology as an ideal, he rejected nearly every Cartesian explanation proffered in his own domain of expertise. Sylvius even held a grudge against Descartes's stubborn adherence to speculations about the motion of the heart, bringing the old controversy into a friendly chat in 1661. As recorded by the Danish polyhistor Ole Borch on a shared trip back from the Hague, Sylvius and Borch first chatted about a recent dissection in which the blood would not coagulate. Then

---

[103] Sylvius 1679, 34: "Ego sanè, qui me non tantùm in rebus naturalibus *Tractûs* possibilitatem (ut Scholasticorum termino utar) mente assequi, at insuper, quod primarium est, plura ejus exempla, sive experimenta observasse puto, haud accedo illis, qui *Motum omnem solo Pulsu absolvi* docent."

[104] Sylvius 1679, 34: "Nec assensum meum coget facilè, qui ex suis praesuppositis, nec probatis, tanquam principiis, modo suo explicabit, nec clarè ac distinctè demonstrabit *Pelli* id, quod *Trahi* testatur Sensus externus non deceptus, sed integer ac ritè agens."

[105] The reference is to Horace's Satire 5. The context is that someone tries to persuade Horace that incense will melt on the temple steps without flame. He withholds belief in what he takes to be flimflam. Horace 2012, 47.

[106] Sylvius 1679, 34–35: "Quid si tubulo ex coriis densissimis & nullum Aëri transitum concidentibus parato, atque non dicam per aquas [*sic*] medias & profundas, sed per transversum parietem undique arctissimè iterum circumclusum transmisso annecteretur similis Ventriculus; dum alterum tubuli extremum asperâ arteriâ strictissimè concluderetur; an diceremus tunc Aërem thoracis, abdomisque expansione pressum per nescio quae longissima & imaginaria itinera vias sibi quaerere ac invenire ad illum Ventriculum, quem tunc tam potenter premat, ut cogatur contenti aëris pars ingredi Pulmonem? Id, si volet, credat Judaeus Apella, non Ego; qui abhorreo & caveo mihi ab omnibus praejudiciis Experientiae evidenter contrariis."

Sylvius recalled that Descartes had not attacked Sylvius in writing, but with his speech: "that Descartes, not indeed in his writings, but with his tongue, had been very oblique, to contradicting himself, and that he had said that Sylvius did not understand mechanics (since in a dissected rabbit not everything answered to the principles of Descartes), and that by mechanics [Descartes] had not understood anything other than the fabrications of his own philosophy."[107]

Once again, we find Sylvius using a distinction between sensory anatomical experiments and Cartesian "fabrications." He occasionally used pore-and-particle mechanical filtering devices in his own explanations, such as his accounts of the spleen and gallbladder, but they did little work for him.[108] Still, the soul did even less explanatory work, and there are no obvious faculties involved in his new medicine. These conceptual affinities and his alliance with Cartesians such as Florentius Schuyl clearly position Sylvius as a mechanistic anatomist and physician.[109]

In Sylvius's view, a genuine mechanistic anatomy required hard work. He even rejected extant explanations of mechanical filtering by pores and particles in the kidneys, a favorite site for mechanistic explanations. Like Marcello Malpighi, Sylvius seemed to despair of finding any empirically adequate explanation of kidney secretion that did not use either the failed Galenic notion of attraction or the fantastic speculations of the Cartesians.[110] He was happier with the mechanists' approach, of course, but not some of the rashness of their speculations: "However, I approach more to those who, as much as possible, *assert that all things in our body are brought about mechanically* [*mechanicé*], and want these things to be demonstrated mechanically: Yet this will never happen, unless *we know the fabric of the body, and we show that it has a similitude with mechanical instruments*."[111] For Sylvius, direct, sensory autoptic witnessing along the lines of Harvey's demonstration of the action of the heart was necessary: "Whence you should also see here that

[107] September 1661, Borch 1983, vol 1, 216: "Cartesium non quidem scriptis, sed linguâ in contradicentes sibi valde fuisse obliquum, dixisse Sylvium non intelexisse mechanicum (quia in secto cuniculo non respondebant omnia Cartesii principiis), per mechanicam a. nil aliud intellexisse quam suae philosophiae commenta."

[108] Sylvius 1679, 14: "*Secretus* sic ab Alvinis foecibus Chylus per *carneam*& spongiosam Intestinorum *Crustam*, veluti per pannum laneum, *transcolatur* quasi, exprimiturque in Venas Lacteas memorato Intestinorum motu peristaltico." "The Chyle thus *secreted* from the Alvinis feces through the *fleshy* and spongy *Crust/surface* of the Intestines, as if through a woolen cloth, as it were, *filtered*, it is squeezed into the Lacteal Veins by the previously mentioned peristaltic motion of the Intestines."

[109] Bertoloni Meli 2011, 12–16 and *passim*.

[110] Bertoloni Meli 2011, 284–289. It is interesting that even by the much later date of Malpighi's *Vita*, mechanistic anatomy had no empirically acceptable account of kidney secretion.

[111] Sylvius 1679, 722: "CCCIV. Quamvis autem vulgò Cipiosae hujus Urinae secretionis, mox & exretionis Causa tribui soleat *Renibus potentiùs serum ad se trahentibus*: non satisfacit tamen mihi haec ratio, quamdiu non constat, vim attractricem inesse ipsis Renibus, quam illi duntaxat tribuere non suffiicit, nî quoque talis evincatur ac probetur. CCCV. Accedo autem illis magis, qui, quantum licet, *omnia mechanicè in corpore nostro perfici autumant*, & mechanicè demonstrari cupiunt: Id autem nunquam fiet, nisi *corporis fabricam cognoscamus, & cum mechanicis organis similitudinem habere ostendamus*."

some of the things that you might wish to add to the Body at your pleasure, I do not know how to construct that fabric of the parts of the body, and the same is not demonstrated by autopsy to actually exist, which yet is wholly necessary."[112] He continued to echo his condemnation of Descartes's anatomical speculations here as throughout his works:

> At any rate, they must not be disjoined one from the other, but the works of the Mind and of the Body must be joined together; nay, rather, as often as a Physician meets with some notable and abstruse effect among his patients, just so often must some probable Causes effective for it be thought out *by the wise Mind*, and it must be examined *by the industrious Hand* or in some other way by experiment. And the solidity and truth of the devised Causes must be weighed, and *a severe Judgment* must be decreed, to be rejected according to experience, since an offspring of the Mind must not be let in.[113]

Sylvius's strong rhetorical empiricism here—rejecting any "offspring of the mind" alone—was an expression of his anatomical training and his chymical work. As we shall see, Sylvius attempted to pass this attitude on to his students, even the dedicated mechanists.

Sylvius's defense of Harvey's account of the heart's motion against the speculative explanations of the Cartesians will now bring us full circle. Sylvius set out to resolve the dispute between Harvey and Descartes, "the two most brilliant Lights of this age" by a recourse to experiment and the personified witnesses of the two premier senses of the anatomist, *Visus et Tactus*.[114] The two faithful witnesses, *Sight* and *Touch*, testified that the arteries pulse and dilate whenever the ventricles of the heart contract— that the blood pours out from the heart with each contraction.[115] This process of muscular contraction of the heart and dilation of the arteries was so well-established that even the first translator of Descartes's *Treatise on Man*, Florentius Schuyl, took it as a matter of obvious fact. His illustrations for the Latin translation of Descartes's text moved toward more empirical pictures—even "virtual dissections" as Rebecca Wilkin has pointed out—and he was forced to accept the observations of the anatomists as well (Fig. 8.1).[116]

[112] Sylvius 1679, 722–723: "CCCVI. Nam, ut hoc obiter moneam, graviter mihi peccare videntur, quotquot mutationes in corpore mechanicas urgent, interim omnen laborem subterfugiunt, ex quo corporis nostri partium cum mechanicis instrumentis similitudinem ac conformitatem deducere licet. CCCVII. Unde videas hîc quoque nonnullos quaevis pro lubitu Corpori affingere, fabricam ipsius partium nescio quam comminisci, non item talem actu existere autopsiâ demonstrare, quod tamen omninò necessarium."

[113] Sylvius 1679, 723: "Utique ab invicem sejungenda non sunt, sed conjungenda Mentis Corporisque opera; quin imò quoties aliquis effectus notabilis & abstrusis occurrit Medico circa suos aegros, toties & *Ingenio sagaci* excogitandae sunt aliquae probabiles illius effectûs Causae, & *Manûs industriâ* vel quovis alio experimento examinanda est, trutinandaque Causarum excogitatarum soliditas & veritas, & *Judicio severo* decernendum, num secundùm experientiam, rejici, admittine debeat Ingenii foetus."

[114] Sylvius 1679, 43: "Binos autem fideles, ac omni exceptione majores hujus Veritatis profero Testes, *Visum & Tactum*." This rhetorical trope of sight and touch as witnesses is also found in Harvey 1653, 415. I am indebted to Karin Ekholm for this reference.

[115] Sylvius 1679, 43.

[116] Wilkin 2003.

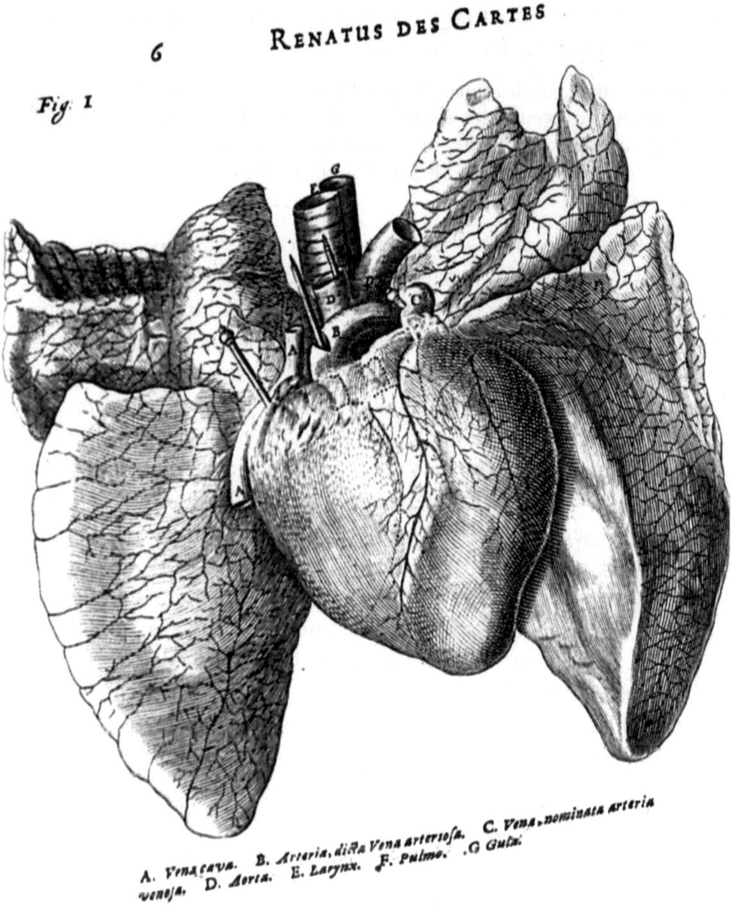

Fig. 8.1 Florentius Schuyl, trans. *De homine*, 1662: Illustration of the heart (Courtesy Hesburgh Library, University of Notre Dame)

For Sylvius, as a leading professor of medicine, the example of Descartes and his wildly inaccurate anatomy became a cautionary tale for his students and fellow physicians. Here, Sylvius's public disputations and published writings emerge as a culmination of his longstanding opposition to Descartes's medical pretensions, and to de Raey's lectures in Cartesian medicine.

> For whatever even the most subtle and sagacious Intellect [*Ingenium*] is able to think up that is most probable and plausible to the human Mind, every such thought, if it looks to the Medical Art, ought to be suspected as False by the Prudent, until Experience the Teacher of Truth has manifested that it is True, that is, that it has actually been observed in the things themselves as it has been imagined [*fingitur*].[117]

[117] Sylvius 1679, 43: "Quicquid enim Ingenium etiam subtilissimum & sagacissimum excogitare potest humanae Menti maximè probabile atque plausabile, id omne, Artem Medicam si spectet, tantisper Falsi suspectum esse Prudentibus debet, donec id ipsum Verum esse, hoc est, tale, quale fingitur, in rebus ipsis actu observari manifestârit Veritatis Magistra Experientia."

Ultimately, Sylvius' frustration with the chimerical Cartesian approach to anatomy spilled out into a denunciation:

> God forbid that among the legitimate Sons of Physicians anyone should be discovered who is so lazy, and who … would rather delight in adhering to his own figments and Chimeras, and those of others, and, since they are ill-pleasing to the sound Mind, he must at length, finally and tardily to exclaim with Medea: "I see better the better way, and I approve; but I follow the worse way."[118]

This charge was not made idly. Shortly after this, Sylvius freely admitted that Descartes was a "famous Mathematician and industrious Philosopher."[119] In Sylvius' view, however, a perspective conditioned by medical tradition and decades of experimentation and practice, Descartes could not be considered an anatomist and physician. Drawing the lines of discipline and identity so that Descartes remained a successful mathematician and philosopher was the gentlest criticism, but one that barricaded him from a place among the "Sons of Physicians." For Sylvius, as for Walaeus and other leading Dutch anatomist-physicians, their discipline was founded on close sensory observation and experimentation. If Descartes could not attend to the witness of his senses, he was not a good anatomist—he was not really doing proper anatomy at all.

We can see the same attitude carried on in one of Sylvius's students, Nicolaus Steno. In Paris in 1665, Steno, one of the foremost mechanist anatomists of his time, gave a lengthy lecture on the brain to an audience well-stocked with Cartesian enthusiasts.[120] He tempered his main objections with praise of Descartes's 'mental construction' of the mechanical man in the *Treatise*, but strongly criticized the Cartesian view of the pineal gland. In Descartes's view, as Steno explained, the pores on the surface of the gland needed to relate directly to the pores opposite, spirits needed to flow from all sides of the gland, and the gland needed to turn about and incline itself variously to the flow of spirits. None of these features is true of the gland observed in anatomical demonstrations, Steno asserted. The gland did not move, was anchored firmly in one inclination, did not appear to be aligned with any pores or major vessels, and was connected to ventricles on only one side. Furthermore, the central system of blood arteries necessary for the separation of spirits upon which Descartes's system depended did not exist, and one could find there only veins. Experience and direct observation proved the case against Descartes. Against the supposed movability of the gland, "experience assures us that it is, in fact, incapable of doing so." The non-existence of the arterial network

[118] Sylvius 1679, 44: "Absit autem, ut inter legitimos Medicorum Filios aliquis reperiatur tam socors, cui (repugnante licet cum Aegrotorum nonnunquam damno Experientiâ) suis, alienisve adhaerere lubeat figmentis ac Chimaeris, quoniam malè sanae placuerunt Menti, aliquando & saepenumerò serò cum Medeâ exclamaturae,—*Video meliora proboque;/Deteriora sequor*." The lamentation of Medea is from Ovid 1972, 59.

[119] Sylvius 1679, 44.

[120] See Steno 1965, 78, for an excellent attempt to reconstruct who might have been in attendance. Compare Bertoloni Meli 2011, 13, 18 and 84, who also mentions that Steno found Descartes to be "no anatomist."

for the elaboration of spirits as Descartes described is immediately known, "if you will believe your eyes."[121]

Steno thus followed the autoptic approach he had learned in Copenhagen and in Leiden, and most of all from Sylvius. Those who think that Descartes's description of the body is a "faithful representation of what lies hidden in the compartments of the human body … do not accept the very clear *demonstrations* of M. Silvius, who has shown many that Descartes' description does not tally with what is revealed by dissection of the body … *I am sure that eyes alone are required to observe and recognize* a great difference between the machine imagined by M. Descartes and what we see when we study the anatomy of the human body."[122] Descartes, of course, occasionally used the rhetorical device of considering bodies of "men resembling us," of "statues" made of earth, but anatomists routinely found his apparent empirical errors to be telling against his anatomical system.[123]

Steno also built on Sylvius' empirical findings on the heart to develop his own critique of the Cartesian model of the heart's function.[124] Later in life, in a 1678 letter to Leibniz, Steno recalled that the empirical observation the heart's muscular nature was a critical blow against his youthful Cartesian sympathies:

> I will tell you, that when I lived in the free country [the Netherlands] and had contact with very free thinking people and read all sorts of books, I had great esteem for Descartes' philosophy … Then a young Swedish friend [Th. Walgensten (1627–81)] brought me the lungs of a cow with the heart attached in order to study the substance of the lungs. That done, we boiled the heart to see if the substance was muscular. Its membrane having been peeled,

---

[121] Steno 1669 in Scherz 1965, 132–133.

[122] *Ibid.*, 129; italics added. Orig.: Ie me serois contenté de l'admirer avec quelque-uns, comme la description d'une belle machine, & toute de son invention; si ie n'avois rencontré beaucoup de gens qui le prennent tout antrement, & qui le veulent faire passer pour une rélation fidele, de ce qu'il y a de plus caché dans les ressors du corps humain. Puis-que ces gens la ne se rendent pas aux démonstrations tres-évidentes de Monsieur Silvius, qui a fait voir souvent que la description de Monsieur des Cartes, ne s'accorde pas avec la dissection des corps qu'elle décrit, il faut que sans raporter icy tout son syste[15]me, ie leur en marque quelques endroits, où ie suis assuré qu'il ne tiendra qu'a eux du voir clair, & de reconnoistre une grande difference entre la machine que Monsieur des Cartes s'est imaginée, & celle que nous voyons lors que nous faisons l'Anatomie des corps humains.

> I should have been prevented from referring to the faults in this treatise by the respect that I feel is owed by everyone, myself included, to intellects of this order, I would have been pleased to admire it, with the rest, as a description of a beautiful machine, invented entirely by him, if I had not met many persons who take it as [129] quite the opposite and who wish to pass it off as a faithful representation of what lies hidden in the compartments of the human body. Since these men do not accept the very clear demonstrations of M. Silvius, who has shown many that Descartes' description does not tally with what is revealed by dissection of the body, it is necessary for me, without describing all of his system here, to note certain parts where I am sure that eyes alone are required to observe and recognize a great difference between the machine imagined by M. Descartes and what we see when we study the anatomy of the human body. [pp. 128–129, trans. by Alexander J. Pollock]

[123] Descartes 1972, 1–2. AT XI 119–120.

[124] As Steno 1664, 5 mentions, both Harvey and Sylvius had observed that the heart was indeed a muscle.

> the first fibres of the heart led me down to the apex, and from apex upwards again, which in fact explained the whole fabric of the heart, till that moment unknown to all and directly contrary to what all the greatest and most dangerous philosophers held for proven truth; in other words they did not listen to things that did not fit into their opinion of the heart and muscles, still holding infallible the system of Descartes. To that end I took the leg of a rabbit previously dissected by me. In the first muscle and by the first incision there became unveiled, what to this moment had been unknown to man, the fabric of the muscles, turning upside-down the system of Descartes ..."[125]

In his 1680 defense of his own conversion to Catholicism, Steno again cited his observation of the muscular structure of the heart, this time as an important means by which God preserved him from the errors of Cartesianism which led to materialism and Spinozism.[126] No doubt, Steno's later recollections appeared in an apologetic framework. But the importance of the critique of Descartes's anatomical claims remained a theme for anatomists at Leiden for decades. In fact, Sylvius' and Steno's objections to Cartesian theory lived on in the work of their critical successors, such as that of Anton Nuck in the late 1680s.[127] Nuck closed his critique with an epitaph for the Cartesian pineal gland in the style of Bartholin's earlier eulogy for the Galenic liver.[128] Ultimately, Harvey's work fostered both the turn from the Galenic liver and the Leiden attacks on Cartesian anatomy. Experimental anatomy cut against old and new alike.

8.5 Conclusions

In the wake of Harvey and Descartes, Dutch physicians worked out complex and varied responses, mixing metaphysical ideals and experiments in varying degrees. Some generalizations, though, are reasonably well-supported.

First, that Descartes's anatomical speculations incited heavy and persistent criticism on empirical grounds. We should not pass over the historical importance of our actors' senses of error, as we tread carefully around anachronistic fault-finding. Perceived errors mattered to the history of early modern philosophy and medicine. The critical response to the speculative anatomy of Descartes, Van Hogelande, and de Raey was very likely not localized to the Low Countries. As mentioned above, Thomas Bartholin's hugely popular anatomical compendia from 1641 on also contained criticism of Descartes's account of the heart, as well as his claims about the

---

[125] Translation in Kardel 1986, 97. Steno 1952, 366–9.

[126] See Steno 1944, vol 1, 388–390.

[127] This was the case even for Cartesians outside of the Netherlands. See Smith 2013.

[128] Nuck 1691: "Traveler, stop a moment and consider with every endeavor the buried conarium as it once was, the first part of your body and the seat of the soul, the PINEAL GLAND, his majesty and splendor in this age born and extinguished—fame established it and opinion preserved it for as long as it lived, until the aura of divine particles completely flew away, and the clear waters/lymph filled its place—go, without the gland, traveler, give up the endeavor/conarion, lest posterity marvels at your ignorance."

pineal gland. Peter Anstey has shown that, regardless of philosophical orientation, no one in England accepted Descartes's account of the heart's motion.[129] I believe the Leiden story might give us a suggestive reason this was so: the anatomical errors were too much for the anatomists to bear.[130] Dennis Des Chene has also commented on the widely negative reception of Descartes's anatomy in the second half of the seventeenth century.[131] The work here shows that his anatomy received a persistent and detailed rejection from its very origin. This response contrasts sharply with Descartes's own self-regard as an anatomist, a favorable view that anatomists in general did not share.[132]

But this does not mean that Descartes's works were of no use to the leading anatomists. The disciplinary bounding performed by these early Dutch anatomists when they ruled Descartes out of the anatomy theater also created a space for their own, improved efforts, often including Cartesian mechanism as an *ideal* of ontology and explanation. The younger generation of Leiden anatomists, in particular, was eager to develop a new, chymico-mechanical physiology properly founded in observation and experiment.

Second, that experimentation was much more constitutive of the anatomical work of physicians who followed Harvey. Though philosophically conservative himself, Harvey's experimental demonstrations of the action of the heart and the directions of blood flow inspired generations of investigative workers at home and abroad.[133] Sylvius and Walaeus modeled much of the rhetoric and practice of their experimentation on Harvey's work. Mechanical explanations in anatomy were not wholly new, as Walaeus pointed out with the example of Erasistratus's model of the heart's action, which depended on material necessity. The physicians discussed here were comfortable with material explanations. Some, such as Van Hogelande and Sylvius, made their metaphysical ideals of a purely quantitative, mechanical medicine and natural philosophy quite explicit. Yet for Sylvius and most other physicians at Leiden, it remained difficult to square these ideals with the testimony of the senses. Thus mechanist anatomy and medicine were better defined by what they avoided—explanations via souls and faculties—and by their insistence on pushing

---

[129] Anstey 2000.

[130] Of course, English solidarity for Harvey's doctrines and anti-French sentiment no doubt played a role.

[131] Des Chene 2001, 153: "Not only did Descartes not manage to complete the science of life: in respect to particulars, he failed even to begin it. Within fifty years of his death, most if not all the mechanisms proposed by him were rejected outright, as were the *feu sans lumière* in the heart, the role of the pineal gland in sensation and memory, and most of his embryology, or substantially modified."

[132] Descartes to Mersenne, 20 January 1639: "In fact, I have taken into consideration not only what Vesalius and others write about anatomy, but also many details unmentioned by them, which I have observed myself while dissecting various animals. I have spent much time on dissection during the last eleven years, and I doubt whether there is any doctor who has made such detailed observations as I." AT II 525; CSMK 134.

[133] For England, see the superb study of Frank 1980.

material explanations as far as possible. Experimentation, often in contrast to speculation, became the primary means for generating these new explanations.

Finally, a suggestion: if we lay the evidence here alongside parallel developments elsewhere in the 1660s and 1670s, a pattern emerges in which self-consciously experimentalist groups in Leiden, Paris, and England reject what they describe as the excessive speculation of some Cartesians.[134] I suspect that the controversy over the action of the heart became a crucial battle, driving polemics about method and perceived errors. Those anatomists committed to mechanical ideals usually held explanations generated from those ideals subordinate to the autoptic evidence of the senses and experiments. This stance, rooted in ancient and early modern anatomical method, contrasts starkly with Van Hogelande's early prioritization of ratiocination over experimentation. As critics and historians noted, Cartesians often held their first principles immune from empirical refutation.[135] This mutual divide between anatomists and some Cartesians resembles the experimental/speculative distinction generating debate among historians of philosophy.[136] A great deal of work remains to be done to articulate the changes in reference and connotation for 'experimental philosophy' in different places and times. Moreover, I suspect the contrasting 'speculative philosophy' was even more variously defined and constituted, and often appeared as the pejorative construction of opponents. In these complexities, learned medicine was clearly a key site for early developments, in which the method of anatomists and chymists drove the rejection of empirically-flawed mechanical explanations, even for those with visions of a fully mechanistic medicine. Disciplinary conventions, pedagogical concerns, manual skill, and metaphysical ideals all met in the heart.

References

Anstey, Peter. 2000. Descartes's cardiology and its reception in English physiology. In *Descartes's natural philosophy*, ed. Stephen Gaukroger, John Schuster, and John Sutton, 420–444. New York: Routledge.

Anstey, Peter, and Alberto Vanzo. 2012. The origins of early modern experimental philosophy. *Intellectual History Review* 22: 499–518.

Aucante, Vincent. 2006. *La philosophie médicale de Descartes*. Paris: Presses Universitaires de France.

Bartholin, Thomas. 1651. *Anatomia, ex Caspari Bartholini Parentis Institutionibus, Omniumque Recentiorum et propriis Observationibus Tertium ad sanguinis Circulationem Reformata*. Leiden: Franciscus Hackius.

Bartholin, Thomas. 1655. *Anatomia, ex Caspari Bartholini Parentis Institutionibus, Omniumque Recentiorum et propriis Observationibus Tertium ad sanguinis Circulationem Reformata*. The Hague: Adrianus Vlacq.

Bartholin, Thomas. 1740. *Thomae Bartholini Epistolarum Medicinalium: Centuria IV*. The Hague: Petrus Gosse.

[134] Roux 2013; Anstey 2000; Frank 1980, 210–213.

[135] Ragland 2014, 135.

[136] E.g., Anstey and Vanzo 2012.

Bartholin, Caspar, and Thomas Bartholin. 1641. *Institutiones anatomicae.* Leiden: Franciscus Hackius.

Bartholin, Caspar, and Thomas Bartholin. 1645. *Institutiones anatomicae.* Leiden: Franciscus Hackius.

Bellis, Delphine. 2013. Empiricism without metaphysics: Regius' Cartesian natural philosophy. In *Cartesian empiricisms*, ed. Mihnea Dobre and Tammy Nyden, 151–184. Dordrecht: Springer.

Bertoloni Meli, Domenico. 2011. *Mechanism, experiment, disease: Marcello Malpighi and seventeenth-century anatomy*. Baltimore: The Johns Hopkins University Press.

Bono, James J. 1995. *The word of god and the languages of man: Interpreting nature in early modern science and medicine, 1: Ficino to Descartes*. Madison: University of Wisconsin Press.

Borch, Ole. 1983. *Olai Borrichii Itinerarium 1660–1665: The Journal of the Danish Polyhistor Ole Borch*, 4 vols., ed. H.D. Schepelern. London: E. J. Brill.

Bos, Jan Jacobus Frederik Maria. 2002. *The correspondence between Descartes and Henricus Regius*. Utrecht: Publications of the Department of Philosophy, Utrecht University.

Buchwald, Jed. 2008. Descartes' experimental journey past the prism and through the invisible world to the rainbow. *Annals of Science* 65: 1–46.

Clarke, Desmond. 1982. *Descartes's philosophy of science*. Manchester: Manchester University Press.

Clarke, Desmond. 1992. Descartes's philosophy of science and the scientific revolution. In *The Cambridge companion to Descartes*, ed. John Cottingham, 258–285. Cambridge: Cambridge University Press.

Clarke, Desmond. 2006. *Descartes: A biography*. New York: Cambridge University Press.

Des Chene, Dennis. 2001. *Spirits and clocks: Machine and organism in* Descartes. Ithaca: Cornell University Press.

Des Chene, Dennis. 2005. Mechanisms of life in the seventeenth century. *Studies in the History and Philosophy of Biology* 36: 245–260.

Descartes, René. 1964–1976. *Oeuvres de Descartes*, ed. C. Adam, and P. Tannery. Paris: Vrin.

Descartes, René. 1972. *Treatise of Man*. Trans. Thomas Steele Hall. Cambridge, MA: Harvard University Press.

Descartes, René. 1984, 1985, 1991. *The Philosophical Writings of Descartes*, 3 vols., edited and translated by John Cottingham, Robert Stoothoff, and Dugald Murdoch, volume 3 including Anthony Kenny. *The Philosophical Writings of Descartes*, 3 vols. Cambridge: Cambridge University Press.

Dobre, Mihnea, and Tammy Nyden (eds.). 2013. *Cartesian empiricisms*. Dordrecht: Springer.

Frank, Robert G. 1980. *Harvey and the Oxford physiologists: A study of scientific ideas and social interaction*. Berkeley/Los Angeles: University of California Press.

French, Roger. 1994. *William Harvey's natural philosophy*. Cambridge: Cambridge University Press.

Fuchs, Thomas. 2001. *The Mechanization of the Heart: Harvey and Descartes*. Trans. Marjorie Grene. Rochester: The University of Rochester Press.

Galen. 1968. *On the Usefulness of the Parts of the Body*, 2 vols. Trans. Margaret T. May. Ithaca: Cornell University Press.

Garber, Dan. 2001. *Descartes embodied: Reading Cartesian philosophy through Cartesian science*. Cambridge: Cambridge University Press.

Gariepy, Thomas. 1990. *Mechanism without metaphysics: Henricus Regius and the establishment of Cartesian medicine*. PhD. dissertation, Yale University.

Goldberg, Benjamin I. 2012. *William Harvey, soul searcher*. Ph.D. dissertation, University of Pittsburgh.

Gorham, Geoffrey. 1994. Mind-body dualism and the Harvey-Descartes controversy. *Journal of the History of Ideas* 55: 211–234.

Grene, Marjorie. 1992. The heart and the blood: Descartes, Plemp and Harvey. In *Essays in the philosophy and science of Descartes*, ed. Stephen Voss, 324–336. Oxford: Oxford University Press.

Harvey, William. 1628. *Exercitatio anatomica de motu cordis et sanguinis in animalibus*. Frankfurt: Guilelmus Fitzerus.
Harvey, William. 1649. *Exercitationes duae anatomicae de circulatione sanguinis*. Rotterdam: Arnoldus Leers.
Harvey, William. 1653. *Anatomical exercitations, concerning the generation of living creatures*. London: James Young.
Harvey, William. 1993. *The Second Anatomical Essay to Jean Riolan on the Circulation of the Blood*. Trans. Kenneth J. Franklin. *The Circulation of the Blood and Other Writings*. London: Everyman.
Harvey, William, et al. 1647. *Recentiorum disceptationes de motu cordis sanguinis, et chyli, in animalibus*. Leiden: Ioannis Maire.
Horace. 2012. *Satires: Book 1*, ed. Emily Gower. Cambridge: Cambridge University Press.
Jorink, Eric. 2010. *Reading the Book of Nature in the Dutch Golden Age, 1575–1715*. Trans. Peter Mason. Leiden: Brill.
Kardel, Troels. 1986. A specimen of observations upon the muscles: Taken from that noble anatomist Nicholas Steno. In *Nicolaus Steno (1638–1686): A re-consideration by Danish Scientists*, ed. J.E. Poulson, and E. Snorrason, 97–134. Gentofte: Nordisk Insulinlaboratorium.
Lennox, James G. 1992. Teleology. In *Keywords in evolutionary biology*, ed. Evelyn Fox Keller and Elizabeth A. Lloyd. Cambridge, MA: Harvard University Press.
Lonie, I.M. 1964. Erasistratus, the Erasistrateans, and Aristotle. *Bulletin of the History of Medicine* 38: 426–443.
McGahagan, Thomas. 1976. *Cartesianism in the Netherlands, 1639–1676; The New science and the Calvinist counter-reformation*. PhD. dissertation, University of Pennsylvania.
Newman, William R., and Lawrence M. Principe. 1998. Alchemy vs. chemistry: The etymological origins of a historiographic mistake. *Early Science and Medicine* 3: 32–65.
Nuck, Anton. 1691. *De Inventis Novis, Epistola Anatomica*. In *Adenographia curiosa et uteri foeminei anatome nova*. Jordanus Luchtmans, Leiden.
Ovid. 1972. *Metamorphoses: Books 6–10*, ed. William S. Anderson. Norman: University of Oklahoma Press.
Pagel, Walter. 1967. *William Harvey's biological ideas*. New York: S. Karger.
Pagel, Walter. 1978. *New light on William Harvey*. Basel: S. Karger.
Petrescu, Lucien. 2013. Descartes on the heartbeat: The Leuven affair. *Perspectives on Science* 21: 397–428.
Ragland, Evan. 2012. Chymistry and taste: Franciscus Dele Boë Sylvius as a chymical physician between Galenism and Cartesianism. *Ambix* 59: 1–21.
Ragland, Evan R. 2014. Between certain metaphysics and the senses: Cataloging and evaluating Cartesian empiricisms. *Journal of Early Modern Studies* 2: 119–139.
Roux, Sophie. 2013. Was there a Cartesian experimentalism in 1660s France? In *Cartesian empiricisms*, ed. Mihnea Dobre and Tammy Nyden, 47–89. Dordrecht: Springer.
Rupp, Jan C.C. 1990. Matters of life and death: The social and cultural conditions of the rise of anatomical theaters, with special reference to seventeenth century Holland. *History of Science* 28: 263–282.
Sakellariadis, Spyros. 1982. Descartes's use of empirical data to test hypotheses. *Isis* 73: 68–76.
Sargent, Rose-Mary. 1995. *The diffident naturalist: Robert Boyle and the philosophy of experiment*. Chicago: The University of Chicago Press.
Schouten, J. 1972. *Johannes Walaeus: zijn betekenis voor de verbreiding van de leer van de bloedsomloop*. Assen: Van Gorcum.
Schouten, J. 1974. Johannes Walaeus (1604–1649) and his experiments on the circulation of the blood. *Journal of the History of Medicine and Allied Sciences* 29: 259–279.
Smith, Justin E.H. 2013. Heat, action, perception: Models of living beings in German medical Cartesianism. In *Cartesian empiricisms*, ed. Mihnea Dobre and Tammy Nyden, 105–124. Dordrecht: Springer.

Steno, Nicolaus. 1664. *De musculis & glandulis observationum specimen*. Copenhagen: Matthiae Godicchenii.

Steno, Nicolaus. 1944–1947. *Opera Theologica*, 2 vols., ed. Knud Larsen, and Gustav Scherz. Copenhagen: Nyt Nordisk Forlag.

Steno, Nicolaus. 1952. *Nicolai Stenonis Epistolae et epistolae ad eum datae*, ed. Gustav Scherz. Copenhagen: Nyt Nordisk Forlag.

Steno, Nicolaus. 1965. *Discours sur l'anatomie de cerveau* (1669). In *Nicolaus Steno's Lecture on the Anatomy of the Brain*, ed. and Trans. Gustav Scherz. Copenhagen: Nyt Nordisk Forlag.

Swammerdam, Jan. 1667. *Tractatus Physico-Anatomico-Medicus De Respiratione Usuque Pulmonum*. Leiden: Adrian à Gaasbeeck.

Sylvius, Franciscus Dele Boë. 1679. *Opera Medica*. Amsterdam: Daniel Elsevier and Abraham Wolfgang.

Toellner, Richard. 1972. The controversy between Descartes and Harvey regarding the nature of the cardiac motions. In *Science, society and medicine in the renaissance*, vol. 2, ed. Allen G. Debus, 73–89. New York: Neale Watson Academic Publications, Inc.

Van Helmont, Joanne [Jan] Baptista. 1644. *Febrium Doctrina Inaudita*. Cologne.

Van Hogelande, Cornelis. 1646. *Cogitationes quibus Dei Existentia item Animae Spiritalitas, et possibilis cum corpore unio, demonstratur: nec non, brevis Historia Oeconomiae Corporis Animalis, proponitur, atque Mechanice explicatur*. Amsterdam: Ludovicus Elzevirius.

Walaeus, Johannes. 1641. Epistolae duae de motu chyli et sanguinis. In *Institutiones anatomicae*, ed. Caspar Bartholin and Thomas Bartholin, 385–408. Leiden: Franciscus Hackius.

Walaeus, Johannes. 1645. Epistolae duae de motu chyli et sanguinis. In *Institutiones anatomicae*, ed. Caspar Bartholin and Thomas Bartholin, 443–488. Leiden: Franciscus Hackius.

Walaeus, Johannes. 1655. Epistola Prima de Motu Chyli et Sanguinis. In *Anatomia Reformata*, ed. Thomas Bartholin, 531–565. The Hague: Adrianus Vlacq.

Wear, Andrew. 1983. William Harvey and the 'Way of the Anatomists'. *History of Science* 21: 223–249.

Wilkin, Rebecca. 2003. Figuring the Dead Descartes: Claude Clerselier's *L'Homme de René Descartes (1664)*. *Representations* 83: 38–66.

Zabarella, Jacopo. 1608. *Opera Logica*. Frankfurt: Lazarus Zetznerus.

Chapter 9
Louis de la Forge and the Development of Cartesian Medical Philosophy

Patricia Easton and Melissa Gholamnejad

Abstract Louis de la Forge (1632–1666) was a medical doctor and an early defender of the Cartesian philosophy. He is best known for his views on causation and his development of occasionalism within the Cartesian school. Commentators such as Balz (1932), Garber (1987), and Nadler (1998) have focused on the consequences of La Forge's views for Cartesian metaphysics and physics, with little consideration of La Forge's medical philosophy. We argue that La Forge provides a sophisticated version of Cartesian mind-body dualism, and he advances a mechanistic account of the animal spirits, corporeal memory, and a host of other topics relevant to Descartes's conception of the human body-machine. We examine La Forge's lengthy *Remarques* in the French edition of Descartes's *L'Homme de Rene Descartes et un Traite de la Formation du Foetus* (1664, 1667) where he advances Descartes's account of the generation and working of the animal spirits and its relevance to the human body-machine. We also examine La Forge's *Traité de l'esprit de l'homme et de ses facultez et fonctions, et de son union avec le corps* (1666), where he explains the functions of the soul while defending dualism and the mechanism of the body-machine against scholasticism. We conclude that La Forge advances Descartes's account of the generation and workings of the animal spirits and their interaction with the human soul, giving us an important vantage point to see the reception and development of the Cartesian medical philosophy in France.

Keywords Louis de la Forge • Rene Descartes • Medical philosophy • Mechanistic • Dualism • Mind-body union • Animal spirits • Aversion • Surprise • Corporeal memory • Habit • Inclination • Fermentation • Muscular movement • Pineal gland

P. Easton (✉) • M. Gholamnejad
Department of Philosophy, Claremont Graduate University,
Harper 152, 150 E. Tenth Street, Claremont, CA 91711, USA
e-mail: patricia.easton@cgu.edu; melissa.gholamnejad@cgu.edu

P. Distelzweig et al. (eds.), *Early Modern Medicine and Natural Philosophy*,
History, Philosophy and Theory of the Life Sciences 14,
DOI 10.1007/978-94-017-7353-9_9

9.1 Introduction

Louis de la Forge (1632–1666) was a medical doctor and an early defender of the Cartesian philosophy. He was educated at La Flèche, the college where Descartes studied several decades earlier, and he practiced medicine in Saumur. He is best known for his views on causation and his development of occasionalism within the Cartesian school. (Clarke 2011) Commentators such as Balz (1932), Garber (1987), and Nadler (1998) have focused on the consequences of La Forge's views for Cartesian metaphysics and physics, with little consideration of La Forge's medical philosophy.

Assessments of his merit as a medical thinker are few, and when offered, various. According to Descartes's biographer, Adrien Baillet, "one can say in praise of this work [*Traité de l'ésprit*] that this disciple surpassed his teacher in his industry."[1] By contrast, historian of medicine Thomas Steele Hall, when commenting on La Forge's *Remarks on Descartes's Treatise on Man*, wrote, "In elaborating upon Cartesian hypotheses, La Forge exceeds his master in the invention of unverified detail."[2] Of the same work Clerselier writes, "So that I could almost say his Commentary is a perfect text, which says everything and assumes nothing, leaves nothing behind, and contains the solution to all the most difficult questions …"[3]

The absence of any comprehensive assessment of La Forge as a medical thinker is what we hope to address here. With the growing interest in Descartes's thought as it applies to medicine, such a study is timely.[4] The main elements of Descartes's medical philosophy that influenced La Forge were Descartes's appeal to the laws of mechanics and the simplicity of his suppositions. More specifically, Descartes's view of matter as extended in three dimensions and bodies as assemblages of divisions of matter moving together provided the framework to explain all physical phenomena, including the functions of the human body. In order to study how the parts of the human body are assembled requires anatomy, and in order to understand how the parts moved in relation to other parts requires an understanding of mechanics. This metaphysical and explanatory framework attracted La Forge to Descartes's posthumously published *Traité de l'homme.* In La Forge's commentary (hereafter, *Remarks*) on the text he writes:

> He [Descartes] supposed nothing except that there are extended bodies, in length, breadth, and depth, that have diverse shapes, and that are moved in diverse ways. This is so simple, and so intelligible, and so much proved by experience or reason, that our adversaries find nothing to say, except that they don't believe that these principles are sufficient to be able to deduce knowledge of all the other things that make up the world.[5]

---

[1] Baillet 1691, 1.vii, ch. xix, 398–99.

[2] Hall 1969, xli.

[3] La Forge 1974, xxxiv–xxxv. Author translation.

[4] See recent studies by Aucante 2006; Carter 1983; Bitbol-Herpéries 1993; 2000; Gaukroger 2002; Easton 2011; Shapin 2000; Verbeek 1993; Wilson 1997; Wright and Potter 2000, and Shapiro 2003.

[5] La Forge 1664, 407. Author Translation.

La Forge took Descartes's theory of the generation and function of animal spirits to be a prime example of this simplicity in reasoning. Descartes describes the generation and function of animal spirits in part V of the *Discourse on Method.* Descartes compares the generation of animal spirits to a fine wind or lively flame that rises upwards, as he imagined that the most agitated and fine particles of blood, which composed the animal spirits, rose to the brain from the heart.[6] The idea is that only the size, shape, and motion of the particles of the blood in relation to the laws of mechanics need be appealed to in order to explain the function of the heart, brain, and the entire human body. As we will see La Forge keeps to this Cartesian framework throughout his *Remarks*, and consistently argues against the scholastic account in its appeals to faculties.

What Descartes and his contemporaries meant by "mechanism" and the "mechanical philosophy" varied greatly. According to Dennis Des Chene, mechanism in the late seventeenth-century could be adopted in two principal ways, as an ontology of nature, and/or a form of explanation.[7] Mechanism as ontology of nature viewed all natural things as having only mechanical properties; and mechanism as form or method of explanation viewed some natural operations as mechanical but not all (such as animal souls). A mechanism of form or method permitted a non-reductionist approach to medicine that was not open to the more restrictive mechanism as ontology of nature. Des Chene argues that Descartes's view that animals are machines derives from a mechanistic and reductionist ontology of nature. We agree with Des Chene with regard to animal souls, but when it comes to human souls, and their effects on the body and vice versa, Descartes, like La Forge, departs from the purely ontological notion of mechanism and adopts a non-reductive mechanism as form of explanation. As we hope to show, this non-reductive mechanism has notable consequences for the Cartesian medical philosophy.

This brings us to the final piece in Descartes's medical philosophy, too often ignored but adopted by La Forge, namely Descartes's view on the union of mind and body, and their causal interaction. Mind-body interaction provides for the effect of ideas, initiated in the mind but embodied in the corporeal imagination, on the functions of the body. It also provides for the effect of the body, through the corporeal imagination, on the functions of the human soul. We see this aspect of Descartes's medical philosophy at work in La Forge's account of aversion, found in his *Treatise on the Human Mind* (1666, 1669). In this work, he provides a sophisticated version of Cartesian mind-body interactionism, of the sort that has some surprises given his dualism and occasionalism. The central question is how the body-machine interacts with the mind and the mind with the body-machine. He dispels the Galenist theory of sympathies and antipathies to explain phenomena such as aversion, and instead explains it in terms of the mechanism of animal spirits and the force of the imagination. (*Treatise*, Chapter 22) La Forge's thesis is that the thoughts of the soul determine the direction of the animal spirits, which in turn produce certain patterns on the pineal gland, which in turn carry out the functions of the soul. As we will see,

[6] Descartes 1984–91. CSM I, 138–139.

[7] Des Chene 2005, 245–260.

the body-machine is primarily responsible for carrying out all the "useful" functions of the soul, with a few mental functions such as intellectual memory left as the province of the mind alone.[8]

In what follows, we hope to show that La Forge, in his *Remarks* (1664), advances Descartes's mechanical model of the body providing a genuine scientific framework for reasoning about and testing the operations of the body—some of which results in falsifying Descartes's account. This advancement made by La Forge will be demonstrated by reviewing his stance on generation and working of the animal spirits and its relevance to the human body-machine and his position on muscular movement.[9] Further, La Forge, in his *Treatise on the Human Mind*, explains the functions of the soul while defending dualism and the mechanism of the body machine against scholasticism thus paving the way for a study of the functions of the soul-body in the form of a non-reductive mechanism. In section I of this paper, we will examine how La Forge's *Remarks* furthers Descartes's account in two important respects: it provides greater anatomical detail along with a robust Cartesian explanation of animal spirits, and second, raises problems for the explanation of muscular movements given the virtually instantaneous communication of motion from the brain to the muscles. Using Descartes's description of animal spirits, La Forge shows how there is insufficient time for the motions of the animal spirits to account for reflex and other muscular motions. In section II, we focus on La Forge's defense of dualism as it relates to the functions of the soul-body.

We hope that a thorough consideration of La Forge's *Remarks* and *Treatise on the Human Mind* will aid in understanding how La Forge develops the Cartesian account of animal spirits, sense perception, appetite, corporeal memory, and soul-body. In doing so, we propose that the Cartesian model of the body-machine provides a strong framework for medical thinking and merits more study than it has received.

---

[8] Descartes 1984–1991 gives only cursory treatment of intellectual memory. See, for example, his exchange with Burman, CSMK III, 336, and his letter to Huygens, CSMK III, 216.

[9] Another example of La Forge advancing Descartes's work, not discussed here, is found in Wilkin 2008. Wilkin discusses how La Forge takes Descartes's explanation of how spirits move through the pores of the brain and the heart to be distributed throughout the body when a passion is felt to explain how some mother's passions can yield a birthmark while others do not. La Forge develops the above Cartesian notion by appealing to the speed by which the animal spirits pass through the mother's body. There must be sufficient exertion of force to "send the animal spirit whizzing through various nerves and arteries so that they pull other spirits along" thus generating the speed to create various stimuli that can generate a birthmark in some cases but not in others. La Forge draws on Descartes account "of how different passions (in the strictly corporeal sense) produce different bodily responses…differences are determined by the pathway taken by the animal spirit" but it is La Forge who attempts to ground the explanation of how birthmarks are generated into a mechanical hypothesis and provide a corporeal explanation for their cause. 552–555.

9.2 La Forge's *Remarks* on Descartes's *Treatise on Man* (1664)

Shortly after Descartes's death in 1650, copies of a French manuscript of what came to be known as *Treatise on Man*, circulated among his friends and correspondents. It was later edited and published in Latin and French (*De homine*, 1662; *Traité de l'homme*, 1664). The 1662 edition was a Latin translation by Florentius Schuyl. Clerselier reportedly found the images in the Latin edition inadequate to illustrating Descartes's descriptions, which were dynamic and often not observable by the naked eye or a microscope. Because Descartes's descriptions required illustration of physiological processes that took place at the minutest physical particles, it was a challenge for the illustrators, Schuyl, and later La Forge, and Gutschoven, to imagine and depict these small structures. The "illustrators were required to interpret the text and create images that rendered its meaning understandable."[10] It was precisely this understanding of Descartes's text that Clerselier was seeking in the 1664 French edition that he felt the 1662 Schuyl edition lacked.

The illustrations found in La Forge's lengthy *Remarks*, appended to the 1664 French edition of Descartes's *L'Homme de René Descartes et un Traité de la Formation du Foetus* (1664), help bring to life Descartes's idea that the body is a machine. La Forge's commentary fills over 200 pages of a 500-page volume, following Clerselier's French translation of *The Treatise on Man*, and *Description of the Human Body.* The translation includes illustrations by La Forge and Gérard van Gutschoven, both asked by Clerselier to produce the illustrations Descartes had desired to accompany the text. Both medical doctors by practice contributed their medical knowledge to the illustrations.[11]

Rebecca Wilkin attests to the importance of "this foundational text of post-Cartesian debate," recounting the oft-cited first encounter by Nicolas Malebranche:

> … he discovered such luminous truths, deduced in such marvellous order, and especially a mechanics [*une mécanique*] of the human body so admirable and divine, that he was ravished by it. … the joy of making so many discoveries caused such violent palpitations in his heart, that he was obliged to put the book down time and again in order to catch his breath.[12]

Wilkin attributes the importance of the text to Post-Cartesians and what struck so calm a figure as Malebranche to the illustrations themselves: "… it seems that he can only be referring to Gutschoven's and La Forge's stark and stylized figures. Single units portraying discrete mechanical systems, they stand out from the mass of print by which Clerselier approximates the unfinished and deformed manuscript of *L'Homme*."[13] While the illustrations help to bring to life Descartes's idea that the body is a machine, the text with its descriptions and metaphors would have sufficed

[10] Zittel 2011, 221.

[11] *Ibid.*

[12] André 1970. Translated and cited in Wilkin 2003. Note that Wilkin attributes Malebranche's attraction to the figures, not the body as machine message.

[13] Wilkin 2003, 60. Thanks to Gideon Manning for drawing our attention to this paper.

to awaken the Oratorian from his dogmatic slumber. The model of the body as a machine is constructed clearly in Descartes's text, accented and developed in La Forge's *Remarks*, and richly illustrated by the figures.

Remarks contains a number of notable discussions: a lengthy description of the nature and generation of animal spirits, how animal spirits produce muscular movements, sensory impressions, appetites and inclinations, a sustained defense of the pineal gland as the seat of the soul, the difference between a brain in a state of wakefulness v. sleep, and the nature of corporeal habit and how it is established. La Forge concludes with a defense of Cartesian physiology and its principle of fermentation by a comparison to Cartesian physics. We take each of these up in turn.

9.2.1 Animal Spirits: Generation and Function

La Forge alerts us to the importance of fermentation in contrast to faculties in the explanation of the movement of the heart: "Most doctors do not attribute the beating of the heart or the blood that flows there without cessation and with each beat, to fermentation, but rather to a Faculty of the soul called the Pulsating faculty."[14] Fermentation is what Descartes attributes to "a continual heat in our heart", which he designated as the first principle of life in the body-machine.[15] Descartes draws on this notion in Part V of the *Discourse on Method*,[16] and in his Letter to Plempius, 15 February 1638, in his discussion of blood circulation and the flow of blood through the heart. Descartes compares the expansion of blood in the heart to the action of a "yeast-like liquid" coming into contact with another liquid and causing it to expand. As Des Chene explains, a culture containing yeast is needed to start the fermentation process in the making of bread dough; in the same manner the blood in the heart requires a starter.[17] The heat created in the heart acts like a starter—a heat that is started without light or fire. For Descartes the heat of the heart is just like "new wine when it ferments…causes some of the particles to collect in a part of the space containing them, and then makes them expand, pressing against the others."[18] This expansion and the repetition of particles resisting and moving away moves the heart to beat, the body to pulse, the muscles to move, and the animal spirits to generate.

[14] La Forge 1664,185.

[15] Descartes, 1984–1991. *Passions of the Soul* I, 8, CSM I: 331.

[16] "I supposed, too, that in the beginning God did not place in this body any rational soul or any other thing to serve as a vegetative or sensitive soul, but rather that he kindled in its heart one of those fires without light which I had already explained, and whose nature I understood to be no different from that of the fire which heats hay when it has been stored before it is dry, or which causes new wine to seethe when it is left to ferment from the crushed grapes." Descartes, 1984–1991. CSM I, 134.

[17] Des Chene 2001, 21.

[18] Descartes 1998,187.

This change is a sudden fermentation that is made possible via "the entire fabric of the heart, the heat in it, the very nature of blood"[19]

The animal spirits are generated from the blood, which upon fermentation in the heart separates the blood into grosser and finer particles. According to La Forge, "By animal spirits we understand not only the first and second elements but also and principally all the smallest parts of the blood, which because of their solidity retain better than any others the agitation they receive in the heart."[20] He likens this process to distillation whereby liquors are separated by heat into their gross particles and finer spirits.[21] The lighter, smaller, and more agitated particles are the animal spirits, which rise and move more easily to the brain while the grosser particles are blood circulated to the rest of the body. La Forge refers his reader to a figure that illustrates the relative routes of the animal spirits depending on their quantity and level of agitation. The finer particles rush up to the brain while the coarser ones descend towards the testicles:

> In my opinion, this dependence and communication between the brain and the testicles is confirmed by experience, which shows that men of study, who exercise their imagination and brain a great deal, are not ordinarily strong in the functions of generation.[22]

La Forge then turns to the explanation of muscular movements, whereby the animal spirits move mechanically through the brain, down the brain stem and into the muscles. According to La Forge, explaining muscular movements by animal spirits raises special problems for the general claim that animal spirits are motions of the body.[23] Muscular movements are not achieved through the power of spirits or faculties: "But in fact, in neither dogs nor man are there any spirits, nor spiritual or corporeal faculties, that can breathe wind into a muscle, without filling the entire body."[24]

The objection La Forge takes especially seriously is how to explain the promptness evident in the movement of a limb—from the command to the execution. Given the relatively small quantity of animal spirits and their location in the brain, La

[19] Fermentation during the early modern period was understood in various ways. Antonio Clericuzio 2012 discusses the view of acid as a fermenter and a cause of digestion and further elaborates the process of digestion as a purely chemical, purely mechanical, or a combination of both chemical and mechanical digestion. 335. Catherine Wilson 1995 discusses how by the nineteenth century, Pasteur differentiated between bad and good infection when the theory of infectious diseases was combined with the chemistry of fermentation. 141.

[20] La Forge 1664, 206.

[21] La Forge 1664, 205–206.

[22] La Forge 1664, 210.

[23] Des Chene 2005, discusses the issue of muscle movement. Des Chene showcases the difference between Perrault and Descartes by their views on the work that animal spirits do in case of muscles. For Perrault animal spirits "operate to relax the muscles and not to tighten or shorten them. They shorten of their own accord after being stretched" while according to Descartes that "the entry of the animal spirits into a muscle shortens it." Perrault recognizes that the animal machine is like a mechanical machine however, the machine requires a "mover" and a pure machine is incapable of providing it.

[24] La Forge 1664, 220.

Forge questions how they traverse the body in sufficient time and quantity to move our extremities.[25] Given what we now know about the electro-chemical transmission of these signals, and the limitations of the mechanical explanation, La Forge's observations are astute. For several pages La Forge painstakingly describes the complex structure of the ventricles, the tiny double canals with valves, the branching pathways—a complex network of spaces and interwoven fibers directing the animal spirits from the brain along the spine to the muscles.

He responds to the promptness problem by supposing that the narrowing and doubling of the canals, the action of valves, the heating of the spirits to increase agitation, are sufficient to explain how the animal spirits could carry out muscle movements in a purely mechanical manner. Given we are not aware of the command as it passes through the body, we merely experience the desire and subsequent action, he reasons that it is again demonstrated that the communication of motion is purely mechanical.[26]

La Forge defends Descartes's reasons for supposing that the pineal gland is the seat of the soul. La Forge likens the pineal gland to a boat floating on water, attached by a few loose ties; relatively free to move about with the motions of the water.[27] The pineal gland is not only a singular structure, as Descartes pointed out, thus making it a natural location for the common sense, it also has the flexibility and mobility to receive a variety of patterns and images on its surface. In beasts, there are three general causes of motion that can affect the images traced on the pineal gland: first, the action of the exterior object on the senses; second, memory—the opening and closing of pores depending on the regularity of the pattern of the animal spirits; and third, the animal spirits themselves, their force, abundance, and course. In men, there is a fourth cause, the force of the soul to determine the movement of the gland, but even here, the pattern of the animal spirits not the soul produces the corporeal image. This mechanical explanation of the animal spirits and the pineal gland is employed to explain a range of phenomena from perception, remembering, dreaming, waking, and the communication and sympathy between a mother and her unborn child. Concerning the latter, La Forge writes, "when the mother has a strong imagination which includes a violent passion, it makes a considerable opening on

---

[25] La Forge 1664, 241.

[26] La Forge 1664, 244. Cf. Descartes 1984–1991; CSMK III: 225–226. Letter to Vorstius, 19 June 1643; "These animal spirits flow from the cavities of the brain through the nerves to all the muscles of the body, where they serve to move the limbs. Finally they leave the body by transpiration that cannot be detected–not merely those which pass along the nerves, but others as well which merely travelled in the arteries and veins. Whatever leaves the animal's body by this undetectable process of transpiration necessarily has the form of spirits … Only the animal spirits are pure; but they vary in strength depending on the differences in the particles which make them up." According to Descartes more can be found on vapours and exhalations and winds in Chapters 1, 2, 3, of his *meteorology*, "… for what I wrote there of vapours, exhalations and winds can easily be applied to spirits …" See also, Des Chene 2001, 37. "The primary difference between blood-particles, and the aereous particles and spirits, is size." Particles that are spirits, which are solid and excited, are unlike the aereous particles; spirit particles do not linger in the lungs but rather make the added journey into the aorta and toward the brain.

[27] La Forge 1664, 302.

the surface of the ventricles of the brain, whose figure is traced on the gland, where it subsists for as long as it takes to become this opening in the network... Because this machine [of the fetus] is similar to that of its mother … the same openings are made in the network of the infant."[28]

La Forge treats memory or recollection as the organ of memory which he illustrates as the space between the interior surface of the ventricles and the exterior surface of the brain in which ideas are retraced on the gland by the animal spirits, the same movements that the object originally produced.[29] La Forge develops his account of corporeal memory in *Treatise on the Human Mind*, treating corporeal memory in mechanical and structural terms:

> By the term 'corporeal memory' here I understand only a certain facility to re-open what remains in those pores of the brain's ventricles which have already been opened by the spirits and in the fibres through which the spirits passed, whatever the cause which had made the opening.[30]

In Fig. 9.1 (below) La Forge shows how the corporeal image is produced by multiple networks beginning with the object perceived, in this case, an apple: the first image is traced on nerves of the eyes; then reflected on the surface of the gland then impressed in the pathways of the brain tissue where it becomes "memory" and capable of retrieval.

La Forge depicts the multiple gateways to the brain that make possible or prevent sensory perception; depending on the quality of the animal spirits along the way, the motions may or may not be communicated to H: the pineal gland. From these examples and illustrations, La Forge concludes that only the pineal gland has the central location, the mobility, and the structure to receive the multiple pathways of motion in the brain.

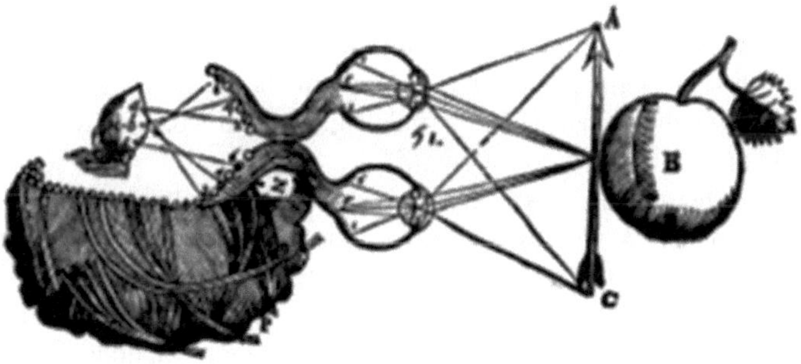

Fig. 9.1 Corporeal image

[28] La Forge 1664, 339–341. The term "reseuil" has been translated from old French as "network"; according to the 1694 *Dictionnaire de l'Académie française*, it is the same term as "réseau" meaning small nets or interlacings; it later acquired the anatomical meaning of interlacing of vessels.

[29] La Forge 1664, 331–332.

[30] La Forge 1997, 178.

Descartes's pineal gland hypothesis received a great deal of criticism by his defenders and critics alike. Willis in England argued that animals have the gland and are "most destitute of imagination."[31] Danish doctor, Niels Steensen, pointed out that Descartes's basic anatomical assumptions were wrong because the pineal gland is not suspended in the middle of the ventricles and is surrounded by veins not arteries.[32]

Perhaps a more biting blow came in 1675 by the Cartesian doctor, François Bayle, who gave detailed anatomical evidence and arguments for why the pineal gland could not perform the functions assigned to it.[33]

> If the trouble is taken to observe thoroughly the substance of the pineal gland, its situation, the disorder and obstruction of the vessels that surround it, the disposition of the parts that form the cavities or ventricles of the brain, the extremity [aboutissement] of the marrow of the backbone surrounding the corpus callosum, the substance that separates them, and the other particularities that have until now escaped the most exact anatomists, and the knowledge of which is very necessary here—by exactly observing all these things, one will then be convinced that it is impossible that the pineal gland could serve the uses that Descartes attributes to it.[34]

Bayle's arguments moved Malebranche's opinion from favorable to unfavorable, such that in 1674 he wrote that, " … for it must be remarked that even if [Descartes] were mistaken, which I do not believe," to a revised comment in 1678 "as appears very likely." Yet Bayle himself adds that Descartes's access to the medical information at the time was limited and one should not fault Descartes for holding the pineal gland hypothesis. The implication is that if Descartes had had the anatomical evidence that Bayle and others had, he would have agreed that the pineal gland could not function as the site of interaction between the mind and the body.

Le Forge admires Descartes's description of the brain, which he likens to a thick forest, Figs. 9.2 and 9.3 (below), "… we compared the brain to a very thick forest; the body of fibers represents the trunk of the tree, and the tiny hairs that advance out from their body, represent the branches and twigs." Moreover, the animal spirits are like a man who enters an opening in the woods and has a tendency to go to the right or left or straight ahead; but when he finds an open and worn pathway, rather than make a new one, he is likely to follow it.[35]

In Fig. 9.4 (below) La Forge illustrates the contrast between the waking brain and the sleeping brain. During sleep the animal spirits are slower, weaker, less abundant and the gland is smaller and its cavities are straightened and less receptive to exterior motions and the generation of muscular movements. By contrast, during wakefulness the animal spirits are faster, stronger, and more abundant. The implication is that there are structural and well as functional differences between the sleeping and the wakeful brain.

---

[31] Willis 1681, chapter 14. Cited in Lokhorst, Gert-Jan 2011.

[32] Steensen 1669.

[33] Bayle 1675; 1677.

[34] Lennon and Easton 1992, 91.

[35] La Forge 1664, 306.

Fig. 9.2 Decartes's Forest-like Brain: Cross-section

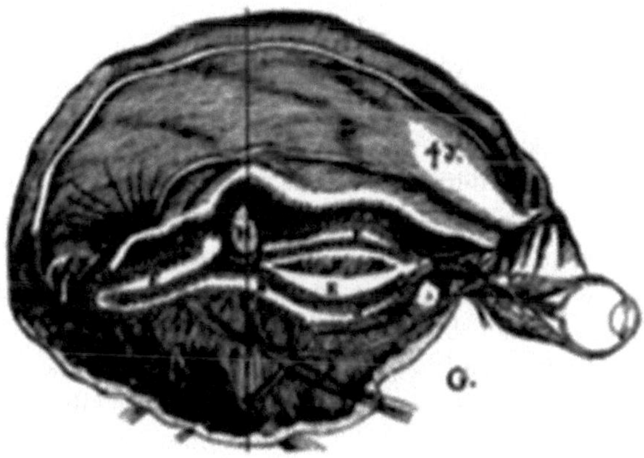

Fig. 9.3 Lateral View

La Forge again examines the hypothesized structural difference of a waking and sleeping brain in *Treatise on Human Mind* when he discusses memory. He reasons that during a wakeful state the animal spirits are plentiful but the brain is occupied with numerous activities when the senses are being used thus, making it very difficult to have a constant flow of the animal spirits. In contrast, during sleep when there are not as many animal spirits, the brain is also not distracted by external objects striking the senses, so that if during sleep animal spirits happened "to come

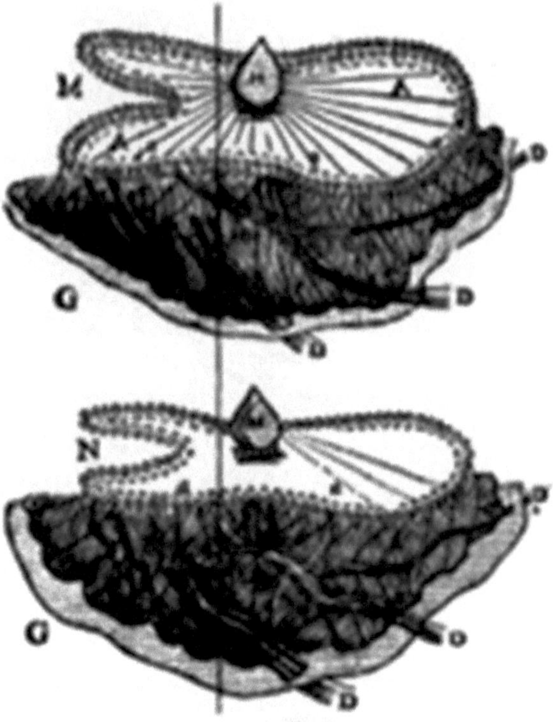

Fig. 9.4 Wakeful brain above; sleeping brain below

close to one of these traces [previous ideas or experience], it can re-open them much more than they were opened by action of the senses because the adjacent fibers are not squeezed as much."[36] This form of animal spirits can sometimes have a more forceful stimulation on the brain than having experienced the object via the senses. This is because, "all the animal spirits, or at least most of those which are in the gland at that time, are not diverted by any other action and can therefore flow towards that particular place."[37] Again, here too the consequence is that there are structural and functional variances between the sleeping and the wakeful brain.

Towards the end of his *Remarks*, La Forge discusses how corporeal habits are formed. He describes how to train a dog to jump for the King by pairing the sound of the words "jump for the King of France" with a piece of bread so that the words will be associated with the reward. Later the bread may be removed and the dog will continue to jump upon hearing the words uttered.[38] La Forge argues that this is because the process of the animal spirits and the images produced is mechanical, causing the physical traces to become physically associated with the images traced

[36] La Forge 1997, 163.

[37] *Ibid.*

[38] La Forge 1664, 385–386.

in the brain; it is not a process involving knowledge but rather corporeal habit/training. He re-visits the making of corporeal habits in the *Treatise* as well, this time in the example of a guard dog. Upon first encounter you chase the dog away with a stick; in subsequent encounters the dog runs away from you even without the stick:

> For as a result of having struck it, the flow of spirits that our presence stimulates became joined with that caused by the blows of the stick. The paths taken by these two flows having met somewhere in the mass of the brain and combined into one, it hardly matters by which of the two it is subsequently re-opened in the brain's ventricles ...[39]

La Forge's description of this mechanical process of the animal spirits and its habitual associations identifies what Pavlov in the twentieth century called "classical conditioning." This likely was not without predecessors prior to La Forge, and certainly other Cartesians, such as François Bayle, held a similar view. However, what seems to differ from the Pavlovian view is the potential role of ideas in the case of humans for countering the corporeal conditioning.

Lastly in this section, we will discuss La Forge's defense of Cartesian physiology and its principle of fermentation by a comparison to Cartesian physics and the application of this useful model and reasoning to medicine:

> Anyway, to explain how the two seeds of man and woman could produce a machine capable of all the bodily functions we have just described, it assumes nothing else, but that they are of such a nature that in coming to mix together they serve as the leaven to each other, and are fermented. Can there be anything simpler? And so if it is possible to judge each part by the aggregate, and if what is sketched in the Second Treatise can be used to speculate about what he might have done had he completed his project, he would have explained the formation of all the parts of the human body, and all its functions, had death not taken him.[40]

Thus fermentation is responsible for the beating of the heart, the production of animal spirits, the movement of blood and the animal spirits, the production of muscular movements, and reproduction itself. In La Forge's *Remarks*, we have seen developments of specific anatomical, physiological, and medical phenomena that are consistent with Descartes's work and developments of it. What La Forge evaluates as the most admirable in Descartes's physiological and medical reasoning is the simplicity of the suppositions and the comprehensiveness of its principles in the explanation of all biological phenomena. What Descartes's mechanical philosophy provided was a unified framework within which to understand the body-machine. Yet, there remains one significant chapter to the story of Cartesian medicine, namely, the role of the human soul and mind-body interaction, and its effects on the functions of the body.

[39] La Forge 1997, 181–182.

[40] La Forge 1664, 407–408.

9.3 La Forge's *Traité de l'esprit de l'homme et de ses facultez et fonctions, et de son union avec le corps* (1666)

Whereas the purpose of La Forge's *Remarks* was to elaborate and clarify Descartes's explanation of the functions of the body and his physiology, the purpose of the *Treatise on the Human Mind* is to explain the functions of the soul. In the opening of Chapter 16 of the *Treatise*, La Forge is concerned with establishing how the mind can move the body or the body can move the mind—which he argues is no more difficult than conceiving how a body has the power to communicate motion to another body. And, he takes it as a given that his audience assumes they "conceive only under a corporeal form and representation."[41] The problem enters in conceiving of mind-body interaction:

> However these people should admit either that they know nothing about most natural phenomena and do not know what they are saying when they explain them according to scholastic principles, or at least that they represent them to themselves in a manner or under an idea which is completely spiritual.[42]

His strategy is to argue that the case of mind-body interaction is as natural to conceive as body-body interaction: "That is why I said at the beginning that it was no more difficult to conceive how the mind moves the body than to know how one moves another, because, in fact, one must have recourse to the same universal cause in both cases."[43] La Forge's description veers clear away from scholastic references to powers and occult qualities of bodies or of minds. God is the universal source of motion in bodies and minds, and thus, conception of all interactions is equally difficult, even though body-body interaction appears more intelligible. As in *Remarks*, the importance of the body-machine in carrying out the functions of the body-soul union remains evident in the *Treatise*:

> You should also know that since no part of our body has power to act on the mind except through the mediation of the gland, likewise the soul has no power to move the body's limbs except by its mediation, by determining the movement of the gland and the flow of animal spirits towards the side of the brain's ventricles …this gland is the principal seat of the soul and the point at which their mutual communication begins and ends.[44]

La Forge saw the interaction of the mind and body as the *power of each to determine the direction of the animal spirits*—downward from the will to the movements of the body, and upwards from the motions in the body to the patterns on the pineal gland. This is illustrated nicely in Chapter 19: *Of Memory and Recollection.* He opens with a discussion of the hydraulic machine that Descartes describes at the beginning of the *Treatise on Man*. He completes the metaphor, which is worth quoting here in full:

---

[41] La Forge 1997, 143.

[42] *Ibid.*

[43] *Ibid.*

[44] *Ibid.*, 150–151.

> To complete the comparison, one could add that there is a mill nearby which provides and moves the water which makes all the parts of the machine work and that it sometimes provides more and sometimes less water. In the middle of the opening where the greatest discharge takes place, there is a little boat which is situated in such a way that, depending on the flow of water coming from the mill or the impulse which the fountaineer gives it (I imagine him lodged inside it and so attached to it that he cannot raise his head above the sides), it pushes the water into some of the tubes which terminate at this opening and thereby moves the figures into which these tubes discharge. It must also be assumed that these tubes are made of leather rather than of wood, lead or earthenware, so that they can expand in proportion to the volume of water which enters them. One will easily see that this fountaineer, thus hidden in the boat—which sometimes leans in one direction and sometimes in another, depending on the impact of the water which flows from the mill or the effect of the tubes which have just opened—should be aware of four very significant things.[45]

The "four very significant things" he outlines are as follows:

1. Actions of patrons walking in the Grotto/actions of external objects on the body-machine
2. Impulses of the fountaineer/effects of the gland itself on the corporeal image
3. Impact of the boat on the flow of water/actions of the animals spirits rising up from the heart
4. Facility of paths formed to be reopened after original path is forged/traces of memory

La Forge notes that all of this could happen without 2 (the impulses of the fountaineer). And, it is worth noting that the effect of the fountaineer/soul is on the direction of the animal spirits, which in turn produce the patterns on the pineal gland. The metaphor of the pineal gland as a boat that is made to move to and fro from above by the fountaineer (mind) and from below by the flow of water from the mill is not uniquely Cartesian. It harkens back to Aristotle's notion of the soul as a captain in the ship in Book II of *De Anima*. What is perhaps unique is the detailed role La Forge gives to the material mechanical function and how it interacts with the human mind. The mind exerts force on the body, causing the animal spirits to trace a corporeal image in the brain, whose corporeal form or structure affects the direction and pattern of animal spirits in the brain, thereby directing the movements of the body.

In Chapter 22, La Forge takes up the question of love and aversion, and offers a fascinating account. According to La Forge, it is not sympathies or antipathies that explain aversion, but rather the cause is surprise. He begins by arguing against the "obscure" terms that mask our ignorance, such as the account of sympathy and antipathy found in the work of Digby:

> The learned and subtle Chevalier Digby, in the treatise he wrote on the power of sympathy, introduces a very remarkable example of these kinds of aversion in the person of King James of Great Britain, who had such an aversion to unsheathed swords that he could not see one without extreme fright. He even says that the origin of the aversion came from the

[45] *Ibid.*, 177.

> fact that his mother the queen, when pregnant with him, was very frightened by the murder of one of her officers who was killed in her presence in her bedroom.[46]

The point about surprise is further explained with an example of a pregnant woman. It is that surprise produces a very noticeable change in the mother's animal spirits, which in turn can disturb the child she carries in her womb because of the intimate connection between the child and mother at the time. Again, the details of the mechanical explanation defend the Cartesian mechanical account:

> To understand this, consider that the principal effect of this surprise is to stop the gland at the place from which the species comes which caused the fright (as Mr Descartes well remarked when he spoke about wonder) and to carry the spirits in that direction so much that they are taken away from all other directions. This species is traced not only on the gland of the mother but also on that of the child (as I explained at length in my *Comments*, pages 335ff.) and even more forcefully than on that of the mother, because the child's gland is more tender and more capable of being disturbed by it. That is why if, after the child's birth, some cause traces the same species on their gland, it produces the same effects and causes one of those aversions for which we cannot find a reason, and many other things, which I have explained in my *Comments*. Indeed, there is something very significant to notice here. These kinds of impression are sometimes so strong that it is not always necessary for the object which usually excites a certain passion in us to be present to the senses. It is often enough to imagine it in order to excite the passion. That happened one day to a friend of mine, who had such an aversion to garlic that he could not eat it without vomiting. One day he had eaten a sauce in which there was no garlic and had found it very good when someone in the group said to him afterwards, as a joke, that there was garlic in the sauce and he vomited up everything he had eaten.[47]

La Forge notes that these impressions need not be caused from external objects, but possibly by the ideas in the mind, the imagination, itself. The mind can call up an image, and that in turn, causes a certain passion in us that is present to the senses. He recounts the experience of his friend whose aversion to garlic was so strong due to the very idea and its associated effects, that the mere mention of it caused him to vomit.

La Forge concludes:

> That, in my opinion, is the most genuine cause one could find for these kinds of natural aversion and inclination. I also think that those who like to have recourse to sympathy and antipathy, which are obscure terms which mean nothing and which are only good for disguising our ignorance under the mask of some fancy words, according to the usual style of peripatetic philosophy, should not be preferred to us.[48]

Inclinations that depend on intellectual knowledge get cursory treatment in La Forge because he believed that intellectual inclinations are better known by us so don't carry with them the obscurity and difficulties of the physical inclinations. Likewise, La Forge grants the existence of intellectual memory but provides no explanation of it beyond what Descartes said. What the *Treatise on the Human Mind* shows is that the Cartesian account of the functions of the mind cannot, in large part,

[46] *Ibid.*, 201.

[47] *Ibid.*, 202.

[48] *Ibid.*, 204.

be carried out without consideration of the nature of the body, and the soul-body union and its interaction. Explanations of the functions of the soul depend, as Descartes argued, on understanding the mind alone, the body alone, and the union of mind and body.[49] Explanations of aversion require understanding of the role of ideas and the imagination, *and* the body and imagination. The animal spirits are essential to the communication of motion from the mind to the body, and from the body to the mind.

9.4 Conclusion

As we have seen in the *Remarks*, and in the *Treatise on the Human Mind*, La Forge develops a detailed mechanical account of the animal spirits, of sense perception and appetite, of corporeal memory and habit, and a host of other topics relevant to the Cartesian conception of the human body-machine. We know that many of the specifics of Descartes's physiology did not stand the test of time. For example, the pineal gland hypothesis did not fare well after Descartes's death. Similar criticisms can be found of Descartes's theory of the heart and its movements, and of muscular movements. Descartes's views fare better in his studies of optics and motor reflexes, and in his descriptions of corporeal habit as an early conception of the cognitive mechanisms behind classical conditioning. There are some grounds for thinking that Descartes's account of corporeal memory and localization of function have relevance if not merit in the development of psycho-physics and later in neuroscience. However, the most significant impact Descartes seems to have had is in offering the model of the body-machine itself, and the mechanical-functional explanations that accompanied it, which promised to yield a science of the body. We have argued that Descartes and La Forge after him adopted an ontological and reductionist sense of mechanism in the development of the body-machine hypothesis, but that humans and the role of the mind and its interaction with the body, demanded a non-reductionist sense of mechanism. Both senses of mechanism inform and shape the Cartesian medical philosophy.

Study of the early Cartesian doctors, long neglected, gives us the opportunity to reassess Descartes's contributions to the biological and medical sciences. In studying La Forge, we hope we have shown that his work provides a rich account of the generation and workings of the animal spirits and their interaction with the human soul, giving us an important vantage point to see the reception and development of the Cartesian medical philosophy in France.

[49] Descartes 1984–1991. Passions, Part I.

References

Andre, Yves Marie. 1970. *La Vie du R. P. Malebranche*. Geneva: Slatkine Reprints.

Aucante, Vincent. 2006. *La Philosophie médicale de Descartes*. Paris: PUF.

Baillet, Adrien. 1691. *La Vie de M. Descartes*, 2 vols. Paris.

Balz, G. Albert. 1932. Louis de la Forge and the critique of substantial forms. *The Philosophical Review* 41(6): 551–576.

Bayle, François. 1675. *Discours sur l'experience et la raison, dans lequel on montre la necessite de les joindre dans la physique, dans la medecine et dans la chirurgie*. Paris: Moette. English trans. Lennon, T.M., and P.A. Easton. 1992. *The Cartesian empiricism of François Bayle*. New York/London: Garland Publishing.

Bayle, François. 1677. *Apoplexia*.

Bitbol-Herpéries, Annie. 1993. Descartes et Regius: leur pensée médicale. In *Descartes et Regius: Leur pensee medicale*, ed. T. Verbeek, 47–68. Amsterdam: Rodopi.

Bitbol-Hespériès, Annie. 2000. Descartes, reader of Harvey: The discovery of the circulation of blood in context. *Graduate Faculty Philosophy Journal* 22(1): 15–40.

Carter, Richard B. 1983. *Descartes' medical philosophy: The organic solution to the mind-body problem*. Baltimore: Johns Hopkins Press.

Clarke, Desmond. 2011. Louis de La Forge. *Stanford encyclopedia of philosophy*. Winter 2011 Edition, ed. E. N. Zalta. http://plato.stanford.edu/entries/la-forge/. Accessed 9 Jan 2014.

Clericuzio, Antonio. 2012. Chemical and mechanical theories of digestion in early modern medicine. *Studies in History and philosophy of Biological and Biomedical Sciences* 43: 329–337.

Des Chene, Dennis. 2001. *Spirits and clocks: Machine and organism in Descartes*. Ithaca/London: Cornell University Press.

Des Chene, Dennis. 2005. Mechanisms of life in the seventeenth century: Borelli, Perrault, Régis. *Studies in History and Philosophy of Biological and Biomedical Sciences* 36: 245–260.

Descartes, René. 1664. *L'Homme de René Descartes et un Traitté de la Formation du Foetus du mesme autheur, Avec les Remarques de Louys de la Forge, Docteur en Medicine, demeurant à La Fleche, sur le Traité de René Descartes; & sur les Figures par luy inventées*, ed. Claude Clerselier. Paris: Charles Angot.

Descartes, René. 1984–1991. The philosophical writings of Descartes, 3 vols. Trans. J. Cottingham, R. Stoothoff, and D. Murdoch (Volume 3 including Anthony Kenny). Cambridge: Cambridge University Press.

Descartes, René. 1998. *The world and other writings*. Trans and Ed. Stephen Gaukroger. Cambridge: University Press.

Dictionnaire de l'Académie Française. Paris: Jean Baptiste Coignard, 1694.

Easton, Patricia. 2011. The Cartesian doctor, François Bayle (1622–1709), on psychosomatic explanation. *Studies in History and Philosophy of Biological and Biomedical Sciences* 42(2): 203–209.

Garber, Daniel. 1987. How god causes motion: Descartes, divine sustenance, and occasionalism. *The Journal of Philosophy* 84(10): 567–580.

Gaukroger, Stephen. 2002. *Descartes' system of natural philosophy*. Cambridge: Cambridge University Press.

Hall, Thomas Steele. 1969. *Ideas of life and matter: Studies in the history of general physiology 600B.C.-1900A.D.*, 2 vols. Chicago: University of Chicago Press. [Editions published 1975ff under the title *A General History of Physiology*].

La Forge, Louis de. 1664. Remarques. In Descartes, R. *L'Homme de René Descartes et un Traitté de la Formation du Foetus du mesme autheur, Avec les Remarques de Louys de la Forge, Docteur en Medicine, demeurant à La Fleche, sur le Traité de René Descartes; & sur les Figures par luy inventées*, ed. C. Clerselier. Paris: Charles Angot.

La Forge, Louis de. 1666. *Traitté de l'Esprit de l'Homme, de ses facultez et fonctions, et de son union avec le corps. Suivant les Principes de René Descartes*. Paris: Theodore Girard, [Another printing the same year, in Paris, by Michel Bobin and Nicolas Le Gras.]

La Forge, Louis de. 1669. *Tractatus de Mente Humana, Ejus Facultatibus &Functionibus, Nec non De ejusdem unione cum corpore; secundum Principia Renati Descartes*. Amsterdam: Daniel Elzevier. [Posthumous Latin translation.]

La Forge, Louis de. 1974. *Œuvres philosophiques, avec une étude bio-bibliographique*, ed. P. Claire. Paris: Presses universitaires de France.

La Forge, Louis de. 1997. *Treatise on the human mind*. Trans. D.M. Clarke. Dordrecht: Kluwer.

Lennon, Thomas M., and Patricia A. Easton. 1992. *The Cartesian empiricism of François Bayle*. New York/London: Garland Publishing.

Lokhorst, Gert-Jan. 2011. Descartes and the pineal gland. *The Stanford encyclopedia of philosophy*. Summer 2011 Edition, ed. E. Zalta. http://plato.stanford.edu/archives/sum2011/entries/pineal-gland/. Accessed 7 Jan 2014.

Nadler, M. Steven. 1998. Louis de La Forge and the development of occasionalism: Continuous creation and the activity of the soul. *Journal of the History of Philosophy* 36(2): 215–231.

Shapin, Steven. 2000. Descartes the doctor: Rationalism and its therapies. *British Journal for the History of Science* 33(2): 131–154.

Shapiro, Lisa. 2003. The health of the body-machine? Or seventeenth century mechanism and the concept of health. *Perspectives on Science* 11(4): 421–442.

Steensen, Niels. 1669. *Discours de Monsieur Stenon sur l'anatomie du cerveau*. Paris: Robert de Ninville.

Verbeek, Theo. 1993. *Descartes et Regius*. Amsterdam: Rodopi.

Wilkin, Rebecca M. 2003. Figuring the dead Descartes: Claude Clerselier's Homme de Rene Descartes 1664. *Representation* 83(1): 38–66.

Wilkin, Rebecca M. 2008. Essaying the mechanical hypothesis: Descartes, La Forge, and Malebranche on the formation of birthmarks. *Early Science and Medicine* 13: 533–567.

Willis, Thomas 1681. The anatomy of the brain and the description and use of the nerves. In *The remaining medical works of that famous and renowned physician Dr. Thomas Willis*. Trans. S. Pordage, London: Printed for T. Dring, C. Harper, J. Leigh and S. Martyn.

Wilson, Catherine. 1995. *The invisible world: Early modern philosophy and the invention of the microscope*. Princeton: Princeton University Press.

Wilson, Catherine. 1997. Between medicina mentis and medical materialism. In *Logic and the workings of the mind*, ed. P. Easton, 237–253. Atascadero: Ridgeview.

Wright, John, and Paul Potter (eds.). 2000. *Psyche and soma: Physicians and metaphysicians on the mind-body problem from antiquity to enlightenment*. Oxford: Clarendon.

Zittel, Clause. 2011. Conflicting pictures: Illustrating Descartes' Traite de l'homme. In *Silent messengers: The circulation of material objects of knowledge in the early modern low countries*, ed. S. Dupré and C. Lüthy, 217–260. New Brunswick/London: Transaction Publishers.

Part III
Matter and Life, Corpuscles and Chymistry

Chapter 10
Transplantation and Corpuscular Identity in Paracelsian Vital Philosophy

Jole Shackelford

Abstract Long recognized as an important antecedent to the development of modern chemistry, Paracelsian chemical philosophy is often left out of historians' reconstruction of "chymical" matter theory in the seventeenth-century scientific revolution, owing to a thematic incommensurability between the new mechanistic theories and the vital philosophy of the Paracelsians. As a result, vital philosophy is more often characterized in terms of "correspondences" and "affinities" than as an explanation for material transformation. This paper explores a key component of vitalist matter theory, Paracelsian *semina* (seeds) as basic organic entities and principles of development, and how the concept of their "transplantation" illuminates both their inner vital nature and their spatialization as material principles. The result is a concept of chemical-mechanical action that is far different from the mechanical matter theory of the Cartesians. By defining temporality as an essential characteristic of seminal matter, the late sixteenth-century Paracelsian theorist Petrus Severinus provided a metaphysically sound basis for explaining internal agency as a foundational property of material being. Severinus' conception of *semina* was widely read and commented on by medical writers and natural philosophers involved in constructing the "new science" of the seventeenth century.

Keywords Paracelsus • Severinus • Chronobiology • Seeds • Mechanical philosophy

Recent studies of mechanical philosophy, corpuscular materialism, and alchemy in sixteenth and seventeenth-century Europe have clouded the once tidy schematic view of the Scientific Revolution as entailing, *inter alia*, a paradigm shift from scholastic medieval Aristotelian matter theory, which was heavily invested in the concept of substantial form, to variants of neo-classical atomism, which formed the basis of Robert Boyle's corpuscular hypothesis and thus the roots of modern chemistry. Rather than sustaining a relatively abrupt, generation-long abandonment of Aristotelian theory and erection of a new mechanical philosophy, historians of

J. Shackelford (✉)
Program in the History of Medicine, University of Minnesota, Minneapolis, MN, USA
e-mail: shack001@umn.edu

P. Distelzweig et al. (eds.), *Early Modern Medicine and Natural Philosophy*, History, Philosophy and Theory of the Life Sciences 14,
DOI 10.1007/978-94-017-7353-9_10

science have revealed a continuity in materialist atomism from medieval alchemy through the seventeenth century and suggested a conceptual fusion with the vitalistic matter theory of the Paracelsians as background to seventeenth-century corpuscular matter theory.[1] Studies of late sixteenth and early seventeenth-century engagement with atomism have revealed an unsettled but fertile matrix of atomist ideas, from Giordano Bruno's monads to Galileo Galilei's point-atoms and Nicholas Hill's gropings toward an atomic hypothesis.[2] If Paracelsian vitalism entered into this turn-of-the-century brew, from which the corpuscular ideas of Pierre Gassendi, Walter Charleton, and Boyle emerged as models for the matter theory of a new materialist philosophy, in what form did it appear and why would it be at all appealing to hard-nosed, anti-Aristotelian theorists, who in general rejected vitalists' claims? The Paracelsian ideas expressed by Petrus Severinus (1540/2–1602) in his 1571 book *Ideal of Philosophical Medicine*, which was read by the influential chemical philosopher Joan Baptista van Helmont, among many others, provides insight into how the principal authors of the corpuscular hypothesis understood Paracelsus. Severinus interpreted Paracelsus quite closely and sought to generalize and explain Paracelsian conceptions in a way that made sense to formally-educated natural philosophers, including Gassendi and Boyle, both of whom knew his work. His vitalist matter theory exemplifies the difficulty that Paracelsus' various views on the nature of matter and its transformations presented to seventeenth-century readers.

Take, for example, Severinus' conception of *transplantatio* (transplantation), which he used pointedly as an alternative to *transformatio* and *transmutatio*. Severinus used this term, which he derived from Paracelsian texts, to explicate his matter theory, which was constructed mainly around his concept of *semina*, another term he took from Paracelsus.[3] Severinus' ideas about *semina* and *transplantatio* provide insight into how he interpreted Paracelsian matter theory in a way that defies easy partitioning of early modern corpuscular matter theory into Aristotelian, Neoplatonic, and Atomist types, and helps explain how opponents of vitalism, such as Gassendi and Boyle, could take his ideas about material transformation seriously. The very title of his book, *An Ideal of Philosophical Medicine, Containing the Foundations of the Entirety of Hippocratic, Galenic, and Paracelsian Teaching*,

[1] This historiography has been clearly propounded by Newman and Principe, who identify Marie Boas Hall as a chief architect of the idea that Boyle was a pivotal figure in discarding "alchemical" matter theory and elaborating a new "chemistry" on the basis of mechanical philosophy and a corpuscular matter theory derived from Pierre Gassendi's resurrection of classical atomism. For a concise statement of this see Newman 2006, 6–8; In chapter three, Newman identifies Andreas Libavius as one link in the incorporation of Paracelsian vitalism into corpuscular matter theory, and it is evident that he was drawing in part on ideas about *semina* discussed below.

[2] For an overview, see Lüthy et al. 2001.

[3] Paracelsus used the term transplantation with a sense much like Severinus explicitly and repeatedly in Book Three of "Von Blatern, Lähmi, Beulen, Löchern und Zitrachten der Franzosen und irs Gleichen, inhaltend zehn schöne Bücher," in Paracelsus 1922, Abt. 1, Bd. 6, 301–479. It is conceivable that this was one of Severinus' sources. On Paracelsus' use of the term, see Shackelford 2014 (forthcoming).

alerts us to his syncretism.[4] Although early modern readers of classical atomism would have identified *semina* in the first instance as the term that Lucretius had used for material atoms in *De rerum natura*, Severinus' *semina* differ significantly from atoms of classical Greek origin in that they are intrinsically temporal, which supports a conception of vital philosophy. Envisioned essentially as ideas of bodies in a state of pure potency, Severinus' *semina* are in actuality embodied in the world and therefore have a subtle materiality, which Severinus expressed through his cautious definition of them as the links between bodies and non-bodies.[5] His use of the term transplantation helps to illuminate the material nature of *semina* and how they fit into a broader theory of organic change.

10.1 Vital Philosophy

Severinus understood his philosophy to be "vital", as did Andreas Libavius, when he characterized it thus in his attack on Johan Hartmann.[6] The term is therefore an actor's category in the period and not a modern construction. Severinus regarded any other kind of philosophy, which failed to take into consideration the active roles of the vital elements of nature, *semina* and *astra*, as empty – a study of the husks with no regard for the kernels; in modern terms, an examination of the anatomy of a body without consideration of its physiology:

> All philosophy that, having neglected the contemplation of these seeds and stars, chases dead privations, formless matters, and dead qualities is deaf and blind. Indeed, we cannot obtain knowledge of the elements without the seeds or stars, for they have unfolded the functions of the elements.[7]

Severinus rejected the philosophy of death, a metaphysics that deprived material reality of internal agency or vitality, which aptly characterizes the most austere forms of later mechanical philosophies of the Scientific Revolution. In these later formulations, associated with Gassendi, Charleton, and Boyle – if one ignores nagging vestigial recourse to self-organizing behavior in specific situations – dead corpuscles were imagined to be moved by external forces and their behavior governed by the dead mechanics of inertia and collision.[8] Motion, which for Aristotelians

---

[4] Severinus 1571.

[5] Ibid., 58. "Semina sunt vincula utriusque naturæ, visibilia invisibilibus coniugentia." He says much the same about *astra* on 46: "Astra sunt vincula rerum."

[6] Libavius 1615, title: "A Harbinger of the Vital Philosophy of the Paracelsians … On Vital Philosophy from Severinus according to Johannes Hartmann."

[7] Severinus 1571, 49. "Surda et coeca est omnis Philosophia, quæ horum contemplatione neglecta, priuationes, informes materias, et mortuas qualitates sectatur. Etenim Elementorum cognitionem, sine his adipisci non possumus. Semina enim uel Astra, officia Elementorum explicauere." The term *explicatio* used in this sense conjurs the metaphysical explanations of identity, multiplicity, and causal relationships that were elaborated by Nicholas of Cusa.

[8] Articles and books treating mechanical philosophy are many. For an overview, see Garber 2006.

connoted change of all sorts, was reduced to local motion imposed on particles from without. In contrast, Severinus' vital philosophy can be seen to invest material corpuscular rudiments with internal efficient and formal causation, constituting an endogenous vitality that was compatible with the Christian theology of creation and providence, but also with the Aristotelian concept of organic teleology. It is precisely when considering the self-organizing behavior that bodies exhibit in some circumstances that we find seventeenth-century mechanists questioning whether an element of internal agency need be assigned and turning to consider Severinus' explanation of *semina* as a possible model for corpuscular matter.[9]

Vital philosophy can be understood as fundamentally a physiological conception, which is dynamic. As dynamic, material units, *semina* are characterized by developmental forms and can be interpreted as organic "vital" corpuscles. These contrast with Epicurean and Democritean atoms, which are extrinsically temporal and support what Severinus deemed *philosophia mortalis* or *anatomia mortis*, a view of material nature as essentially devoid of life.[10] Such a view is not characteristic of physiology, as we understand the term today, but rather morphology, with the assumption that forms are stable, even static. Such a static conception underlies the ideal of materialist corpuscles that were part of what has been called mechanical philosophy in early modern science. Such corpuscles were supposed to possess size, shape, and impenetrability and not to "behave" in any characteristic manner, which would suggest that soul or volition was an essential feature. This limited set of fundamental properties lies at the root of the insufficiency of seventeenth-century mechanical philosophy to fulfill the explanatory demands that many early modern medical theorists placed on a matter theory, namely that it explain coordinated systemic functions, orderly growth and development, healing, and other activities that seemed to distinguish the living from the dead.[11] Behavior is change over time and logically presupposes a temporality, a sequence or plan for material change. The fact that even the relatively austere corpuscular hypotheses of Gassendi and Boyle attributed some measure of autonomous developmental possibility to fundamental material particles reflects the unsuitability of a totally de-vitalized conception of nature to explain observed phenomena.[12] A key distinction between Paracelsian

---

[9] There have been several studies of the persistence of active principles in early modern corpuscular philosophy in recent years. See, for example, Clericuzio 2000; Henry 1986.

[10] For example, Severinus 1571, 138: "De corruptibilibus Radicibus, Elementis, Principijs, earumque Mixtione cum puris, copiosius agemus in morborum Generationibus: mortis enim Anatomiam, et morborum continent." [We will more abundantly treat the corruptible roots, elements, and principles, and their mixture with the pure in [the chapter on] the generation of diseases, for they contain the anatomy of death and diseases.]

[11] Terms like biology and botany are anachronistic in the late Renaissance and early modern periods, and I am subsuming under the term "medical theorists" all writers who used natural-philosophical ideas that were associated with medical education and medical writing, whether they were using them to elaborate a medical theory *per se* or more broadly in what would later be called biological studies.

[12] Robert Boyle's matter theory has been examined by several authors. See, for example, Kaplan 1993, especially 56–68.

semina, which were studied and admired by Gassendi and Boyle, and the *atomi* of the ancients was their intrinsic temporality. *Semina* were not subject to development through the actions of external agents, but contained their efficient and final causes within them. This intrinsic temporality speaks to a fundamentally distinct way of looking at the world that hearks back to Hippocratic and Aristotelian conceptions of nature rather than looking forward to the soul-less conceptions of modern materialism, and yet carries within it the seeds of modern materialist-mechanist conceptions of change as development that is directed by an endogenous agency or programming.[13] Severinus, of course, did not use this modern-sounding term, embedded today in computer science, but rather the terms *scientia* and *lithurgia* (*liturgia*).

For Severinus, *scientia* was the Paracelsian conception of the knowledge of things that is embedded in them, which determine their characteristics, functions, and behaviors and which the adept physician could apprehend and use therapeutically. *Scientia* was the information that guided development, which we can liken to a program. This knowledge or information gave *semina* their ontological significance. But *scientia* is programming in a latent state, comparable to a program for a play that is handed out when those who have congregated enter the premises of the play – or a concert or a funerary service. The program is not the play's progress or the play *in actu*, but a plan for what might unfold in time, the temporal sequencing of development. The play itself is the acting out of this program or, at another level, of a script, which is the exercise or application of the *scientia*. Severinus evoked this idea of program in action with the term *liturgia*, which conjures in the reader's mind the ritualized, programmed performances of the Church.[14] The term 'programming', though anachronistic, thus aptly captures Severinus' conception of planned sequence of development both *in potentia* and *in actu*, which is fundamental to his development of *semina* in vital philosophy. The insistence on the indivisibility of potency and act evident in this formulation reflects the Paracelsians' understanding that diseases as causes and diseases as processes were inseparable and essential aspects of diseases, a point of contention with Galenists' distinction of the three contranaturals. The *scientia* and *lithurgia* of *semina* are what make nature vital and thus make Paracelsian philosophy more 'biological' than 'chemical' in modern terms. This point is worth making, inasmuch as Paracelsian theory is traditionally placed by post-Enlightenment historians of science in a narrative of the history of chemistry rather than history of biology, and this historiographical choice has underpinned an enduring misconception of the essentially 'medical' and 'biological' context of some of Paracelsus' most novel ideas.[15] Inasmuch as we form our initial assess-

[13] The difficulty that self-actualization, directed growth, and coordinated activities presented for mechanical reductionists in the modern period is a recurrent theme in Hall 1969.

[14] See Shackelford 2004, 178–80 for an analysis of Severinus' use of the terms *scientia* and *lithurgia* and the identification of the latter with *liturgia* by seventeenth-century readers.

[15] Despite Kurt Sprengel's placement of Paracelsus and the Paracelsians prominently in the narrative of the development of medicine in volume 3 of his mammoth history of Western medicine (Sprengel 1794), where he devoted 231 pages to Paracelsus and the Paracelsians, twentieth-century historiography of Paracelsus has been strongly influenced by Allen G. Debus, who situated Paracelsus as the impetus behind early-modern "chemical philosophy" in his influential two-vol-

ments of past ideas by comparing them to ones we understand, it is important to point to the 'biological' nature of Paracelsian thought as a corrective to the existing bias of approaching it in a 'chemical' context. Inasmuch as modern readers cannot wholly distance themselves from anachronistic terms and ideas, which are intrinsic to the process of understanding through analogy, it is important to examine these concepts in a thoughtful way.[16]

10.2 Form and Development

Embryology was for Aristotle a study of the generation or development of individuals over time, from their rudimentary beginnings to their mature, final states, and supported his concept of organic teleology. Embryological studies aptly confront and problematize the Scholastic concept of form as an essential component of hylomorphism: The developing chicken has a distinct form and matter at any particular moment, but this hylomorphic identity changes from day to day as the embryo grows and changes shape, aiming to realize the form of the adult chicken. Clearly the form of the total animal throughout its development, namely its species, must then be dynamic, and yet if it is continuous, it cannot be a random succession of momentary, static forms or stages of development. This is worth reflecting on, inasmuch as Bill Newman's recent attention to the metaphysics of mixture in chemical reactions, which underwent profound reinterpretation in the Scientific Revolution, gives the impression that early modern chemists understood the Aristotelian substantial form of reagents as if it is a static component of a hylomorphic substance, and that material change implied a discontinuity, as one static form vanished and

ume *The Chemical Philosophy* (Debus 1977) and several other books. The term "chemical philosophy" is early modern, but to modern ears it places Paracelsus in a chemical context. An exception is Walter Pagel (Pagel 1967, 336–37), who linked the vital philosophy of Paracelsus and Severinus to the ideas of William Harvey. In his later development of these ideas (Pagel 1976, 172) he appears to have understood the Paracelsians in just this light when he wrote that "parallels can be found between Harvey and early vitalists, notably among chemical philosophers such as Paracelsus, Van Helmont, Quercetanus and Fludd. Such parallels do not necessarily spell an 'influence' of one savant on the other, but are due to their common vitalist sources and persuasion." This perspective builds on Pagel's earlier suggestion that William Harvey may have been familiar with the support given to Aristotle's theory of *epigenesis* by Paracelsians Petrus Severinus and Johannes Marcus Marci of Kronland.

[16] The temptation to refer to Severinus' Paracelsian ideas as "medical" rather than "biological" merely exchanges one anachronism for another, unless the reader is careful to disassociate the modern conception of the scope, identity, and purposes of "medicine" from a sixteenth-century conception, which, for Severinus, included much of what we today would call agricultural biology and animal husbandry. Was Paracelsian thought "medical" rather than "biological"? Inasmuch as vital philosophy extended to plants and minerals, and even planets and stars, it was more than medical or even veterinary, in both the classical and the modern senses. The classical roots of "medicine" refer to healing, and clearly Paracelsian "medicine" was more broad than healing – it was vital philosophy, philosophy of life.

was replaced by another.[17] This portrayal is quite suitable when considering what happens when reagent A and reagent B are mixed and result in a product C with a new substantial identity, since the reagents are themselves treated as stable chemicals. The forms of substances A and B cease to exist and a new substantial form C is adduced in the product or imposed on it from outside. As Newman argues persuasively, experiments and analysis by Daniel Sennert and other seventeenth-century chemists, who could not reconcile the supposed destruction of reagent forms with the manifest recovery of constituent substances in reversible reactions, implied that the reagents' forms and thus their substances must persist in the product. The preservation of substantial forms through such reactions supported their understanding that durable, corpuscular chemical parts with substantial identities were hidden in the product and could be recovered in a reverse reaction.[18]

The scholastic Aristotelian conception of a substantial form as a static substance is suitable for explaining inorganic chemical reactions, but it is misleading when considering organic development. Aristotle's concept of form, an inextricable component of an individual that accounts for its resemblance to others of its species (in the modern sense), contains within it its range of possible development to its end or *telos*, sometimes termed *energeia*.[19] The medical or vital philosopher is confronted with the choice between viewing the developing individual as subject to a continuous succession of infinite forms, much like the perfect cinematic filmstrip that might be made of an infinite number of still frames running infinitely fast, or else defining the concept of form to include the range of development within a single substantial entity. The latter fits better with Aristotle's concept of form as the functional organization of a body throughout its lifespan.[20] I believe that it is just this problem that

[17] Newman 2006, various places, esp. 120: "This left the other alternative, that the gold and silver existed together as heterogeneities in a compound, and that their substantial forms had not really been supplanted by a *forma mixti*. To admit this, however, … was equivalent to saying that the gold and silver were composed of semipermanent particles having their own unchanged substantial forms." Newman does not explicitly identify these forms as essentially unchanging, but gives the impression that they are in this case by saying that they are unchanged and resulting in the particles' "semipermanence."

[18] On the importance of reversible reactions to Bill Newman's assessment of revolutionary changes in seventeenth-century matter theory, see his comments in Newman 2006, xiii: "My use of 'reversible reaction' has nothing to do with the modern notion of chemical equilibrium but rather with the alchemists' rebuttal of the strict Aristotelian concept of 'perfect mixture', according to which (at least in the eyes of the major scholastic schools of thought) there was no possibility of reversing the process that we now refer to as a chemical reaction." Also, 224: "The theory of perfect mixture and the concomitant denial of its reversibility were iconic features of a conventional scholasticism whose overthrow was genuinely epoch-making."

[19] Aristotle, *Metaphysics* IX 1050a-b (McKeon 1941, 830–31): "Where, then, the result is something apart from the exercise, the actuality is in the thing that is being made, e.g. the act of building is in the thing that is being built … Obviously, therefore, the substance or form is actuality"; "If, then, there are any entities or substances such as the dialecticians say the Ideas are, there must be something much more scientific than science-itself and something more mobile than movement-itself; for these will be more of the nature of actualities …"

[20] Nussbaum 1978, 71: "But in the case of living things, it is very clear that to explain behavior we must refer, not to surface configuration, but to *the functional organization that the individuals*

Paracelsus was confronting when he refused the Galenic distinction between the three contra-naturals, namely disease, its cause, and its *sequellæ*. For Paracelsus, these three were merely stages of development of a single entity, leading historians of medicine to distinguish Paracelsus' aetiology from Galenic aetiology by labeling the former an "ontological" concept and the latter a "physiological" one.[21] I will not comment on the aptness of this characterization here, but rather emphasize that Paracelsus' conception of disease as both cause and effect willy-nilly embraced the Aristotelian notion of the disease's formal cause containing its range of development or actualization within it. Understanding this is helpful for understanding Severinus' idea of seminal development, which is the essential foundation for his entire biological metaphysics and why it was taken seriously by a number of readers who otherwise rejected Paracelsus' philosophy.

10.3 Seeds as Endogenous Agents of Generation and Corruption

It was principally Severinus' doctrine of seminal development – the idea that material bodies sprout forth from *semina* or seeds – that drew comment from his readers.[22] The precise ontological status of these seeds is difficult to understand from Severinus' main publication, *The Ideal of Philosophical Medicine* (1571), in which he assigned them a liminal existence on the threshold of materiality – not really bodies but not really non-bodies. Following Walter Pagel's interpretation, I have previously emphasized that these *semina* were in essence *not* material entities, but rather were *loci* or places at which material bodies come into being through a Neoplatonic emanation of ideas into materiated individuals.[23] This reading prioritizes *semina* as seminal reasons, which is their historical root and true essence. However, the very fact that *semina* are *somewhere*, providentially disseminated in the material world, gives them a liminal materiality (at the very least in a

share with other members of their species. This is the form; this, and not the shape, remains the same as long as the creature is the same creature." My emphasis.

[21] Pagel 1982, 325 *et passim*; Niebyl 1971.

[22] The salience of Severinus' *semina* theory for late sixteenth and seventeenth-century readers is explained in Shackelford 2004.

[23] See Pagel 1967, 239–247. One is reminded of William Harvey's *De Motu Cordis* (1653), in Harvey 1995, p. 34: "In a Hen's egg I shewed the first beginning of the Chick, like a little cloud … in the midst of which cloud there was a point of blood which did beat, so little, that when it was contracted it disappeared, and vanish'd out of our sight, and in its dilation, shew'd it self again red and small, as the point of a needle; insomuch as *betwixt being seen, and not being seen, as it were betwixt being and not being it did represent a beating, and the beginning of life.*" Emphasis added. I have puzzled over the metaphysical implications for Harvey of the explicit image he presented of the embryo arising from an invisible seed: "Ita ut inter ipsum videri, et non videri quasi inter esse et non esse, palpitationem et vitæ principium ageret."

mathematical sense) that becomes evident when they sprout.[24] Contemplation of the Paracelsian and Severinian concept of transplantation and the temporal characteristics that Severinus attributed to *semina* further complicates their ontological status in his work and even calls into question their meaning for Paracelsus.

To begin with, it seems likely that Severinus used the term *semina* with full knowledge that Lucretius had used this name for an Epicurean atom and that Fracastoro had used the term to denote a material particle. Nevertheless, the term also evoked the notion of *rationes seminales*, the *logoi spermatikoi* of Neoplatonic and Stoic derivation, notable in medieval natural philosophy influenced by St. Augustine, a tradition that informed Paracelsus' ideas.[25] It makes sense, then, to posit the idea that Severinus developed his medical philosophy around the concept of *semina* with the understanding that it had both materialistic and formalistic nuances.[26]

The term *semina* refers in the first instance to the physical objects that, when sown into the earth, sprout when conditions are fitting and begin the process of vegetative growth. It is a biological or agricultural metaphor that any sixteenth-century farmer or gardener would grasp as readily as the student of Aristotelian metaphysics. For Severinus, the term had a much broader meaning. The seed is not merely a material vehicle for growth, but the special location at which a body begins to generate and will ultimately degenerate through corruption. It is the location that bears in it the programming for that generation and corruption. Generation is therefore a key function of *semina*:

> And so, we claim that generation is the progression of seeds, in which they go forth from their sources, abysses, and vital principles, onto the world stage, in an orderly unfolding of bodies and at determined times. By the renewal of individuals, they safeguard the continuity of species. In this process, the visible is made from the invisible, the corporeal from the incorporeal, by the power of the vital and immortal knowledge flourishing in all nature.[27]

[24] Giordano Bruno grappled with this idea of locating material atoms with no parts, and thus no extension, in *De minimo*. On Bruno's vitalist atomism, see Gatti 2001.

[25] Shackelford 2004, 175–76. Pagel 1982, 8–9, admitted that little is known of Paracelsus' early biography, but is persuaded by Kurt Goldammer's opinion that he was educated chiefly by his father and churchmen. This certainly explains the echoes of medieval scholastic natural philosophical and theological debates that one hears in his ideas, which often have more of a medieval character than that of Renaissance humanism. Webster 2008, 10–11, is more agnostic about Paracelsus' academic formation, suggesting that Paracelsus' early critics, who believed he had gathered his ideas from peasants, might have been more accurate.

[26] Indeed, I believe Severinus used this duality to place his conception of *semina* as bearers of intrinsically incorporeal developmental progress within a corpuscular framework and, more generally, to characterize his conception of Paracelsian "vital philosophy" as a reaction against materialist tendencies in natural philosophy and medicine.

[27] Severinus 1571, 131. Itaque generationem esse progressionem seminum asserimus: in qua, ex fontibus, abyssis, et vitalibus principiis ordinata corporum explicatione, in hanc mundanam scenam, definitis temporibus, progredientia. Individuorum renovatione, specierum perpetuitatem custodiunt. Fierique in hac lithurgia. Ex invisibilibus visibilia, ex incorporeis corporea, potestate vitalis immortalisque scientiæ in universa natura vigentis.

We can see from this point that Severinus' basic conception of the generation of all bodies from *semina* is fundamentally Aristotelian in conception, modeled on formal cause guiding efficient epigenetic embryological development acting on a material substrate. Drawing on Paracelsus, Severinus plainly referred to the material elements of this substrate as the wombs or *matrices* in which the *semina* are lodged, making the embryological context explicit. However, he embedded this idea in a fundamentally Neoplatonist worldview, where the *semina* are themselves rudimentary units containing a seminal form or idea, and the bodies that develop from them are therefore material emanations from the seminal plans or reasons that were planted into the physical world at the original creation, as manifestations of the divine mind. The seed as metaphor captures a moment in the developmental cycle when an entity exists in a state of potency, waiting to spring into actuality with the passage of time:

> All things seem to be made by means of seeds, and in them are contained the regulation of all creation. … For whatever alterations, mutations, and motions might appear, do not exist, except as a flowing of seeds showing distinct changes through the destructions, renewals, and transplantations of bodies.[28]

Given appropriate environmental conditions, the seed sprouts at a pre-ordained time and proceeds through a developmental pattern of maturity, decay, and eventual oblivion, which Severinus interpreted as a return to latency, chaos, or abyss. It is a powerful metaphor that embraces the process by which species renew themselves, the teleological order of Aristotelian nature, and the Neoplatonic cycle of bodies going forth from the dark night of non-being to the efflorescence of being and eventual return to non-being.[29]

From this definition it is obvious that Severinus meant the concept of seed to apply not merely to visible, palpable vegetable husks and kernels, but also to invisible sources of generation for all animal, vegetable, and mineral bodies. Severinus used *semina* in his book to explain sexual, a-sexual, and spontaneous generation and even extended the concept to the generation of diseases from *semina morborum* – the seeds of diseases. Note that in the above quotation Severinus said that these

[28] Ibid., 79–80, in chapter eight, *De generatione rerum naturalium et seminum mechanica lithurgia.* "Itaque omnia videntur propter semina facta esse, et in iis contineri totius Creaturæ dispensationem, ut antea quoque admonivimus. Quæcunque enim hic alterationes, mutationes, motus apparent, non sunt nisi fluxus seminum, corporum consumtionibus, restaurationibus, transplantationibus, vicissitudines definitas ostendentium."

[29] Ibid. "Hoc est: 'Nihil fit quod prius non erat, neque quicquam perit, sed permixta et discreta alterantur. Homines vero opinantur hoc ex Orco in lucem auctum generari, illud vero ex luce in Orcum imminutum perire ac corrumpi.' … Ita Orpheus quoque et prisci Theologi consueverunt, Tenebras, Noctem, Quietem, Orcum, eadem significatione accipientes." [That is [quoting and translating Hippocrates' Greek]: 'nothing comes into being that did not exist before, and nothing perishes, but mixed and separated, things are changed. But in fact, men supposed that the increased thing is brought forth from death into light, that the diminished passes away from light into death and is destroyed.' … So Orpheus, too, and the *prisci theologi* were accustomed, understanding darkness, night, sleep, and death in that same sense.]

seminal processes show distinct changes through the transplantation of bodies. These generations, as processes, were also subject to transplantation.

Severinus developed the concept of transplantation in the context of his theory of generation and corruption, which is the metaphysical core of vital philosophy, namely an explanation for change in the organic, material world. He regarded his theory as an elaboration of Paracelsus' ideas about living things, which were the basis of his medical revolution:

> Theophrastus Paracelsus has changed the whole of medicine in our time. Explaining the natures and circumstances of the elements in a far different manner, he proposed the doctrine of generation and transplantation clearly and fully. Recalling the methods of vital astronomy, he rejected the family of humors and attributed the generation and transplantation of diseases to other mechanisms.[30]

This statement, which appears in the first few pages of Severinus' *Ideal of Philosophical Medicine*, alerts the reader to the far reaching consequences of this medical reform: Not only did Paracelsus replace the basic tenets of Galenic pathology and therapeutics, but he restaged medicine on a new theoretical foundation, a new theory of generation and transplantation that is connected with something called vital astronomy.

The term vital astronomy gives the reader pause. What does astronomy have to do with organic processes? Medieval medicine took the influence of the planets and stars on human health and disease for granted, and Marsilio Ficino brought this idea into mainstream Renaissance Platonism in his *De triplici vita*.[31] But a directly applicable explanation can be apprehended in Severinus' conception of *astra* – literally stars – as entities existing throughout the cosmos, to be found in celestial as well as in terrestrial matter and thus within us, too. Vital astronomy, then, is basically another term for vital philosophy, but one that implicates Paracelsus' ideas about human generation from celestial dust as well as from terrestrial *limbus* in a broadly theologized biology.[32]

10.4 *Semina* and *astra*

Severinus used the terms *semina* and *astra* promiscuously in his book, emphasizing the astral nature of *semina*. They explained the appearances of things in time – the movement from potential to actual that defined organic movement (change) in the sublunary world for Aristotle. The continuity of cosmic nature implied by this view

---

[30] Ibid: 4. "Horum vestigiis insistens Theophrastus ille Paracelsus, nostris temporibus universam Medicinam immutavit. Elementorum naturas et conditiones longe diverso modo exponens, Generationum ac Transplantationum doctrinam cumulate clareque proposuit. Astronomiæ vitalis Methodum revocans morborum Generationes et Transplantationes aliis Mechanicis ascripsit, repudiata humorum familia."

[31] Mebane 1989 lucidly puts Ficino's astrological vision into a broad context of Renaissance Platonist thought.

[32] On Paracelsus' anthropology see Daniel 2002; Rudolph 1998.

was ably caricatured by Severinus' one-time friend Tycho Brahe, who sent a pair of emblematic images in Renaissance neo-classical style to his friend Falche Gøye in the late 1580s, which he later used in his published books. These represent complementary sciences that Tycho termed "celestial astronomy" and "terrestrial astronomy," but inspection of the latter, which features a chemical laboratory beneath an idealized Greek man bearing a handful of herbs and the snake Asklepios, identifies "terrestrial astronomy" as chemical medicine.[33] Tycho Brahe, the Renaissance man, sought to know the *astra* of both places, observing their behaviors above with his various astronomical devices and their behaviors below via chemical separations in his cellar laboratory. The correspondence between these two was affirmed by the Emerald Tablet of Hermes, a familiar text held to be replete with alchemical wisdom.[34] According to Severinus, these *astra* become manifest as actual visible stars only in the celestial region, owing to conditions there, which enable them to develop fully; elsewhere their development stops short of perfection, limited by the elemental matter that nourishes their growth and narrows the scope of their maturation.[35] Severinus thought that the causes of things throughout nature are in this sense astral, but that early philosophers had erroneously assigned such causation only to celestial stars because they did not understand that terrestrial ones also exist.[36] About these terrestrial *astra* with their seminal virtues he wrote: "From these proceed actions. In

[33] On these images and Tycho Brahe's involvement with alchemy and his relationship with Severinus, see Shackelford 1993.

[34] On the Emerald Tablet, see Ruska 1926.

[35] Severinus 1571, 96. The temporal manifestations of seeds are in part affected by the elemental bodies they take on; terrestrial ones are slowed down by their earthly matter, but even among these there are variations determined by the seminal programming: "In ipsis quoque procreationibus terrenis, ratione seminum, quædam celeres habent periodos ad Exaltationem et insequentem declinationem subito festinant. Aliæ longioribus intervallis circuitus absoluunt: habent enim, ut magnitudines, ita durationis quoque innatam legem" [In those very terrestrial procreations, too, certain [things] have swifter periods, by reason of [their] seeds; they suddenly hasten to exaltation and ensuing decline. Others complete [their] cycles in longer intervals, for they have an innate law of duration as well as of magnitude.] It is worth noting that the limitations that material nature placed on divine seminal expression as development, which Severinus built into his Paracelsian biology, is reminiscent of Galen's insistence that the divine creator's power is not total, but rather limited by material nature, and was therefore an idea that was readily available to students of medicine. For a discussion of Galen's ideas on omnipotence, see French 1994, 158–59.

[36] Severinus 1571, 52–53: "Quot, quanti, Philosophi, totam Astronomiam soli Firmamento ascripserunt: reliquorum Elementorum vicissitudines, non ab innatis astris, sed a coelestibus absolui prodiderunt: et, si Diis placet, etiam seminarias virtutes et foecunditates, ex coelo inconsiderate deduxerunt. At certe, quæcunque in coelo explicata conspiciuntur, in coeteris quoque Elementis, virtute et vitali potestate continentur." [What numerous and even great philosophers have attributed the whole of astronomy to the firmament alone! They have reported that the vicissitudes of the rest of the elements are carried out not by innate stars, but by celestial ones. And, if you will, they have even rashly deduced seminal virtues and fertilities from the heavens. But surely, whatever things are perceived to be unfolded in the heavens are also contained in the rest of the elements, by a virtue and vital power.]

these the knowledge and mechanical processes of things flourish. In these the determined moments of timing are kept."[37]

This sentence conveys two basic concepts that Severinus attributed to these "astral" *semina*, namely that they contain knowledge of some sort (*scientia*) and command something he variously called *lithurgia mechanica*, or *mechanicus processus*, which I have translated simply as mechanical process. These fascinating terms go right to the heart of Severinus' *semina*-theory and how he thought nature was constructed. I have explained what I think Severinus meant by mechanical liturgy or mechanical process in detail elsewhere.[38] But only recently have I begun to understand what the Paracelsian biologist meant by the phrase "in these the determined moments of timing are kept." Even my choice to translate *temporum decreta momenta* and similar phrases as 'timings' or 'moments of timing' struck me as speculative, inasmuch as I was unsure what concept underlay these terms and phrases, since I had little experience with anything like this in early medicine and had no context in which to understand it.[39] I have come to suspect that Severinus was working with a conception of the mechanical that was quite different from the one that was reified in the Mechanical Philosophy of the late seventeenth century[40]; that he was groping toward an expression of an inner temporality in things that grasped time and motion in a sense that was not wholly Aristotelian, but certainly was unlike temporality in the post-Cartesian meaning that was assimilated in Enlightenment thought.

The triumph of Cartesian philosophy and other philosophical variations encompassed by the term "mechanical philosophy" imposed a sense of temporality that privileged local motion, that associated time with the measurement of motion within a dimensional space. This mechanization of the world picture displaced the notion that motion is fundamentally internal, formal change – teleological change – which had characterized Aristotelian metaphysics and natural philosophy.[41] Indeed, histo-

[37] Ibid., 53. Following hard upon the previous quotation: "A quibus [innatis astris] actiones proficiscuntur: in quibus Scientiæ mechanicique rerum processus vigent: in quibus temporum decreta momenta custodiuntur."

[38] For a discussion of the meanings of these terms and how semina function in Severinus' biological philosophy, see Shackelford 2004, 176–85.

[39] Severinus' *Idea medicinæ philosophicæ* (Severinus 1571) is replete with such phrases, for example: predestined timings as characteristics compared to movements, "motuumque leges et temporum prædestinationes" (51), fixed or determined moments, "temporum decreta momenta (53), specific intervals of times or timings, "certis temporum intervallis" (54) and "definito temporum intervallo" (56), periods or timings of rest and work, "quietis et laborum tempora" (57), inborn laws of timings "contra innatas temporum leges" (280), timings expressing seasonality, "pro natura seminum, tempus morborum, vel vernum, vel autumnale, vel hybernum, vel solstitiale obtinebunt" (288), timings as one characteristic of the celestial stars, "visibilium astrorum distinctas proprietates, officia, tempora" (288–89), and as internal characteristics of bodies, "in quibus temporum, saporum, odorum, colorum, caliditatis, frigiditatis, coagulationis, sed non in mundanæ, figuræ, sed non externæ, proprietates constantes custodiuntur" (319–20).

[40] I developed this idea in Shackelford 1998 and further in Shackelford 2007.

[41] Hall 1969, v. 2, 67 hints at this when he writes that "What happened in the eighteenth century was, in part, a continuation of the Galenic mode of interpretation but with the difference that the

rians of science tell us that it was the main intention of the "mechanical philosophers" to eliminate these elements of Aristotelian explanation of material transformation.[42] The subsequent history of vitalism in the European Enlightenment can be in a coarse generalization viewed as scholars' attempts to account for phenomena that did not conform to the mechanical paradigm or did so only with great violence to the integrity of the mechanical models that were conceived. Such phenomena included the apparent ability of some kinds of matter to self-organize – phenomena of generation and growth – as well as tissue irritability and sensitivity and the coordination of processes that are evident in complex organisms. Such phenomena were amenable to the postulation of a supervening animating principle or soul or to the hypothesis of an inner vital agency in material organs. In all cases, it was the seeming capacity for interior motion, development in time, that was under consideration.[43]

The essential temporal nature of *semina*, by which I mean that *semina* contain characteristic timings for development as well as more familiar morphological properties such as size, color, flavor, etc., is borne out in Severinus' explanations of generation and transplantation. Transplantation is of special interest, not only because, when applied to *semina*, it implies their existence as subtle material bodies, i.e. corpuscles, but also because seminal programming (*rationes seminales*) is evident in the transplantation of behaviors.

10.5 Transplantations, Biological Variation, and the Appearance of "New" Forms

Severinus' description of mechanical liturgy, the process by which the *semina* carry out their developmental programming, makes it clear that he regarded transplantation as one aspect of the seminal developmental process at work in the quotidian operations of nature: "And so by their help the world province is governed; the

latent equivalents [faculties, etc.] were now generally regarded not as Aristotelian-Galenic 'qualitative movements' but, usually, as insensible contractions, vibrations, or other displacements of the microconstituents … of the living system. This shift was undoubtedly in part an outgrowth of the seventeenth-century transition to a predominantly mechanical … world picture." That is, the concept of motion as broadly-conceived change shifted from qualitative alteration to local motion with the formation of mechanical philosophy in the Scientific Revolution, which explained changes in quality, like quality itself, in terms of local motions of and collisions of corpuscles.

[42] Newman 2006, 153 locates the transition from reliance on Aristotelian substantial forms to explain change to the local motion of corpuscles in the work of Daniel Sennert, who himself did not quite abandon the concept of substantial form, but deprived it of its explanatory power; Robert Boyle and posterity would complete the job – see Newman and Principe 2002.

[43] Consider some of the Cartesian mechanisms, such as his explanation for terrestrial magnetism, that even contemporaries found implausible. For discussion of Descartes' images and their place in his projection of natural philosophy, see Lüthy 2006, esp. 120–26. On the enduring legacy of vitalism in post-Cartesian physiology, see Hall 1969.

predestinations of timings, the laws of motion, the constant processes of generation and transplantation are carried out."[44] While programmed generation implied a rule-bound, determined development, transplantation accounted for the variability observed in nature. It is a process that supervenes on normal generation.[45] A seed can in actuality develop into something other than what its mechanical knowledge programmed it to become, owing to contingent elemental and astral conditions: "If the [astral] impressions will have been more moderate, the places not wholly hostile, then when the hope of generation has been abandoned, nature will undertake a plan of transplantation, and what it did not wish will become, as Hippocrates would say."[46] Transplantation explained pathological deviations from normal development, but also spontaneous generation, a-sexual reproduction, which was widely practice in horticulture, and hybridization.

Severinus clearly thought about transplantation in terms of fundamental material transformation, which could happen at the phenotypic level of the individual body, without affecting its core identity, or it could reach into the fundamental genetic type:

> I call the transplantations of individuals, those in which the characteristics are altered, while the root remains, such as colors, flavors, magnitudes, shapes, etc., however, so that the authority still remains in the root and it can reproduce certain signs of its family, although vehemently altered. … But I understand the transplantations of species to be when all the characteristics have been changed, and the distinguishing marks of a new family are manifestly reproduced.[47]

Described in anachronistic terms, this passage is readily understood in a modern biological context as an early statement about observed heritability of characteristics. But when viewed within the sixteenth-century context of Paracelsian

[44] Severinus 1571, 48–9: "Horum itaque ministerio mundana provincia administratur, temporum prædestinationes, motuum leges, generationum et transplantationum constantes lithurgiæ absolvuntur."

[45] Ibid., Chap. 10 ("De Generatione humana, et Transplantationibus Generationi supervenientibus") applies this idea of transplantation specifically to human generation.

[46] Ibid., 99. "Si vero moderatiores fuerint impressiones, loca non plane inimica, Generationis relicta spe, Transplantationis consilium aggredietur Natura, fietque quod non vult, ut cum Hippocrate loquamur." Severinus did not quote or cite Hippocrates here, but his comment refers back to his earlier statement (92), "Validioribus vero tincturis confluentibus, mixtione signaturæ mutabuntur, fietque quod non volunt, seminum generatione in sterilem transplantationem conversa." His analysis appears to be based on his Paracelsian reading of *Regimen I.v,14-17* (Hippocrates 1931, 236–37), which he quoted in Greek and translated (Severinus 1571, 92) "Hoc est: Omnibus inter se ultro citroque commeantibus et conspirantibus unumquodque destinatum fatum explore limitibus suæ naturæ custoditis; fierique a seminibus Omnia per divinam necessitate asserit, et quæ volunt et quæ non volunt." It therefore appears that the account of generation in *Regimen I,* read through the lens of the Paracelsian *Philosophia ad Athenienses*, provided an influential non-Aristotelian classical legitimation for Severinus' vital philosophy.

[47] Ibid., 140. "Individuorum Transplantationes dico, in quibus manente Radice signaturæ mutantur, ut colores, sapores, magnitudines, figuræ, etc. ita tamen ut maneat etiamnum imperium. Radicis et familiæ suæ certa signa, quamvis vehementer immutata referre possit. … Specierum vero Transplantationes intelligo, ubi omnibus signaturis immutatis novæ familiæ insignia manifeste repræsentantur."

philosophy, which had a much broader, hylozoic conception of "organic" that extended to planets and minerals, Severinus' explanation clearly speaks to matter theory more generally.

Severinus adduced examples of transplantations in plants, animals, minerals, and humans, implicating the influence of the stars in these. Farmers know that despite their efforts to select wheat seed to breed true for next year's planting, darnel can appear in the field, as if spontaneously. This is because the wheat seed contains within it the potential to become transplanted to darnel. Likewise, mustard seed can produce mint, turnip seed radishes, and so on.[48] As long as the seminal power within the seed remains strong, the seed will breed true, but if it becomes weak, it will undergo transplantation and be taken over by external influences, such as come from the stars.[49] Transplantations of this sort are quite common in nature and also employed intentionally by breeders to alter the characteristics of things.[50]

Animals that are characterized by distinct sexes (perfected animals) are less liable to transplantations than lower forms, owing to the perfection and strength of their *semina*.[51] But even here, transplantations can occur when the seeds are weak

[48] Ibid., 141: "Thus we have observed that mustard has degenerated into mint, the turnip into the radish, imperatoria into angelica, wheat into darnel, basil into thyme, and many transplantations of this kind. In all these the seeds seem equivocal, which can be the principles of many differing individuals with almost all properties. Thus in the seed of wheat, the form of darnel lies hidden, but as an attendant, equivocal and accidental." [Sic Sysimbrium in Mentam, Rapam in Raphanum, Imperatoriam in Angelicam, Triticum in Lolium, Ocimum in Serpyllum degenerasse et multas huiusmodi Transplantationes conspeximus. In quibus omnibus semina videntur æquivoca, quæ multorum Individuorum fere omnibus proprietatibus dissidentium Principia esse possunt. Sic in semine Tritici Forma Lolii delitescit, sed ministra, æquivoca, accidens.]

[49] Ibid, 141–42: "In the same way the rudiments of mint lie concealed and be subordinate in sysimbrium, as long as the elements and principles of mustard maintain their authority by a strong and vital power. But if, with the rays of other stars flowing together, the beginnings of mint will have obtained powerful increases, then the mint will dominate, and mustard will be subordinate. The situation is similar in the turnip and radish, and all the rest." [Eodemque modo in Sysimbrio Menthæ rudimenta latent ministrantque, quousque Sysymbrii Elementa, Principia, imperium robore vitalique potestate obtinuerint. Quod si aliorum Astrorum radijs confluentibus, Menthæ exordia incrementa sumserint robusta, dominabitur Mentha, servietque sysymbrium. In Rapa et Raphano similiter, et cæteris omnibus.]

[50] Ibid., 142: "Transplantations of individuals in the vegetables are so frequent that philosophers have taken the use of the term from this. Thus wild things are tamed, flavors are made mild, colors are altered, the times of maturation are accelerated or retarded, and all nature submits itself to the judgment of mortals, and refreshes us with a useful and delightful variety." [Individuorum Transplantationes in Vegetabilibus usqueadeo sunt frequentes, uthinc nominis usarum receperint Philosophi. Ita cicurantur agrestes, mitigantur sapores, mutantur colores, maturitatis tempora accelerantur vel retardantur, totaque Natura arbitriis Mortalium se submittens, utili delectabilique varietate nos recreat.]

[51] Ibid., 139: "In the category of animals, the perfect, which are distinguished by the distinction of their sexes, admit transplantation with difficulty, and not all of them do this: those in which there is a great affinity between their seeds and nature, and that not unless the seeds have been mixed, such as wolves and dogs, horses and asses, partridges and chickens, etc. Animals of this sort have roots connected in nature's great society." [In genere Animalium perfecta quæ sexuum distinctione separata sunt difficulter Transplantationem admittunt: et hoc non omnia: solummodo illa in quibus

and mixed, since mixing can be attended by transplantation, the subject of Chap. 9 of his book.[52] When this happens, hybrids arise:

> Certain kinds of animals are generated not from their own seed, but from the confused and mixed up seeds of diverse animals, which, however, harmonize in the society of nature, and have the impulse to reproduce at the same times, such as the horse and the ass, which produce the mule, dogs and wolves, and partridges and chickens. These matters are more fully treated in the doctrine of transplantations, which we will soon add.[53]

Among the less perfected animals, those which do not clearly reproduce sexually, transplantations are more frequent. The *semina* of insects and other lower forms contain the programming for many species, making transplantation more likely:

> For insects and things similar to them admit frequent transplantations, and warm foreign seeds in their tents. Paracelsus often speaks in this manner. For he [Hippocrates][54] says that from the same aliment the dog produces canine substance, flesh, seed, root, [but] the man produces human substance, the pig produces porcine substance, and the lion produces leonine substance, on account of the power of the mechanical spirits, which can furnish all aliments with characteristics.[55]

From this it is clear that Severinus built his conception of generation and development on Aristotelian epigenesis, which was supported by Hippocratic theory of nutrition and growth, despite the fact that he made a point of saying that this process is not as explained by Aristotle and Galen; that seminal development (involving generation and transplantation) is not caused by a mixing of the four Aristotelian elements or the corresponding four primary qualities.[56] The same process by which

---

magna est seminum et Naturæ affinitas, idque non nisi seminibus permixtis, ut lupi, canes: equi, asinæ: perdices, gallinæ, etc. Huiusmodi Animalia Radices habent magna Naturæ societate coniunctas.]

[52] Ibid., Chapter Nine: "De Mixtione, et huius comite Transplantatione."

[53] Ibid., 109. "Generantur et Animalia quædam non ex proprio semine, sed ex seminibus diversorum Animalium confusis commixtisque, quæ tamen naturæ societate conspirantia, iisdem temporibus ad Generationem impetus habent, ut equus et asina mulum, canes et lupi, perdices et gallinæ: de quibus commodius in Transplantationum doctrina, quam mox subiungemus, agetur."

[54] The logical antecedent is Paracelsus, but earlier and also just below this passage we see "him" expressing himself in Greek, suggesting that "he" is Hippocrates, whom Severinus occasionally quotes in Greek in the *Idea medicinæ philosophicæ*, and that Paracelsus is mentioned as an aside.

[55] Ibid.,154. "Insecta enim et his finitima frequentes admittunt Transplantationes, ac peregrina semina in suis tentoriis fovent. Ad hunc modum sæpe loquitur Paracelsus. Dicit enim ex eodem alimento canem caninam producere naturam, carnem, semen, radicem: hominem humanam, suem suillam, leonem leoninam, ob potestatem Mechanicorum spirituum, qui alimenta omnia suis signaturis ornare possunt." Also 139–40: "In the class of the insects, and the rest of the animals in which no evident distinction of the sexes appears, transplantations occur more frequently and more readily. For their seminal principles, furnished with the knowledge of many species, admit equivocal generations." [In genere Insectorum, coeterorumque Animalium, in quibus sexuum discrimina manifesta non apparent, frequentiores et faciliores contingunt Transplantationes. Seminalia enim eorum Principia, multarum specierum Scientia instructa, æquivocas Generationes admittunt.]

[56] Ibid., 45: "But, they will say, the four qualities carry out generations by means of mixing and meeting. Right; those things that are mixed and compounded produce generations and fruits. But we firmly assert that the elements, which are the present concern, are not mixed. Indeed it has been

an organic body assimilates diverse nutrients and transforms them into specific kinds of flesh (i.e. proper to its species) accounts for transplanted developmental processes; the new seminal program uses this process of assimilation to carry out its function by epigenesis and thus produces a "new" material body instead of the one expected. We can extrapolate from this to the even less perfected forms of animal life: invisible seeds sprouting in diverse material fields (e.g. a cow's carcass) appear to generate animals such as bees and maggots spontaneously.

Minerals also are subject to transplantation, giving rise to variant kinds, most obvious in the case of gemstones, which have their roots in the metals. Through transplantation, lead can become steel, silver can become sapphire, copper can become emerald, and so on.[57] And, much as horticulturalists make use of transplantation to produce useful variations, the adept alchemist can take advantage of these wonderful powers of nature to produce salts, mineral waters, sulphurs, and other chemical products by means of artifice, as Paracelsus taught.[58] That Paracelsus used the term transplantation in this and other instances, where one might expect the more familiar alchemical term transmutation, might be attributed to the quirkiness of his writing style and penchant for peculiar neologisms, but it could also be the case that he intentionally avoided the term transmutation, which was so closely bound up with scholastic Aristotelian metaphysics and rife with sacramental implications. Transplantation was much more agricultural, more biological. Such anthropological and Christological nuances would have been even more apparent when applied to human physiology and especially pathological processes.[59]

---

accepted on the authority of the philosophers that the elements are mixed. But yet when they examine the sources of generation, the development of seeds, and the ways of transplantation, they make use of the other elements, and they do not at all reach those first and greatest foundations of nature." [At dicent, Mistione et congressu generationes absolvent. Recte: quæ miscentur et componuntur, generationes et fructus producunt: Elementa vero, de quibus nunc agimus, constanter asseveramus non misceri. Philosophorum quidem authoritate receptum est, Elementa misceri: at vero Generationum fontes, Seminum progressiones, et Transplantationum modos, dum scrutantur, aliorum Elementorum ministerio utuntur, prima illa et summa Naturæ fundamenta nequaquam attingunt.] Pagel 1967 noted Severinus' Aristotelian conception of epigenesis, where he used it to illuminate the intellectual background to Harvey's embryology.

[57] Severinus 1571, 143: "Indeed, gems are produced by specific transplantation, the prime and common root of the metals having been occupied by more powerful tinctures, and that happens when the individual characteristics of the metals have been received before. Thus, steel is transplanted from the roots of lead, sapphire from the roots of silver, emerald from the roots of copper, beryl from the roots of iron, etc." [Gemmæ etenim specifica Transplantatione producuntur, validioribus Tincturis occupata prima communique Metallorum Radice idque receptis antea Individuis Metallorum signaturis. Ita ex radicibus plumbi transplantatur Adamas, ex radicibus Lunæ saphyrus, ex radicibus Veneris smaragdus, ex radice Martis Berillus, etc.]

[58] Ibid., 143: "Individuæ uero Transplantationes Metallorum, ubique conspiciuntur in Metallis, Marchasitis, Sulphuribus, Salibus, Thermis. Quinetiam artificiosæ Transplantationes Salium, ut aluminis, vitrioli, salis nitri, ammoniaci, Sulphurum, antimonialium, cupri, ferri, plumbi, stanni, naturæ mirabilem potestatem satis ostendunt. De quibus copiose et docte in multis locis disserit Paracelsus."

[59] Redondi 1987, 203–226. The corpuscular hypothesis, like godless atomism, projected the idea that formal characteristics were essential manifestations of material identity and therefore did not

10.6 Transplantation in Human Generation and Pathology

Severinus described generation and transplantation most expansively in Chaps. 10 and 12, where he applied *semina* theory to human biology and pathology. While one might have expected a fuller exploration of monstrous births as an example of mixing *semina* of different kinds and transplantations resulting from impressions, such as one sees on display in Ambroise Paré's work, Severinus' thematic development is oriented more toward internal medicine and physiology, perhaps reflecting his education as a physician rather than as a surgeon.[60] He mainly discusses transplantation in the context of internal pathology and used it to explain inherent diseases as well as chronic and acute diseases.

Disease is closely connected with decay, itself an aspect of material transformations and thus within the general scope of generation and corruption. Severinus noted that some kinds of things are more readily subject to decay than others, owing to a constitution that is more mercurial than saline.[61] This explanation clearly reflects Paracelsian matter theory, which embodied the experience that salts tended to preserve organic matter from natural decay.

All beings, whether plants, animals, or minerals, are subject to decay, but the temporal nature of that process is specific to kind. That is, each species has its own natural temporal pattern of development and hence a natural life span. However, an animal is more vulnerable than a plant or mineral to diseases, which result in premature decay, owing to its consumption of both minerals and plants, which incorporates their liabilities into its own. Humans, eating from all three kingdoms, are especially exposed. This is because with the diverse consumed materials come their

support the idea that the Eucharistic host could actually be Christ while retaining the characteristics of the wafer. Thomas Aquinas' interpretation of Aristotelian substantial form obviated this difficulty and became approved doctrine. Paracelsus was writing during the thick of the confessional controversies in German lands and was himself called in for questioning in connection with popular rebellion in 1525, so it is unlikely that he was unaware of the importance of terminology and ideas associated with transubstantiation, even if he may have been unaware of its relation to debates over the plurality of forms and suppression of corpuscular alchemy in thirteenth-century Europe.

[60] The surgeon Ambroise Paré devoted a monograph to monstrous births and other portents, Paré 1982 (original French edition 1573), in which he explained monstrous animal births as influenced by the relative strengths (and origins) of the parental seed as well as the effects of visual and celestial impressions, but Severinus' *Ideal of Philosophical Medicine*, as its title suggests, belongs to a more academic medical genre. The idea that characteristics of the child diversely resemble those of the mother or father, arising from the relative strengths of the parents' seed, was commonplace in medieval Galenic medical theory, but Paré extends this to the mixing of the seed of different species to produce monsters (e.g. human crosses with farm animals), perhaps drawing on folk traditions that may also have inspired the Paracelsian ideas about seminal mixing.

[61] Severinus 1571, 219: "In quibus vero impuritates Mercuriales non sunt ita copiosæ, et salium robustior natura, ἄβαυσιν uel cariem expectant, ut arbores plurimæ fruticesque nonnullæ." [In those [creatures] in which mercurial impurities are not so abundant, and the nature of the salts is stronger, they await "abausin" or decay, as do most trees and some fruits.]

semina and consequently new physical (corporeal) and temporal characteristics that might become transplanted to the host's anatomy:

> When seeds have been transplanted and newly introduced, so too are their characteristics, colors, flavors, hotnesses, coldnesses, et cetera. Thus parts are rendered discolored, fouled with strange flavors, intemperate, slow, dull, immobile, or are contaminated by disagreeing changes and actions. These are all symptoms and fruits of seeds causing transplantation.[62]

The tendency to decay earlier can in this way become transplanted from the body of ingested material into the *mumia* or core essential matter (the root) of the animal and alter its nature.[63]

Contagious (acute) diseases are by their very nature short lived and therefore less challenging to the physician than are chronic diseases, especially those deemed hereditary. Paracelsian physicians were notorious for taking on cases that were deemed incurable by Galenic practitioners, so it is not surprising that Severinus showed particular interest in using his theory to explain hereditary and chronic diseases and thus provide guidance to curing them. Applying his general biological theory, Severinus regarded such diseases as particularly intractable because they are well rooted in the depth of the organic being. Fortunately, the physician who is well-informed about this theory can both expel disease-seeds using powerful purgatives, diuretics, and sudorifics, or else he can use the power of transplantation to correct seminal developments, by employing medications with powerful tinctures to effect a new transplantation and in effect reprogram an organic function, restoring health: "And when the torments of symptoms have been removed, a friendly peace has been returned to nature, and the general cure will be seen to overcome the hostile impurities by means of the strengthening of the balsam, soothing, and peaceful transplantation."[64] The main message, however, is that the physician must be knowledgeable about vital philosophy, because the transplantations that bring about diseases are similar in nature to biological transplantations of all sorts.[65]

---

[62] Ibid., 245. "Transplantatis seminibus novisque introductis, transplantantur quoque signaturæ, colores sapores, caliditates, frigiditates, etc. Ita discoloratæ, peregrinis saporibus inquinatæ, intemperatæ, segnes, stupidæ, immobiles redduntur partes, vel dissentaneis motibus et actionibus contaminantur. Suntque hæc omnia symptomata et fructus seminum Transplantationem causantium."

[63] Ibid., 219: "In the root or mummy of animals, because they are nourished by the resolutions of vegetables and minerals, similar impurities are found, yet differing according to the nature of the fields [in which they lodge], for transplantation is made from the province of vegetable balsam into the animal republic, consequently they [animals] experience similar kinds of destruction and death." [In radice vel Mumia animalium, quia Vegetabilibus et Mineralium resolutionibus nutriuntur, similes impuritates reperiuntur differentes tamen secundum agrorum naturam: facta est enim Transplantatio ex provincia Balsami vegetabilis, in Rempub.[licam] animalis: dissolutiones proinde et mortes similes experiuntur.]

[64] Ibid., 354. "… et ablatis symptomatum cruciatibus, amica quies naturæ reddebatur, videbaturque Universalis curatio corroboratione Balsami, mitigatione, et quieta Transplantatione, hostiles impuritates superare."

[65] Ibid., 412: "Transplantation happens to generation by means of an intervening [interveniente] mixture, and it is common to the orders of all generations, whether the properties of the individual substance [naturæ] are altered by slighter tinctures, or with stronger [tinctures] the transplanted characteristics of species demonstrate signs of a new family." [Transplantatio, Generationi, inter-

Recall that Severinus mentioned before that he regarded transplantations as occurring to individuals and also to species, depending on whether they became heritable. What this amounts to is a kind of theory of the inheritance of acquired characteristics, an idea he elaborated in connection with diseases:

> For whatever is contained in the parents, whether healthy or diseased, which can affect the balsam and root of the human nature by means of a powerful and vital impression, becomes hereditary by transplantation. Impressions of this sort are plainly astral, incorporeal, spiritual, and invisible, but imitating the efficient natures of the seminal reasons, developing by the virtue of [their] knowledge and predestinations, in which a corporeal mass is not needed for the formation of bodies.[66]

In this particular example that Severinus presents, the ontological liminality of *semina* and the process of transplantation shine forth in the description of impressions: The impressions are the means by which seminal programming is transferred from the parent or other agent to whatever they impress, in which this programming then formatively acts. Like the seminal reasons or mechanical knowledge they imitate, these impressions are not corporeal and therefore not material, but they are connected with subtly material *semina*, which enables them to be spatially relocated, i.e. transplanted.[67]

10.7 Semina as the Principles (Rudiments) of Vital Corpuscles

Returning to the problem of defining the ontological status of Severinus' *semina*, that is, whether they are formal or material, and whether we can consider them to be corpuscular in any sense, we must balance what Severinus wrote about how they behave, namely their role in the generation of bodies from seminal reasons lodged in elementary matrices – a plainly Aristotelian embryology – with the liminal material status he ascribed to them and what he wrote about the role of transplantation in organic development. The fact that Severinus and Paracelsus pointedly used the term transplantation to mean a process distinct from transmutation or

veniente Mixtione accidit, communisque est Generationum omnium ordinibus, sive levioribus Tincturis, Individuæ naturæ proprietates immutentur, sive validioribus, Specierum signaturæ transplantatæ novæ familiæ insignia demonstrent.]

[66] Ibid., 220–21. "Quicquid enim in parentibus continetur, sive sanum sive morbidum, quod valida vitalique impressione, Balsamum et radicem humanæ naturæ afficere possit, Transplantatione hæreditarium evadit. Huiusmodi impressiones plane sunt astrales, incorporeæ, spirituales, invisibiles: sed efficaces seminalium Rationum naturas imitantes, Scientiæ et prædestinationum virtute procedentes, in quibus mole corporea non opus est ad corporum constitutiones." Severinus' theory that transplantation can occur either phenotypically or at the heritable level implies that therapeutic transplantation can, in principle, effect a cure of hereditary disease.

[67] The distinction between seminal reason and impression thus seems to be instrumental rather than formal: the former is bound to the seed as a hylomorphic unit, while the latter is the transfer of the formal information or seminal programming (*rationes*) from agent to patient, where it induces its "impression" on the native seminal reason and thus alters it through transplantation.

transformation implies that transplantation was not simply a change of substantial form, but that, true to its etymology and use in medieval and Renaissance Europe, it must involve the relocation of a material thing from one place to another.[68] However, as Severinus used the term, transplantation is clearly a process that can result in the programing or entelechy of one form or species becoming modified, hijacked, or replaced by another, with the result that a body of different form or species matures instead. Or, to paraphrase Hippocrates as quoted by Severinus, what turns out is not what was planned.

One way to reconcile these formulations while preserving Severinus' fundamental adherence to a Neoplatonic view of the world of material *corpora* as an emanation from the immaterial world of seminal ideas is to impose Aristotle's requirement that the substantial form and constituent matter of a body be physically inseparable, if logically distinct. Another way of stating this is that all bodies are hylomorphic entities in their actuality: whatever undergoes transplantation undergoes translocation and consequently must be a dimensioned hylomorphic entity, a body.

Severinus' requirement that *semina* be the links between body and non-body, at once both *corpora* and *incorpora*, was an attempt to surmount this problem of giving an intrinsically formal entity a spatial identity, while at the same time constituting a third metaphysical category that has the properties of both ideas and bodies and can therefore interface with both. This was a role traditionally assigned to spirit in Renaissance Platonism or, as Hiro Hirai has shown, to some sort of vehicle for the soul.[69] But what Severinus did by associating this spirit with *semina*, a term that for Lucretius, Fracastoro, and also in common parlance carried a sense of spatially-confined and granular material things, was to reify it and indeed to atomize it. His *semina* are ideally and intrinsically immaterial, nondimensional locations which, at the instant they emerge from pure potentiality, pure chaos and abyss, become dimensioned and begin to grow materially and epigenetically. Or, to use Severinus' language, they put on husks and vestments and enter the world stage to perform the planned comedy, a mechanical liturgy. He likened this process to actors donning vestments for their appearance on the stage, playing certain roles, and retiring from the scene. Generation, mixture, and transplantation are components of this emergence process.[70] Offstage, *semina* exist only as ideas, *rationes*, but once that first

---

[68] The term transplantation derives from the preposition "across" and the verb "to plant," which itself derives from the Latin *plantar*, the bottom of the foot used to plant seeds or seedlings. It therefore carries with it a strong physical and spatial deep sense. In medieval use the term could be more abstract.

[69] On this matter see Hirai 2011.

[70] Severinus made occasional references to the phenomenological world as a theatrical production unfolding on a stage, but did not explicitly develop the analogy. For example, at the end of chapter four of *Idea medicinæ philosophicæ* (Severinus 1571, 39) Severinus explains how the rest of the book will unfold: "Third, we will establish the principles of all bodies, which are the vestments and dwellings of the seeds going forth onto this world stage. [Tertio, Principia constituemus omnium corporum, quæ sunt vestimenta et domicilia Seminum, in hanc mundanam scenam progredientium.]; This stage is the scene, literally, for coming to be and passing away (80–81): "There are two famous questions, which very much obscure the knowledge of natural things: where they have

instant of sprouting has passed, the seminal reason, the seed's mechanical knowledge, becomes part of a hylomorphic substance, an informed corpuscle. The seeds themselves are, then, the rudiments of these vital corpuscles, the liminal existences of corpuscles as they become active and manifest on the world stage. They have only a dynamic existence. And if there is a seminal transplantation, and the actor changes his role, then what results is not what was intended, and the comedy might become a tragedy. Either way, the biological theory that Severinus built around these seminal rudiments presented a Platonized reconciliation of Aristotelian, Galenic, and Paracelsian metaphysics, which provided a thoroughgoing explanatory system that brought together everyday agricultural, alchemical, and medical experience. His vision of hylozoic corpuscular rudiments – not the inert *semina* of Lucretian atomism, but the vital, informed *semina* of Paracelsian chemical philosophy – offered readers a middle way between the unsatisfactory scholastic Aristotelian theory of material change as a qualitative mixing and imposition of new form and the austere, agentless world of pagan atomism. Protestant Christians like Severinus, who saw nature as a providential creation, found this latter view difficult to reconcile with the theology and cosmogony revealed in holy scripture.

References

Aristotle 1941. *The basic works of Aristotle*, ed. Richard McKeon. New York: Random House.

Clericuzio, Antonio. 2000. *Elements, principles and corpuscles: A study of atomism and chemistry in the seventeenth century*. Dordrecht: Kluwer Academic Publishers.

Daniel, D.T., and D.T. Daniel. 2002. Paracelsus on Baptism and the acquiring of the eternal body. In *Paracelsian moments: Science, medicine, and astrology in early modern perspective*, ed. S.W. Gerhild and C. Gunnoe, 117–134. Kirksville: Truman State University Press.

Debus, Allen G. 1977. *The chemical philosophy*. New York: Science History Publications. Reprint, Mineola: Dover, 2002.

French, Roger K. 1994. *Ancient natural history*. London: Routledge.

Garber, Daniel. 2006. Physics and foundations. In *Early modern science, vol. 3 of the Cambridge history of science*, ed. Katharine Park and Loraine Daston, 21–69. Cambridge: Cambridge University Press.

Gatti, Hilary. 2001. Giordano Bruno's soul-powered atoms: From ancient sources towards modern science. In *Late Medieval and early modern corpuscular matter theories*, ed. Christoph Lüthy, John E. Murdoch, and William R. Newman, 163–180. Leiden: Brill.

Hall, Thomas Steele. 1969. *Ideas of life and matter: Studies in the history of general physiology 600 B.C.–1900 A.D*, vol. 2. Chicago: University of Chicago.

Harvey, William. 1995. De Motu Cordis (1653). In *The anatomical exercises. De Motu Cordis and De Circulatione Sanguinis in English Translation*. New York: Dover.

come from, and where they are proceeding to, those forms and species affecting us at defined intervals of time that carry out the functions of this world comedy, through the service of generations, transplantations, and mixtures." [Duæ sunt famosæ dubitationes, quæ naturalium rerum scientiam vehementer obscurant: unde profectæ, quo pergant Formæ, Species, quæ nobiscum definitis temporum intervallis negotiaturæ, Generationum, Transplantationum et Mixtionum ministerio mundanæ Comoediæ lithurgiam peragunt.]

Henry, John. 1986. Occult qualities and the experimental philosophy: Active principles in pre-Newtonian matter theory. *History of Science* 24: 335–381.

Hippocrates, 1931. *Hippocrates IV*. Trans. W.H.S. Jones. Loeb Classical Library 150. Cambridge, MA: Harvard University Press.

Hirai, Hiro. 2011. *Medical humanism and natural philosophy. Renaissance debates on matter, life and soul*. Leiden: Brill.

Kaplan, Barbara Beigun. 1993. *"Divulging of useful truths in physic": The medical agenda of Robert Boyle*. Baltimore: Johns Hopkins University Press.

Libavius, Andreas. 1615. *Prodromus vitalis philosophiæ Paracelsistarum Examen philosophiæ novæ, quæ veteri abrogandæ opponitur: in quo agitur de modo discendi novo: De veterum autoritate; De magia Paracelsi ex Crollio; De philosophia vivente ex Seuerino per Johannem Hartmannum; De philosophia harmonica magica Fraternitatis de Rosea Cruce*, in his *Syntagmatis selectorum undiquaque et perspicue traditorum alchymiæ arcanorum, tomus primus*. Frankfurt.

Lüthy, Christoph. 2006. Where logical necessity becomes visual persuasion: Descartes' clear and distinct illustrations. In *Transmitting knowledge: Words, images, and instruments in early modern Europe*, ed. Sachiko Kusukawa and Ian Maclean, 97–133. Oxford: Oxford University Press.

Lüthy, Christoph, John E. Murdoch, and William R. Newman (eds.). 2001. *Late medieval and early modern corpuscular matter theories*. Leiden: Brill.

Mebane, John S. 1989. *Renaissance magic and the return of the Golden Age: The occult tradition and Marlowe, Jonson, and Shakespeare*. Lincoln: University of Nebraska Press.

Newman, William R. 2006. *Atoms and alchemy: Chymistry and the experimental origins of the scientific revolution*. Chicago: University of Chicago Press.

Newman, William R., and Lawrence M. Principe. 2002. *Alchemy tried in the fire: Starkey, Boyle and the Fate of Helmontian Chymistry*. Chicago: University of Chicago Press.

Niebyl, Peter. 1971. Sennert, Van Helmont, and medical ontology. *Bulletin of the History of Medicine* 45: 115–137.

Nussbaum, Martha Craven. 1978. *Aristotle's De Motu Animalium*. Princeton: Princeton University Press.

Pagel, Walter. 1967. *William Harvey's biological ideas; Selected aspects and historical background*. New York: Hafner.

Pagel, Walter. 1976. *New light on William Harvey*. Basel: Karger.

Pagel, Walter. 1982. *Paracelsus: An introduction to philosophical medicine in the era of the renaissance*. Basel: Karger.

Paracelsus. 1922. *Paracelsus. Sämtliche Werke*, ed. Karl Sudhoff. München: R. Oldenbourg.

Paré, Ambroise. 1982. *On Monsters and Marvels*. Trans. Janis L. Pallister. Chicago: University of Chicago Press. Original French edition 1573.

Redondi, Pietro. 1987. *Galileo Heretic*. Trans. Raymond Rosenthal. Princeton: Princeton University Press.

Rudolph, Hartmut. 1998. Hohenheim's anthropology in the light of his writings on the Eucharist. In *Paracelsus. The man and his reputation, his ideas and their transformation*, ed. Ole Peter Grell, 187–206. Leiden: Brill.

Ruska, Julius F. 1926. *Tabula Smaragdina: Ein Beitrag zur Geschichte der Hermetischen Literatur*. Heidelberg: Carl Winter's Universitatsbuchhandlung.

Severinus, Petrus. 1571. *Idea medicinæ philosophicæ fundamenta continens totius doctrinæ Paracelsicæ, Hippocraticæ et Galenicæ*. Basel: Henric Petri.

Shackelford, Jole. 1993. Tycho Brahe, laboratory design, and the aim of science: Reading plans in context. *Isis* 84: 211–230.

Shackelford, Jole. 1998. Seeds with a mechanical purpose: Severinus' *Semina* and seventeenth-century matter theory. In *Reading the book of nature: The other side of the scientific revolution*, Sixteenth century essays and studies 41, ed. Allen G. Debus and Michael T. Walton, 15–44. Kirksville: Sixteenth Century Journal Publishers.

Shackelford, Jole. 2004. *A philosophical path for Paracelsian medicine. The ideas, intellectual context, and influence of Petrus Severinus (1540/2-1602)*. Copenhagen: Museum Tusculanum Press.

Shackelford, Jole. 2007. "Severinus' Paracelsian Seeds: Mechanical Actors on the World Stage," an unpublished presentation at the History of Science Society in Washington DC, 2 November, 2007.

Shackelford, Jole. 2015. La transplantation comme transformation chez Paracelse: de la structure à la fonction. In *Transplanter: Une approche transdisciplinaire : art, médecine, histoire et biologie*, ed. François Delaporte, Bernard Devauchelle, and Emmanuel Fournier, 35–40. Paris: Éditions Hermann.

Sprengel, Kurt Polycarp Joachim. 1794. *Versuch einer pragmatischen Geschichte der Arzneikunde*, III. Halle: Johann Jacob Gebauer.

Webster, Charles. 2008. *Paracelsus: Medicine, magic and mission at the end of time*. New Haven: Yale University Press.

Chapter 11
Mysteries of Living Corpuscles: Atomism and the Origin of Life in Sennert, Gassendi and Kircher

Hiro Hirai

Abstract This paper aims to spotlight some important, but neglected, aspects of early modern interactions between matter theories and the life sciences. It will trace the ways in which atomistic or corpuscular modes of reasoning were adopted to explain the origin of life. To that end this paper will examine three seventeenth-century natural philosophers: Daniel Sennert (1572–1637), Pierre Gassendi (1592–1655) and Athanasius Kircher (1602–1680). Through the analysis of their discussions on the minute constitutive parts of living beings (plants, animals and human beings) as living corpuscles, it will inquire into the exchange of ideas among those who advocated "non-mechanist" or "vitalistic" types of corpuscular philosophy (Here I am using the term "vitalistic" broadly construed as the currents that emphasized the role of the life principle or vital principle). This paper's ultimate goal is to shed light on the role of bio-medical ideas in seventeenth-century natural philosophy.

Keywords Atomism • Vital principle • Seminal principle • Spirit • Soul • Matter • Life • Sennert • Gassendi • Kircher • Redi

11.1 Introduction

Despite the alleged empirical proofs and counterarguments against the postulate of ultimate particles, atomistic or corpuscular matter theories triumphed by associating themselves with a mechanistic philosophy of nature in the seventeenth century. This

---

I would like to thank Peter Distelzweig, Benjamin Goldberg, Evan Ragland and James Lennox for the organization of the Pittsburgh conference and the preparation of its consequent volume.

H. Hirai (✉)
Center for the History of Philosophy and Science, Radboud University, Nijmegen, Netherlands
e-mail: hhirai2@gmail.com

P. Distelzweig et al. (eds.), *Early Modern Medicine and Natural Philosophy*, History, Philosophy and Theory of the Life Sciences 14,
DOI 10.1007/978-94-017-7353-9_11

alliance has often been acknowledged as the foundation of modern science.[1] Atomism, well adopted in such domains as physics, optics and microscopy, however, became too mechanistic to avoid the conventional objections reiterated in medieval university teachings. "Vitalistic" or non-mechanistic types of corpuscular reasoning could circumvent many of these objections, explaining various phenomena especially in biological and medical fields. Recovering and reconstituting their forgotten role will surely help elucidate the evolution of early modern corpuscular philosophy.

This study is only a first step of such an attempt, and, by way of examining some "vitalistic" or non-mechanistic corpuscular ideas, will address seventeenth-century interactions between philosophy and medicine, or more precisely, biological and medical ideas. To this end I will focus on three authors: Daniel Sennert of Wittenberg, French atomist Pierre Gassendi and Jesuit polymath Athanasius Kircher. They addressed the emergence of life in terms of living or "ensouled" corpuscles. This idea might be regarded as contrary to a mechanistic view of nature. Although these figures were highly influential in their own time and their works widely read, their ideas have not attracted their due attention in the traditional narrative of the history of science and philosophy. They do not seem to belong to a single current. By analyzing their divergent ideas, I hope to shed light on their common sources and eventual interactions.

11.2 Sennert, Corpuscles and Spontaneous Generation

A professor of medicine at the University of Wittenberg, Daniel Sennert (1572–1637) has recently drawn renewed interest from scholars. His work encompasses a cluster of issues raised by the intersection of matter theories and the life sciences in the early seventeenth century.[2] In this field, the origin of life was one of the most important questions, and a belief in spontaneous generation (*i.e.*, abiogenesis, or the emergence of life from lifeless matter) played a key role. To address this very issue, Sennert wrote *On the Spontaneous Generation of Living Beings* (*De spontaneo viventium ortu*), a treatise published at the end of his philosophical masterpiece, *Physical Memories* (*Hypomnemata physica*) (Sennert 1636). In this treatise, important in its own time but unfairly overlooked by historians, he developed a corpuscular interpretation of the emergence of life.[3]

Sennert's *Hypomnemata physica* are composed of five books treating, respectively: (1) the principles of natural things; (2) occult qualities; (3) atoms and mix-

[1] On the history of atomism, see among others Lasswitz 1890/1926; Hooykaas 1983[orig. 1933]; van Melsen 1952; Emerton 1984; Clericuzio 2000; Lüthy et al. 2001.

[2] See especially Newman 2006, 85–153; Michael 1997, 2001; Stolberg 2003; Blank 2010a, 167–205. This section is partially based on Hirai 2011, 151–172. For Sennert's influence on Leibniz, see Arthur 2006.

[3] I have used the text of Sennert 1650, 132–242.

tures; (4) the generation of living beings; and (5) spontaneous generation. In the fifth and last book, after dismissing the series of causes for spontaneous generation advanced before him, Sennert argues that this phenomenon occurs by an internal principle lying hidden in matter. Living beings, he adds, that do not reproduce themselves through seeds still possess something analogous. He calls this entity "seminal principle" (*principium seminale*) or "soul" (*anima*).[4]

To describe the presence of this hidden principle in matter, Sennert first tries to explain two modes by which the soul resides in the body: (1) the soul acts as a form in the body; and (2) the soul performs the functions of life through bodily organs. The former corresponds to the first and essential actuality and the latter to the second and accidental actuality. To these Sennert adds a third mode:

> But besides these two modes there is yet a third and the soul can be in some matter after yet another way without informing or vivifying this matter or providing the actions proper to this living being. Thus the seeds of plants and animals can reside in water and earth and the soul [can reside] in these [seeds] without informing or vivifying water or earth.[5]

Thus the soul is contained in a thing as if it were placed in a container. The soul in this state remains dormant without informing or vivifying the container. Once this latent soul is placed under suitable conditions, it starts to execute its functions: informing the body as its form in the first actuality and vivifying it in the second actuality.

As every seed contains a soul or an analogous principle in the eyes of Sennert, the omnipresence of seeds in the world, earth and water, directly signifies that of souls. That is why, he argues, Aristotle taught in his *Generation of Animals*, 3.11: "There is water in earth, and *pneuma* in water, and in all *pneuma* is soul-heat, so that all things are in a sense full of soul."[6] According to Sennert, Aristotle did not mean that all things are animate, but that there is in all things such a hidden entity that becomes manifest and executes the functions of life when it encounters suitable conditions.

Accounting further for the omnipresence of this special entity, Sennert introduces a corpuscular interpretation. According to him, minute corpuscles, coming from the bodies of living beings and their cadavers, are scattered and diffused everywhere in the world by such factors as wind and rain. Each of these corpuscles is endowed with a soul, and their wide diffusion explains the omnipresence of souls in the natural world.[7]

---

[4] Sennert 1650, 5.2, 214–215.

[5] Sennert 1650, 5.2, 216. "Verum praeter hos duos modos datur adhuc tertius, et potest adhuc alio modo anima esse in materia aliqua, ita ut neque eam informet, et vivificet, neque etiam operationes viventis illius proprias edat. Ita in aqua et terra semina plantarum et animalium, et in iis anima inesse possunt, ut tamen neque aquam, neque terram informent et vivificent."

[6] Aristotle 1984, 3.11, 762a18–21. On Aristotle's notion of *pneuma*, see Solmsen 1957, 119–123; Freudenthal 1995.

[7] Sennert 1650, 5.2, 216. For Sennert this is the real meaning of Aristotle's words. For other interpretations in the Renaissance, see Hirai 2005, 143, 147; Hirai 2011, 42, 95, 112, 124, 148.

To illustrate the presence of a soul in these corpuscles, Sennert chooses the example of the "chymical" (chemical/alchemical) dissolution of metals by qualifying this phenomenon as the clearest explanation of all.[8] According to him, gold and silver are dissolved into *minima*, that is, minute corpuscles or atoms, by special acids. Even dissolved, the forms of gold and silver are entirely preserved, although they do not inform the acids in which they swim. These forms exist in the acids and remain dormant as if contained in a vessel. Although gold and silver are divided into atoms or minute corpuscles, they retain entire their essence in these atoms. Sennert applies the same reasoning to the soul of living beings.[9] In his view the seed's matter, provided as a suitable subject or carrier of the soul, is so well disposed that it can retain the soul within itself even if it is divided into the state of *minima*. Without being altered in their essence, the souls can reside in these *minima* and remain dormant. In the case of living beings, concludes Sennert, their seminal force can persist with the soul even down to the level of atoms. As he applies the example taken from the mineral kingdom to that of plants and animals, the traditional clear-cut distinction between living beings and non-living things advanced by the Aristotelians is blurred in his system.

Relying on this account of metallic dissolution, Sennert explains the way spontaneous generation occurs. According to him, whether it is called "seed," "seminal principle" or "soul," there must first be some special entity that comes from the bodies and cadavers of living beings and lies hidden in the water and the earth. When this entity is placed under suitable conditions and stimulated by ambient heat, it begins to perform the functions of life. This special entity, made of minute corpuscles endowed with a soul, is the real origin of the generation of inferior living beings such as bugs and insects, which were believed to be born spontaneously.

Sennert's atom is the carrier of a form in the case of minerals and the carrier of a soul in the case of living beings. The soul can be conceived as a higher type of form. Now let us take a closer look at the relationship between the soul and its carrier atom. First of all, in Sennert's hierarchy of corpuscles, the "atoms" of living beings are corpuscles composed of primordial atoms. In this sense they may correspond better to the particles that would be called "molecules" (*moleculae*) or "seeds of things" (*semina rerum*) by Gassendi.[10]

Sennert then argues that although the soul residing in one atom is weak, several atoms can be united, allowing the souls contained therein to gather and become more powerful. He reports this idea under the authority of Fortunio Liceti (1577–1657), a Paduan professor of philosophy and a friend of Galileo Galilei (1564–1642).[11] Although barely known to historians, Liceti was respected and influential in his own time. He was interested not only in natural philosophy, but also in biological and medical issues, especially those related to generation, embryology, and teratology or monstrous births. Among other writings Liceti published a treatise entitled *On*

[8] I have adopted the term "chymistry" to avoid any arbitrary distinction between chemistry and alchemy which did not exist in Sennert's time. See Newman and Principe 1998.

[9] Sennert 1650, 5.2, 216.

[10] On Gassendi's idea, see Sect. 11.3 below.

[11] On Liceti, see Ongaro 2005; Marangio 1973; Blank 2010b; Hirai 2011, 123–150. I have used the following edition: Liceti 1618.

the Spontaneous Generation of Living Beings (*De spontaneo viventium ortu*) in Vicenza near Padua in 1618. Entirely devoted to the problem of spontaneous generation, this treatise collects and studies the ideas formulated on the phenomenon from antiquity to the sixteenth century, followed by Liceti's own theory. It can be qualified as an "encyclopedia" of spontaneous generation. Sennert entitled his own treatise after Liceti's. If we turn to the work of Liceti, we can observe him claiming almost the same point as Sennert, even with the use of the term "atom," which is not typical for him. Thus these two ideas: (1) the residence of a soul in one atom; and (2) the gathering of the souls of many atoms, are not original to Sennert, but must be attributed to Liceti himself.[12]

What is more striking comes next. Like Liceti Sennert accepts the idea that the seeds of living beings do not always appear as visible bodies but sometimes as minute corpuscles enclosing a soul and lying hidden in the water and the earth. Although Liceti admitted that these corpuscles act as seeds and correspond by analogy to them, he was not willing to call them "seeds." By contrast Sennert finds no reason to refuse them this title. Arguing that such an entity, if not "seeds," can at least be called "seminal principle," he adds:

> Indeed what primarily constitutes a seed is neither its external figure nor its formation in a definite way, but the soul latent in it, with that implanted spirit which is said to correspond to the element of the stars and makes the seeds fertile. Since they, with their subject, can even reside in minute corpuscles, there is no reason why they cannot be called "seeds" in their own way, or "seminal principle." The soul lying hidden in such corpuscles does not inform the earth or the water which contains it. But lying hidden in these [elements] as in a vessel, it nevertheless informs these corpuscles in which it resides as if it were in its proper subject, and exists in them under the first actuality. It attains the second actuality when it finds a suitable place.[13]

The part "that implanted spirit which is said to correspond to the element of the stars and makes the seeds fertile" is an allusion to a passage in Aristotle's *Generation of Animals*, 2.3, which was repeatedly commented on by embryological writers during the Renaissance. Natural philosophers and physicians fueled the debate, which culminated in the emergence of a new "astral medicine."[14] Thus Sennert is paying attention to this tradition. But what is more important for the purpose of the present study is that, according to Sennert, the soul of one atom of a living being does not inform or animate the earth or the water which surrounds this atom, since they are only its recipients, but rather this same soul well and truly animates the atom which

[12] Liceti 1618, 3.10, 203. Liceti's main argument (1618, 3.13, 206–207) is reproduced in Hirai 2011, Appendix 4, 191–193.

[13] Sennert 1650, 5.7, 226. "Non enim externa figura et certo modo facta formatio, semen constituit primario, sed anima in eo latens, cum spiritu illo insito, qui elemento stellarum respondere dicitur, et foecunda facit semina: quae quia cum illo suo subiecto in minimis etiam corpusculis esse possunt, nulla causa est, cur non et illa semina suo modo, aut seminale principium dici possint. Et licet anima, quae in talibus corpusculis latet, terram vel aquam, in qua continetur, non informet, sed in iis ut in vase lateat: tamen corpuscula illa, in quibus ut proprio subiecto est, informat, et in iis actu primo est, ad actum vero secundum accedit idoneum locum nacta." Cf. Michael 1997, 351; Stolberg 2003, 181.

[14] Aristotle 1984, 2.3, 736b33–737a7. On its Renaissance interpretations, see Hirai 2011, 25–30, 69–72, 92–94, 111–112; Hirai 2012, 2014.

carries it. Thus Sennert's atom or molecular corpuscle is not a mere vehicle of the soul. Being itself animated, it provides the basis for living matter. He coins the name of the "seminal principle" for this special soul. Sennert built his philosophical reflection on this particular conception of the soul. For him the soul informs and vivifies its vector corpuscle to guarantee the permanence of the species. This particular conception thus holds the key to understanding Sennert's notion of the seminal principle. In the mid-seventeenth century, the young Robert Boyle (1627–1691), as an assiduous reader of Sennert, was to be keenly interested in the notion of seminal principle and pay it considerable attention.[15]

In a previous study, I showed that Sennert was well informed of the theory of the invisible and spiritual seeds of things, advanced by the Danish Paracelsian Petrus Severinus (1540/42–1602).[16] In his treatise *On the Agreement and Disagreement of the Chymists with the Aristotelians and the Galenists* (*De chymicorum cum Aristotelicis et Galenicis consensu ac dissensu*) (Sennert 1619), Sennert rejected this theory by identifying these spiritual seeds as substantial forms or souls themselves.[17] To his eyes, Severinus and the other chymists who followed him unnecessarily multiplied terms to refer to the same thing. It is true that Sennert's conceptions of the seminal principle and living corpuscles hint at certain ideas of Severinus. As we have seen, however, the fundamental inspiration for his corpuscular interpretation of the origin of life derived from his reading of Liceti's work.

11.3 Gassendi and Seminal Molecules

Like Sennert, Pierre Gassendi (1592–1655) was one of the most fervent defenders of the newly revived atomism.[18] He was also quite familiar with the chymical philosophy of his time. In his masterpiece *Syntagma philosophicum* (Lyon 1658), he supports the idea that minerals and metals result from "seeds" (*semina*) created and disseminated in the earth by God at the moment of Creation. To explain the nature of these seeds, he acknowledges the existence of a certain internal power, called a "seminal force" (*vis seminalis*). According to him, this force, residing in the seeds of minerals and metals, guarantees the regularity of their structure, figures, colors and geographical distribution. Gassendi then introduces the notion of an "elaborator spirit" (*spiritus elaborator*) as the vector of this force. The formation of minerals and metals is thus explained in a similar way to living beings by the invisible seeds conceived in the form of a spirit. He further integrates an atomistic interpretation for the structure of these seeds.

[15] See Clericuzio 1990, 583–587; Anstey 2002, 597–630; Hirai and Yoshimoto 2005. For Sennert's notion, see also Clericuzio 2000, 24–25.

[16] On Severinus, see Shackelford 2004; Hirai 2005, 217–265 (and 401–403 on Sennert's debt to Severinus).

[17] I have used the following edition: Sennert 1633, here 1.9, 88.

[18] On Gassendi, see among others Bloch 1971; Osler 1994; Murr 1997; Fisher 2005; LoLordo 2006; Taussig 2009. The present section is based on Hirai 2003 and 2005, 463–491. I have used the text of Gassendi 1658. His *Syntagma philosophicum* is found in its first and second volumes (hereafter *SP* I and *SP* II, followed by the page number with column a or b).

In the field of biology, the determined geographic distribution of plants leads Gassendi to appeal again to the idea of a seminal force carried by invisible seeds, just as in the case of minerals and metals. The visible seeds of plants are, according to him, constructed through the union of these invisible seeds. Gassendi then argues that plants and their visible seeds are endowed with a soul. For him this soul is nothing but a corporeal substance diffused in the whole body of a plant. He compares it to a "spirit" (*spiritus*) or "miniature flame" (*flammula*), which is, to him, extremely subtle, pure and active. Because of their perfection, visible seeds can preserve this special substance for a long period. Gassendi then refers back to the idea of seminal force:

> Certainly, you will not consider that the grain of wheat, when it is preserved in a storehouse, is deprived of this kind of substance or soul as well as life (it is rather called "seminal force," but it is almost the same thing). For it lies dormant as long as there is no heat or humidity outside [...].[19]

The seminal force is conceived here as almost synonymous to the soul, which is for Gassendi a corporeal substance in the form of spirit or miniature flame. He adds that the soul of a new plant is already present in the visible seed. This soul, called "miniature soul" (*animula*), stems from the soul of the parent plant. The French atomist likens the departure of the *animula* from the parent soul as the emancipation of a daughter from her family.

As for the soul of animals, Gassendi also conceives it as a corporeal substance like "a flame composed of very fine, mobile and active corpuscles."[20] As in the case of plants, the visible seeds of beasts are animated by their internal *animula* conceived as a "lighted torch." Calling upon the image of the soul as burning particles, depicted by Democritus, Gassendi explains that these particles go here and there to produce "molecules" (*moleculae*), which are the invisible "seeds of things" (*semina rerum*). These seeds are further united to form the visible seeds in the generation of animals. In Gassendi the spontaneous generation of plants and animals also occurs because of these invisible seeds or their internal *animulae*. To conclude his discussion, he reveals what is contained in these invisible seeds:

> [We must] confess that nothing has been done to produce the true knowledge of that internal and invisible economy and to expose to our mind's eye that artist or craftsman, so to speak, who skillfully serves tiny instruments so sophisticated in elaborating matter into so proportionate a work [...]. That is why there remains to us, having admired works inimitable and exceeding all human understanding, to sing a hymn to that divine and incomparable Architect who created and established these craftsmen, so to speak, endowed with such great providence, diligence and faculty in the seeds of things [...].[21]

---

[19] Gassendi 1658, *SP* II, 172b. "Ne putes certe tritici granum, cum asservatur in horreo esse orbatum tali substantia, sive anima, atque vita (vim seminalem potius vocant, sed perinde est) ea quippe duntaxat consopita manet, donec deest, humor, calorque exterior [...]."

[20] See Rosenfield 1941, 111–120; Canguilhem 1955, 79–88; Spink 1960, 85–102; Roger 1963, 135–140; Bloch 1971, 229–230, 268, 364–366; Murr 1991; Osler 1994, 64–67.

[21] Gassendi 1658, *SP* II, 267a. "[...] fatendumque est nihil esse actum, quod germanam notitiam creet internae illius, occultaeque œconomiae, quod obiiciat mentis obtutui artificem illum, sive quasi fabrum scite organulis adeo exquisitis utentem ad elaborandum materiam in opificium adeo concinnum [...]. Quare superest, ut mirati opera inimitabilia, captumque omnem superantia hymnum canamus divino illi, ac incomparabili Architecto, qui intra rerum semina creavit, constituitque hosce quasi fabros tanta providentia, industria, atque facultate instructos."

In Gassendi these invisible seeds possess a corpuscular structure within the framework of a mechanistic system. But they are endowed with striking properties that guarantee the regularity and stability of each species. These properties exceed the limit of pure and simple mechanism. They result from the contents of the invisible seeds: internal craftsmen and their "knowledge" or "science" (*scientia*). Gassendi explains that there is nothing more admirable than this kind of *scientia* assigned to invisible seeds for the careful and constant elaboration of the bodies of natural things. According to him, God the Creator imprinted this *scientia* on these invisible seeds. What does Gassendi mean by this special *scientia*?

In a lesser-known passage in his discussion devoted to the chymical principles, Gassendi tries to show that the extraordinary regularity of natural things does not result from the simple combination of the four traditional elements (fire, air, water and earth) of the Aristotelians or the three principles (Sulfur, Mercury and Salt) of the Paracelsian chymical philosophers. He insists on the existence of a seminal power that disposes these elements and principles, and discloses the source of his ideas:

> Only that famous Severinus clearly perceived the truth, and the others like Quercetanus who followed him, when he posited numerous invisible seeds beyond the four elements and the three principles.[22] These seeds can also be called "principles" and "elements" while the grosser elements are, so to speak, just their vestments, matrices and receptacles, so that everything receives from them not only its vigor and action but also its art and *scientia* by which the mechanical spirits contained in these seeds have a power to form the body of minerals, plants and animals as well as their parts. Let us consider them those elaborator spirits [...]. Severinus would also say, as he will, that these principles are the mechanical spirits endowed with the *scientia* and vigor to act. When he said, he said at once everything that he would never repeat. For he would never tell in general how this idea or *scientia* elaborating the work can settle in a certain spirit which is a thing so thin, invisible and impalpable; the consideration of the goal to which [the spirit] prepares the work; the knowledge of matter and necessary conditions, by which matter is made appropriate; the distinction of the ratio or mode, by which [matter] must be reduced, turned, returned, mixed, separated, fashioned and perfected; and how settles the vigor or energy now to use convenient instruments and now to execute everything prescribed by this *scientia*.[23]

---

[22] On French Paracelsian Joseph Du Chesne (1546–1609), *alias* Quercetanus, see Hirai 2010.

[23] Gassendi 1658, *SP* II, 558b–559a. "Unus praeclare rem agnovit memoratus iam Severinus, et qui illum sunt, ut Quercetanus, aliique sequuti, cum praeter quatuor elementa, et tria principia, innumera posuit invisibilia semina, quae dici etiam principia, elementaque valeant, quorumque haec crassiora sint solum quasi vestimenta, matrices receptacula; idque ut ipsis omnem non modo vigorem, actionemque acceptam ferat, sed etiam artem, et scientiam, qua contenti in ipsis mechanici spiritus polleant ad efformandum, ut mineralium, sic vegetabilium, animaliumque corpora, ipsorumque parteis, ut puta elaboratores isti [...]. Et dicat Severinus, ut volet, esse haec principia mechanicos spiritus scientia, et vigore agendi pollenteis; cum id dixerit, semel dixerit quicquid dicturus unquam est. Neque enim unquam praeterea universe manifestabit, quemadmodum cuiquam spiritui, rei tam tenui, rei tam inuisae, tamque intactili insidere possit idea, ac scientia elaborandi operis; consideratio finis, ad quem comparare illud debeat; perspectio materiae, conditionumque necessariarum, ob quas idonea efficitur; dignotio rationis, seu modi, quo eam subigere, versare, reversare, concernere, deligere, fingere, perficere, oporteat: quomodo item possit insidere vigor, ac energia tum usurpandi instrumenta congrua, tum exsequendi omnia, quae talis scientia praescripserit."

This is the true source of Gassendi's theory of the *scientia* imprinted on the invisible seeds by God the Creator. It is the doctrine of the invisible and spiritual seeds of things advanced by Severinus. However, Gassendi himself is not satisfied with simply reproducing the Danish Paracelsian's ideas but tries to interpret them further from an atomistic perspective.

Before concluding this section, let us briefly note an echo of Gassendi's theory in Italian natural philosopher Francesco Redi (1626–1697).[24] At the beginning of his treatise, *Esperienze intorno alla generazione degli insetti* (Florence 1668), he takes up Gassendi's opinion about spontaneous generation without mentioning his name and introduces the idea of the invisible seeds of things:

> There is still another group of wise people who held and hold as true that this generation derives from certain small groups or aggregates of atoms, such aggregates being the seeds of all things, and all things being full of these seeds. That [all things] are full of them, many others admit it by saying that God created these [seeds] at the beginning of the world and scattered them everywhere to render the elements fertile, not of a momentary nor incomplete fertility but as durable as [the elements] themselves. One must understand, they say, in this manner what is written in the Holy Scripture: "God created all things together."[25]

Surprisingly enough, Redi links this argument with the theory of William Harvey (1578–1657), according to which every generation takes place through a seed, conceived as an egg, which encloses a motive principle. He argues that Harvey regards these invisible seeds as atoms that fly through the air and are distributed here and there by winds.[26] Thus Redi closely connects Gassendi's idea with the famous theory of *omne vivum ex ovo*. Harvey's egg is most strikingly interpreted in terms of atoms and corpuscles.

Later Redi explicitly refers to Gassendi and refutes his idea of the invisible seeds. What is important in his discussion is, as he acknowledges, that Gassendi's opinion was popular and widely diffused in his time.[27] It was exactly the period when the theory of the preexistence of germs was spreading in embryology, but the true role played by Gassendi's ideas in this field still remains to be explored thoroughly.[28]

---

[24] On Redi, see especially Findlen 1993; Bernardi and Guerrini 1999; Hirai 2003, 220–221; Duris 2010, 1–25. I have used the text of Redi 1996.

[25] Redi 1996, 78. "Egli c'è ancora un'altra maniera di savie genti, le quali tennero e tengono per vero che tal generazione derivi da certi minimi gruppetti ed aggregamenti di atomi, i quali aggregamenti sieno i semi di tutte quante le cose, e di essi semi le cose tutte sien piene. E che ne sieno piene lo confessano ancora molti altri dicendo che sì fatte semenze nel principio del mondo furono create da Dio, e da lui per tutto disseminate e sparse, per render gli elementi fecondi, non già d'una fecondità momentanea e mancante, ma bensì durevole al pari degli elementi stessi; ed in questa maniera dicono potersi intendere quello che ne' Sacri Libri si legge, *avere Iddio create tutte le cose insieme*." Cf. Ecclesiasticus (Sirach), 18.1.

[26] Redi 1996, 78–79. Cf. Harvey 1650, exercitatio 57. On Harvey, see Roger 1963, 112–121; Pagel 1967; Foote 1969; Duchesneau 1997, 29–42.

[27] Redi 1996, 138.

[28] On the preexistence of germs, see Roger 1963, 325–384; Bowler 1971, 221–244; Duchesneau 1997, 229–237. On the influence of Gassendi's biology, where the concept of seeds as "living atoms" or "living molecules" played an important role, see Rey 1997.

11.4 Kircher and the Corpuscular Origin of Life

A friend of Gassendi, Athanasius Kircher (1602–1680) amassed considerable fame for his intellectual activity carried out during several decades at the Jesuit Collegio Romano. To support his arguments, he constantly appealed to the experiments that, he claims, were performed by him or were reported by others. His work was the object of both admiration and suspicion among his contemporaries and provoked passionate debates all over Europe.[29] The question of spontaneous generation attracted the particular attention of those who tried to reproduce the experiments reported by Kircher. In England Royal Society members such as Robert Boyle and Henry Oldenburg (1615?–1677) inquired into the issue.[30] In Italy the work of Redi publicly accused the Jesuit Father. Many historians have interpreted these reactions as manifestations of the new culture of experimental science. However, very few of them seem to have examined what Kircher really taught and what stimulated the scientific activity of his admirers and adversaries.

In Kircher the problem of spontaneous generation, or more precisely, the origin of life, was intimately connected to diverse important and difficult issues debated among scholars: the Creation of the world, the origin of contagious diseases, the formation of figures and colors in minerals, plants and animals, the origin of fossils and so on.[31] Although Kircher repeatedly addressed these matters in his entire corpus, a good example can be found in the twelfth and last book of his famous geocosmic encyclopedia, *Mundus subterraneus* (Amsterdam, 1664–1665).[32]

As for the generation of living beings, which seems to occur spontaneously from putrefied materials, Kircher first argues that the four elements themselves cannot produce living beings. For him there must be something that plays the role of seeds. He calls these seeds "separated seeds" (*semina decisa*). In his view, when one part of the body of a living being or its corpse is separated from the body, the invisible seeds residing in this part are separated and diffused everywhere in nature. For Kircher these seeds can produce inferior beings that are considerably degenerated from the original living beings; the cause of this degeneration is the weakening of the heat that supports the seminal power of these seeds.

In this context Kircher addresses the emergence of life. Following Thomas Aquinas, the intellectual guide of the Society, he first argues that the substantial form of living beings is drawn from the potentiality of matter. However, he regards these forms or souls as being "material" and divisible because they are drawn from the bosom of matter. Kircher then affirms that there is something formal lying hidden in the body parts of living beings and their cadavers. He identifies this entity with a spirit which resides in the "saline-sulfurous-mercurial" core of the visible

[29] On Kircher, see among others Kangro 1973; Leinkauf 1993; Findlen 2004; Fletcher 2012.

[30] See Hunter and Davis 1999–2000 13, 273–288. For Redi, see Redi 1996, *passim*.

[31] See Singer 1913, 9–11; Belloni 1985; Wilson 1995, 155–159.

[32] I have used the text of Kircher 1664–1665. See also Strasser 1996; Hirai 2007a.

seed and which vivifies matter.[33] This spirit is the real identity of his invisible seeds for Kircher. Thus the secret of the origin of life is intimately linked to the notion of this material spirit, which he certainly borrowed from the tradition of chymical philosophy.[34]

Like Sennert and Gassendi before him, Kircher goes even further to formulate a corpuscular interpretation of the invisible seeds by identifying them as minute corpuscles. These corpuscles can be easily transported by winds and rains and dispersed everywhere in the world. When these corpuscles find an appropriate matrix, thanks to their internal seminal power, they form a "web of life" (*tela vitae*) which provides the basis of a new living being.[35] To reinforce his theory, Kircher argues that living beings cannot be born spontaneously but only from materials which have previously been alive and animated. For him something of the soul, which once resided in a living being, can survive with its seminal power after the death of that living being. Thus he says:

> That is why the most immediate matter of the generation of beings born spontaneously is that seed of ours, in which lies hidden a spirit, so to speak, a certain soul separated from a living being (as Fortunio Liceti teaches skillfully) and remaining in its cadaver, not as a form but as spirituous corpuscles of this living being. A soul lies in these [corpuscles], as if it were placed in a vessel, after the death of the living being.[36]

Kircher clearly connects the origin of life with a certain material spirit that lies hidden in the cadaver of living beings. This spirit is conceived almost as a soul under the form of spirituous corpuscles. Because of this spirit, certain living beings are born from cadavers. But these newly born beings will not be of the same species as the cadavers. According to Kircher, at the death of a living being, the nature of its soul is weakened by the loss of original heat and degenerates into an inferior essence.

Kircher explains the emergence of life with these seminal corpuscles dispersed in the world. When these corpuscles are united under the form of a viscous mass and digested by ambient heat, they acquire an appropriate mixture and heat which allow the material soul, lying hidden in them as if they were placed in a vessel, to manifest itself in the form of life. Kircher argues that these corpuscles play the role of seeds only by analogy since they are not the seeds properly speaking, but certain "envelopes of the seminal reason-principles" (*involucra seminalium rationum*).[37] Coming from the body of beings that once lived, these corpuscles retain within themselves a

---

[33] Kircher 1664–1665, 12.1.6, 336–337.

[34] On the quest for a material spirit of life in chymical philosophy, see Debus 1984; Clericuzio 1994. On Kircher and chymical philosophy, see Baldwin 1990 and Baldwin 1993.

[35] Kircher 1664–1665, 12.1.6, 337.

[36] Kircher 1664–1665, 12.1.6, 337. "Quare materia proxima sponte nascentium generationis est semen illud nostrum, in quo spiritus latet, veluti anima quaedam a vivente decisa (uti Fortunius Licetus scite docet) et in cadavere remanens, non ut forma, sed veluti corpuscula spirituosa istius uiuentis, in quibus anima consistit, veluti in vase post mortem viventis relicta."

[37] On the doctrine of the seminal reason-principle in the Renaissance, see Hirai 2002; Hirai 2005, *passim*.

small, weakened part of the material soul. This soul, which is clearly identified with a certain spirit, comes to act as a substantial form which vivifies the mass. In this way, according to Kircher, living beings that seem to be born spontaneously are produced. The birth of such beings is not really spontaneous, nor abiogenetic; in other words, their life is not generated from a purely lifeless matter.[38]

11.5 A Brief Conclusion

Kircher's idea of seminal spirituous corpuscles carrying a material soul seems to come close to Gassendi's theory: seminal molecules are endowed with a "miniature soul" (*animula*), identified with a "miniature flame" (*flammula*), which is explicitly construed as a corporeal substance, namely a material spirit of life. Both authors thus defended kinds of living corpuscles or ensouled particles. This is not surprising, since Kircher was well informed of the contents of his friend's work. But a very similar idea can also be found in Sennert. As we have seen, Sennert was inspired by the work of Paduan natural philosopher Fortunio Liceti, and Kircher himself once refers to this name. As I have shown elsewhere, Kircher actually absorbed the text of Liceti at length in the course of his discussion on the corpuscular interpretation of spontaneous generation. The fame of Kircher overshadowed that of Liceti in this regard.[39]

One of the most striking observations set forth in this study of Sennert, Gassendi and Kircher is the gradually increasing importance of a material spirit of life and its identification with the soul or the seminal principle. Corpuscular interpretations played a crucial role in this evolution of identification. All three figures explained the origin of life, and especially spontaneous generation, in terms of living corpuscles which are internally animated by this material spirit of life.

References

Anstey, Peter R. 2002. Boyle on seminal principles. *Studies in the History and Philosophy of Biology* 33: 597–630.

Aristotle. 1984. Generation of animals. In *The complete works of Aristotle*, vol. 1, ed. Jonathan Barnes. Princeton: Princeton University Press.

Arthur, Richard T.W. 2006. Animal generation and substance in Sennert and Leibniz. In *The problem of animal generation in early modern philosophy*, ed. Justin E.H. Smith, 147–174. Cambridge: Cambridge University Press.

Baldwin, Martha R. 1990. Alchemy in the Society of Jesus. In *Alchemy revisited*, ed. Z.R.W.M. von Martels, 182–187. Leiden: Brill.

---

[38] On spontaneous generation, see among others von Lippmann 1933; Mendelsohn 1976, 37–65; Farley 1974; Van Der Lugt 2004.

[39] See Hirai 2007b.

Baldwin, Martha R. 1993. Alchemy and the Society of Jesus in the seventeenth century: Strange bedfellows? *Ambix* 40: 41–64.

Belloni, Luigi. 1985. Athanasius Kircher: Seine Mikroskopie, die Animalcula und die Pestwürmer. *Medizinhistorisches Journal* 20: 58–65.

Bernardi, Walter, and Luigi Guerrini (eds.). 1999. *Francesco Redi: un protagonista della scienza moderna*. Florence: Olschki.

Blank, Andreas. 2010a. *Biomedical ontology and the metaphysics of composite substances. 1540–1670*. Munich: Philosophia.

Blank, Andreas. 2010b. Material souls and imagination in late Aristotelian embryology. *Annals of Science* 67: 1–18.

Bloch, Olivier R. 1971. *La philosophie de Gassendi*. The Hague: Nijhoff.

Bowler, Peter J. 1971. Preformation and pre-existence of in the seventeenth century: A brief analysis. *Journal of the History of Biology* 4: 221–244.

Canguilhem, Georges. 1955. *La formation du concept de réflexe aux XVII*^e^*et XVIII*^e^*siècles*. Paris: PUF.

Clericuzio, Antonio. 1990. A redefinition of Boyle's chemistry and corpuscular philosophy. *Annals of Science* 47: 561–589.

Clericuzio, Antonio. 1994. The internal laboratory: The chemical reinterpretation of medical spirits in England (1650–1680). In *Alchemy and chemistry in the 16th and 17th centuries*, ed. Piyo Rattansi and Antonio Clericuzio, 51–83. Dordrecht: Kluwer.

Clericuzio, Antonio. 2000. *Elements, principles and corpuscles: A study of atomism and chemistry in the seventeenth century*. Dordrecht: Kluwer.

Debus, Allen G. 1984. Chemistry and the quest for a material spirit of life in the seventeenth century. In *Spiritus*, ed. Marta Fattori and Massimo L. Bianchi, 245–263. Rome: Ateneo.

Duchesneau, François. 1997. *Les modèles du vivant de Descartes à Leibniz*. Paris: Vrin.

Duris, Pascal. 2010. L'introuvable révolution scientifique: Francesco Redi et la génération spontanée. *Annals of Science* 67: 1–25.

Emerton, Norma E. 1984. *The scientific reinterpretation of form*. Ithaca: Cornell University Press.

Farley, John. 1974. *The spontaneous generation controversy from Descartes to Oparin*. Baltimore: Johns Hopkins University Press.

Findlen, Paula. 1993. Controlling the experiment: Rhetoric, court patronage and the experimental method of Francesco Redi. *History of Science* 31: 35–64.

Findlen, Paula (ed.). 2004. *Athanasius Kircher: The last man who knew everything*. London: Routledge.

Fisher, Saul. 2005. *Pierre Gassendi's philosophy and science*. Leiden: Brill.

Fletcher, John Edward. 2012. *A study of the life and works of Athanasius Kircher, "Germanus Incredibilis"*. Leiden: Brill.

Foote, Edward T. 1969. Harvey: Spontaneous generation and the egg. *Annals of Science* 25: 139–163.

Freudenthal, Gad. 1995. *Aristotle's theory of material substance: Heat and pneuma, form and soul*. Oxford: Clarendon.

Gassendi, Pierre. 1658. *Opera omnia*. 6 vols. Lyon.

Harvey, William. 1650. *De generatione animalium*. London.

Hirai, Hiro. 2002. Concepts of seeds and nature in the work of Marsilio Ficino. In *Marsilio Ficino: His theology, his philosophy, his legacy*, ed. Michael J.B. Allen and Valery Rees, 257–284. Leiden: Brill.

Hirai, Hiro. 2003. Le concept de semence de Pierre Gassendi entre les théories de la matière et les sciences de la vie au XVIIe siècle. *Medicina nei Secoli* 15: 205–226.

Hirai, Hiro. 2005. *Le concept de semence dans les théories de la matière à la Renaissance: de Marsile Ficin à Pierre Gassendi*. Turnhout: Brepols.

Hirai, Hiro. 2007a. Athanasius Kircher's chymical interpretation of the creation and spontaneous generation. In *Chymists and chymistry: Studies in the history of alchemy and early modern chemistry*, ed. Lawrence Principe, 77–87. New York: Science History Publications.

Hirai, Hiro. 2007b. Interprétation chymique de la création et origine corpusculaire de la vie chez Athanasius Kircher. *Annals of Science* 64: 217–234.

Hirai, Hiro. 2010. The world-spirit and quintessence in the chymical philosophy of Joseph Du Chesne. In *Chymia: Science and nature in medieval and early modern Europe (1450–1750)*, ed. Miguel Lopez-Perez et al., 247–261. Cambridge: Cambridge Scholars Press.

Hirai, Hiro. 2011. *Medical humanism and natural philosophy: Renaissance debates on matter, life and the soul*. Leiden: Brill.

Hirai, Hiro. 2012. Il calore cosmico in Telesio fra il *De generatione animalium* di Aristotele e il *De carnibus* di Ippocrate. In *Bernardino Telesio tra filosofia naturale e scienza moderna*, ed. Giuliana Mocchi et al., 71–83. Roma: Serra.

Hirai, Hiro. 2014. The new Astral medicine. In *A companion to astrology in the Renaissance*, ed. Brendan Dooley, 267–286. Leiden: Brill.

Hirai, Hiro, and Hideyuki Yoshimoto. 2005. Anatomizing the sceptical chymist: Robert Boyle and the secret of his early sources on the growth of metals. *Early Science and Medicine* 10: 453–477.

Hooykaas, Reijer. 1983. *The concept of element: Its historical-philosophical development* [orig. Ph.D. diss.: *Het Begrip Element in zijn historisch-wijsgeerige Ontwikkeling*. University of Utrecht, 1933].

Hunter, Michael, and Edward B. Davis, eds. 1999–2000. *The works of Robert Boyle*. 14 vols. London: Pickerling & Chatto.

Kangro, Hans. 1973. Kircher, Athansius. *Dictionary of Scientific Biography* 7: 374–378.

Kircher, Athanasius. 1664–1665. *Mundus subterraneus*. 2 vols. Amsterdam.

Lasswitz, Kurd. 1890/1926. *Geschichte der Atomistik*. Leipzig: Voss.

Leinkauf, Thomas. 1993. *Mundus combinatus: Studien zur Struktur der barocken Universalwissenschaft am Beispiel Athanasius Kirchers SJ. (1602–1680)*. Berlin: Akademie Verlag.

Liceti, Fortunio. 1618. *De spontaneo viventium ortu*. Vicenza.

LoLordo, Antonia. 2006. *Pierre Gassendi and the birth of early modern philosophy*. Cambridge: Cambridge University Press.

Lüthy, Christoph, et al. (eds.). 2001. *Late medieval and early modern corpuscular matter theories*. Leiden: Brill.

Marangio, Marilena. 1973. I Problemi della scienza nel carteggio Liceti-Galilei. *Bollettino di storia della filosofia* 1: 333–350.

Mendelsohn, Everett. 1976. Philosophical biology *vs* experimental biology: Spontaneous generation in the seventeenth century. In *Topics in the philosophy of biology*, ed. Marjorie Grene, 37–65. Dordrecht: Reidel.

Michael, Emily. 1997. Daniel Sennert on matter and form: At the juncture of the old and the new. *Early Science and Medicine* 2: 272–299.

Michael, Emily. 2001. Sennert's sea change: Atoms and causes. In *Late medieval and early modern corpuscular matter theories*, ed. Christoph Lüthy et al., 331–362. Leiden: Brill.

Murr, Sylvia. 1991. L'âme des bêtes chez Gassendi. *Corpus: Revue de philosophie* 16–17: 37–63.

Murr, Sylvia (ed.). 1997. *Gassendi et l'Europe (1592–1792)*. Paris: Vrin.

Newman, William R. 2006. *Atoms and alchemy: Chymistry and the experimental origins of the scientific revolution*. Chicago: University of Chicago Press.

Newman, William R., and Lawrence M. Principe. 1998. Alchemy vs. chemistry: The etymological origins of a historiographic mistake. *Early Science and Medicine* 3: 32–65.

Ongaro, Giuseppe. 2005. Liceto, Fortunio. *Dizionario biografico degli italiani* 65: 69–73.

Osler, Margaret J. 1994. *Divine will and the mechanical philosophy: Gassendi and Descartes on contingency and necessity in the created world*. Cambridge: Cambridge University Press.

Pagel, Walter. 1967. *William Harvey's biological ideas: Selected aspects and historical background*. Basel: Karger.

Redi, Francesco. 1996. *Esperienze intorno alla generazione degli insetti*, ed. Walter Bernardi. Florence: Giunti.

Rey, Roselyne. 1997. Gassendi et les sciences de la vie au XVIIIe siècle. In *Gassendi et l'Europe*, ed. Murr, 189–201. Paris: Vrin.
Roger, Jacques. 1963. *Les sciences de la vie dans la pensée française du XVIII^{e}siècle*. Paris: Colin.
Rosenfield, Leonora C. 1941. *From beast-machine to man-machine*. Oxford: Oxford University Press.
Sennert, Daniel. 1619. *De chymicorum cum Aristotelicis et Galenicis consensu ac dissendu*. Wittenberg.
Sennert, Daniel. 1636. *Hypomnemata physica*. Frankfurt.
Sennert, Daniel. 1633. *De chymicorum cum Aristotelicis et Galenicis consensu ac dissensu liber*. Paris: Societas.
Sennert, Daniel. 1650. *Opera omnia*. Lyon.
Shackelford, Jole. 2004. *A philosophical path for Paracelsian medicine: The ideas, intellectual context, and influence of Petrus Severinus, 1540–1602*. Copenhagen: Museum Tusculanum Press.
Singer, Charles. 1913. *The Development of the Doctrine of Contagium Vivum, 1500–1750*. London: Singer.
Solmsen, Friedrich. 1957. The vital heat, the inborn pneuma, and the aether. *Journal of Hellenic Studies* 77: 119–123.
Spink, Joseph S. 1960. *French free thought from Gassendi to Voltaire*. London: Athlone.
Stolberg, Michael. 2003. Particles of the soul: The medical and Lutheran context of Daniel Sennert's atomism. *Medicina nei Secoli* 15: 177–203.
Strasser, Gerhard F. 1996. Science and pseudoscience: Athanasius Kircher's *Mundus subterraneus* and his *Scrvtinivm... pestis*. In *Knowledge, science, and literature in early modern Germany*, ed. Gerhild S. Williams and Stephan K. Schindler, 219–240. Chapel Hill: University of North Carolina Press.
Taussig, Sylvie. 2009. L'Examen de la philosophie de Fludd de Pierre Gassendi par ses hors-texte. Rome: Serra.
Van Der Lugt, Maaike. 2004. *Le ver, le démon et la vierge: les théories médiévales de la génération extraordinaire*. Paris: Les Belles Lettres.
Van Melsen, Andreas. 1952. *From atomos to atom*. Pittsburgh: Duquesne University Press.
von Lippmann, Edmund O. 1933. *Urzeugung und Lebenskraft: Zur Geschichte dieser Probleme von den ältesten Zeiten an bis zu den Anfängen des 20. Jahrhunderts*. Berlin: Springer.
Wilson, Catherine. 1995. *The invisible world: Early modern philosophy and the invention of the microscope*. Princeton: Princeton University Press.

Chapter 12
Mechanism and Chemical Medicine in Seventeenth-Century England: Boyle's Investigation of Ferments and Fermentation

Antonio Clericuzio

Abstract In this paper I take into account Boyle's explanation of vital phenomena, paying special attention to his work on fermentation. Boyle never published a specific work on ferments and fermentation, yet, this subject played a central part in his medical agenda. He pointed out that the understanding of ferments and fermentation would throw new light on physiological phenomena, notably on digestion. He was not isolated in his quest for the knowledge of fermentation: most early modern natural philosophers and physicians thoroughly investigated this topic providing different accounts of the fermentative process. The research on fermentation became an integrant part of the Oxford physiologists' work on blood and respiration. In the first part of the paper, I examine the alchemical and Paracelsian roots of early modern research on fermentation, in the second, I investigate the chemical and medical work on fermentation carried out by the English physiologists (including Thomas Willis and the English Helmontians), as well as by Boyle and Newton.

Keywords Fermantation • Chemistry • Atomism • Matter • Boyle • Physiology • Blood • Respiration

12.1 Introduction

In *A Free Inquiry into the Vulgarly Receiv'd Notion of Nature* (1686), Boyle summed up his view of the human body along the following lines:

Parts of this paper have been presented at a range of seminars and conferences. I would like to thank all the audiences for their questions and suggestions, as well as the anonymous reviewers for their comments.

A. Clericuzio (✉)
Dipartimento Studi Umanistici, Roma Tre University, Rome, Italy
e-mail: antonio.clericuzio@libero.it

P. Distelzweig et al. (eds.), *Early Modern Medicine and Natural Philosophy*, History, Philosophy and Theory of the Life Sciences 14,
DOI 10.1007/978-94-017-7353-9_12

> …I look not on a human body as on a watch or a hand-mill, i.e., as a machine made up only of solid, or at least consistent parts; but as an hydraulical, or rather hydraulo-pneumatical engine, that consists not only of solid, and stable parts; but of fluids, and those in organical motion. And not only so, but I consider that these fluids, and the liquors and spirits, are in a living man so constituted, that in certain circumstances the liquors are dispos'd to be put into a fermentation or commotion, whereby either some depuration of themselves, or some discharge of hurtful matter by excretion, or both, are produc'd…[1]

Boyle included the above statement in a work he sent to the press at the end of his career, several years after the publication of *The usefulness of experimental philosophy* (1663), where he adopted a range of iatrochemical theories.[2] The quotation from *Notion of Nature*, testifying to Boyle's continual commitment to some central themes of iatrochemistry, deserves special attention, notably the statement that body fluids are "in organical motion." He was not explicit about the meaning of "organical motion," yet one can argue that he referred to a kind of motion that he saw as not reducible to the mechanical laws. Boyle by no means jettisoned mechanical explanations; he argued that in the investigation of living organism they were to be integrated by taking into account agents and processes following rules other than those of the impact of corpuscles.[3]

In order to assess Boyle's view of human body it seems worthwhile to provide a brief account of his mechanical philosophy – a topic that is still matter of contention among Boyle scholars. My first point is that a variety of versions of the 'mechanical philosophy' were to be found in seventeenth-century science. Those who followed Descartes aimed at reducing all natural phenomena to the motion and impact of particles of inert matter. Those who adopted Epicurus' atomism, like Pierre Gassendi, Walter Charleton and Robert Hooke, explained the physical world by means of matter and motion, yet they maintained that matter was not inert, stating that motion was a property of corpuscles.[4] Boyle was concerned about the renaissance of Epicurean philosophy and criticized the modern atomists' theory of the origin of motion. He rejected their tenet that motion was innate to matter: God bestowed motion to corpuscles and guided their movements. For Boyle, the origin and the determination of motion (velocity and direction) depend on God.[5] In Boyle's view, matter and motion cannot "constitute this beautiful and orderly world" if they are not directed by an intelligent agent. To Boyle, matter was homogeneous and inert: "The great mass of lazy matter was created by God at the beginning, and by him put into a swift and various motion, whereby it was actually divided into small parts of several sizes and figures, whose motion and crossing of each other were so guided by God, as to constitute, by their occursions and coalitions, the great

[1] Boyle 1999–2000, vol. 10, 540. On Boyle's *Notion of Nature*, see Hunter and Davis 1996.

[2] See Clericuzio 1993.

[3] Boyle 1999–2000, The Usefulness of Experimental Natural Philosophy (1663), vol. 3, 310–311.

[4] See Clericuzio 1998 and Clericuzio 2001.

[5] Boyle 1999–2000, *Reason and Religion* (1675), vol. 8, 259–261.

inanimate parts of the universe."[6] Although he subscribed to Descartes' view of inert matter, Boyle differed with Descartes over the contents and method of the mechanical philosophy. I am here focusing on the classification of corpuscles, a relevant aspect of Boyle's matter theory as it enabled him to provide corpuscular explanations of chemical change. Simple corpuscles, which are endowed with mechanical properties (i.e., shape, bulk, motion, or rest), form compound corpuscles that Boyle called "corpuscles of the second order." The latter are very seldom broken as their texture remains unchanged in several chemical reactions. Boyle wrote:

> That there are multitudes of corpuscles, which are made up of the coalition of of several of the former *minima naturalia*, and whose bulk is so small, and their adhesion so close and strict, that each of these little primitive concretions or clusters (if I may so call them) of particles is singly below the discernment of sense, and though not absolutely indivisible by nature into the prima naturalia that composed it, or perhaps into other little fragments, yet, for the reasons freshly intimated, they very rarely happen to be actually dissolved or broken, but remain entire in a great variety of sensible bodies, and under various forms or disguises. As not to repeat what we lately mentioned of the undestroyed purging corpuscles of milk, we see that even grosser and more compounded corpuscles may have such a permanent texture: for quicksilver, for instance, may be turned into a red powder for a fusible and malleable body, or a fugitive smoke, and disguised I know not how many other ways, and yet remain true and recoverable mercury.[7]

This topic crops up again in the 1685 tract on specific medicines, where Boyle argues: "the particles of divers bodies may retain their nature in all the digestion and strainers they pass through."[8] Compound corpuscles are endowed with mechanical properties as any parcel of matter, but in addition to the more fundamental properties, they have specific chemical ones, ultimately depending on their texture. What is to be stressed here is that for Boyle compound corpuscles operate according to their chemical properties. Boyle's effort to reform the chemists' classification of chemical substances is in fact grounded on experimental arguments as well as on the notion of chemical corpuscles.[9] The importance of chemical corpuscles can hardly be overestimated, as it bridged the gap between Boyle's corpuscular theory of matter and the chemical and medical investigations. Indeed, Boyle's classification of corpuscles according to their complexity marks a sharp difference between Boyle's

[6] Boyle 1999–2000, The Usefulnesse of Experimental Natural Philosophy (1663), vol, V, 253. In *The Christian Virtuoso* Boyle stated that the mind/ body union is not supernatural, but natural, though it is not mechanical. He styled it "supra-mechanical." Boyle 1999–2000, vol. XII, 478. See Anstey 2000, 190–197. Boyle's manuscripts on this subject were published in MacIntosh 2005, 246–255.

[7] Boyle 1999–2000, *The Origine of Formes and Qualities* (1666), vol. V, 326. Boyle articulated this view by stating that the corpuscles of second order have "their particles so minute and strongly coherent, that nature of her self does scarce ever tear them asunder, as we see, that Mercury and Gold may be successively made to put on a multitude of disguises, and yet so retain their nature, as to be reducible to their pristine forms." Boyle 1999–2000, About The Excellency and Grounds of the Mechanical Hypothesis (1674), vol, VIII, 113. On the reduction to the pristine state, see Meinel 1988; Clericuzio 2000:135–148; Newman 2006, 112–123; 190–198.

[8] Boyle 1999–2000, vol. X, 366.

[9] See Clericuzio 2000, 133–135.

and Descartes' version of the mechanical philosophy, in that Boyle refrained from adopting a reductionist approach to chemical and medical phenomena.[10] Boyle did not have recourse to imaginary sizes and shapes of corpuscles and was very cautious in framing hypotheses about micro-mechanisms operating in the human body.

In his recent book on Malpighi, Bertoloni Meli has thoroughly investigated the importance of mechanical views in early modern medicine, providing a broad definition of the term 'mechanical,' more inclusive than the strict version adopted by Descartes and his followers.

Bertoloni Meli argued that "By mechanical he [e.g. Steno] and other anatomists understood 'machine-like' rather than based on the laws of mechanics: this interpretation goes hand in hand with a view of seventeenth-century mechanics according to which objects take center stage and embody more abstract relations. As in mechanics, in anatomy too understanding a complex structure meant decomposing it and recognizing in it elements associated with simpler, known objects that could be understood and handled separately. The notion of machine at the time was quite a complex and heterogeneous one…" [11] This is a very clarifying statement, especially as it points to the importance of considering complex machines for understanding early modern anatomists' cognitive practice. As it is apparent from the quotation at the beginning of this paper, Boyle saw the human body as a complex machine, i.e., as a "hydraulo-pneumatical engine." The analogy between living bodies and machines occurs in a number of writings, including *The Christian Virtuoso*, where Boyle described the egg as "a wonderful mechanical contrivance… so fine and difficult a piece of mechanism, that the most skillful artists, (whether mathematicians or chemists) that have attempted to detain one liquor in the midst of another, every way ambient, have found their industry defeated."[12] It is to be noticed that Boyle imposed some restrictions to the use of the mechanical analogies. He maintained that perfect devices as living organisms "cannot be satisfactory explicated after the manner of the acting of meer corporeal agents," as they are produced under the "superintendence and guidance" of God's wisdom.[13] He therefore stressed the gap between human artifacts and the works of the divine workman.

Despite Boyle's use of mechanisms for understanding organisms, it is apparent that his view of living bodies was more nuanced than the mechanical one, as he often merged chemical theories and experiments with mechanical notions. I believe that Boyle subscribed to the view that "understanding a complex structure meant decomposing it and recognizing in it elements associated with simpler, known objects that could be understood and handled separately," yet such a knowledge, for Boyle, was useful as far as the structure of parts was concerned, but could not

[10] See Boyle 1999–2000, The Usefulnesse of Experimental Natural Philosophy (1663), vol. III, 256–257. On Boyle's intermediate causes, see Clericuzio 2000, 129–148 and Anstey 2014, 118–119. I disagree with Alan Chalmers 2012, 561, who maintains that Boyle "lacked appropriate notions of intermediate causes in chemistry.".

[11] Bertoloni Meli 2011, 13–14.

[12] Boyle 1999–2000, vol. XII, 447.

[13] Boyle 1999–2000, vol. III, 247–8.

explain *per se* the functions of living bodies.[14] According to Boyle, the study of size, shape and structure of body parts is useful to anatomy, but is inadequate for the understanding of living organisms. This is spelled out in *The Christian Virtuoso*, where Boyle aimed at showing "the religious use a Christian Virtuoso may make of the contemplation of the Microcosm."[15] He split the section devoted to the study of man into two sub-sections, the first dealing with the anatomy of a dead man, the second with the living human body. A living human body performs functions that cannot be reduced to the number, size, shape, fabrick of parts and to the "contrivance and symmetry of the whole machine as such."[16] He listed a series of topics to be taken into account for the understanding of living organisms, including "the chemical operations that nature exercises in a living body, especially in the liquors and other fluids it contains," notably, the invisible fluids (i.e. spirits), made of "active corpuscles," and "the ferments, or principles analogous to them, that are supposed to be lodged in particular parts."[17] He articulated his view of anatomy as follows:

> For all anatomy can do, is to manifest or display the structure of the *consistent parts*, such as the bones, cartilages, nerves, arteries, veins etc., and expose to our senses the visible liquors of the body, such as blood, gall, the concreted juices, urine, etc. But it cannot show us either of the two sorts of invisible parts, viz. the *animal* and other *spirits*, and the ferments, (or principles analogous to them) that may reasonably be supposed to lodge in the stomach, kidneys, and other particular parts. And yet the influences and operations of these are so considerable, that I am apt to think, that most of the parts of the grosser body seem intended by nature, but as a kind of kitchens to dress the aliment, and make its finer parts pure and subtil enough to become animal; or if you please, hormetick or impulsive spirits, fit to actuate the brain and nerves, and thereby to become the grand instruments of sense, motion, and imagination.[18]

As attested by this quotation, Boyle adopted a relevant notion of Paracelsian and Helmontian medicine, namely, the so-called living anatomy – which was evidently at odds with the strict version of the mechanical view of human body. Unlike most Paracelsians and Helmontians, Boyle (who had practiced anatomy throughout his career), did not rule out anatomical knowledge, he rather aimed at integrating it with the chemical analysis of body fluids.[19]

The study of ferments and fermentation was meant to fulfil such a task. Boyle maintained that fermentation was responsible "for some depuration of themselves [e.g., body fluids]," and it produced "some discharge of hurtful matter by excretion."[20] Boyle maintained that physiological phenomena such as digestion, assimilation and

[14] Bertloni Meli 2011, 14.

[15] Boyle 1999–2000, vol. XII, 449.

[16] Boyle 1999–2000, XII, 450.

[17] Boyle 1999–2000, XII, 471–2.

[18] Boyle 1999–2000, vol.XII, 473. Italics are Boyle's. *Hormetic* (from the Greek ὁρμητικός) means having the property of exciting.

[19] On the living anatomy, see Pagel and Rattansi 1964. On Boyle's medical agenda, see Kaplan 1993 and Hunter 1997. Kaplan and Hunter paid little or no attention to Boyle's use of chemistry in medicine

[20] Boyle 1999–2000, vol. X, 540.

excretion were not the outcome of mechanism only (as in Cartesian medicine), being the result of mechanisms and chemical reactions – the latter being activated by ferments.

In this paper I set out to investigate Boyle's explanation of vital phenomena, paying special attention to his work on fermentation, in the context of seventeenth-century English medicine. An important part of Boyle's study of fermentation was aimed at exploring its practical uses, i.e., the preparation of medicines and food preservation.[21] As we shall see, most seventeenth-century physicians had recourse to fermentation to account for physiological phenomena, notably, respiration, digestion, hematopoiesis and the motion of blood, as well as for a range of diseases.

In order to locate Boyle's and the English physiologists' research on fermentation in historical perspective, in the following section I take into account the alchemical and Paracelsian views of ferments and fermentation. As we shall see, they paved the way to a research tradition that reached a peak in the second half of the seventeenth century.[22]

12.2 Fermentation in Alchemy and in Paracelsian Medicine (to 1650)

The recent reassessment of the history of alchemy has highlighted the experimental contributions the alchemists gave to modern science and has rightly stressed the importance of quantitative approaches in alchemy, long deemed spiritual and esoteric, and having little or no impact on early modern science.[23] Yet, the so-called new historiography of alchemy has often downplayed a significant aspect of alchemy, i.e., the interpretation of the *opus* (i.e., the preparation of the philosophers' stone) as a qualitative process involving a series of changes expressed in terms of purification, fermentation, digestion, maturation. It is my cotention that alchemical operations were often described by means of analogies with processes occurring in living organisms. The alchemical *opus* was described as a process triggered by active principles and formative powers, not as the outcome of mechanical change, namely, aggregation and separation of corpuscles of matter. Metals were supposed to undergo generation, vegetation and ripening – processes to be reproduced by the alchemist in the laboratory in order to achieve transmutation.

---

[21] On early modern English medicine see Davis 1973; Webster 1975; Frank 1980; and Wear 2000. According to Davis 1973: 212, Harvey's (limited) use of fermentation to explain the motion of blood and the heart diastolic motion encouraged English physicians to pursue the study of blood fermentation.

[22] For the Paracelsian and Helmontian theories of fermentation see Pagel 1982, 79–87.

[23] Newman and Principe, 1998, 2001 and 2002; Newman 2006. Newman's and Principe's interpretation has been criticized in Vickers 2008. Newman's response is found in Newman 2009. See also Moran 2005; and Principe 2012.

Volatile substances and formative powers played a central part in alchemy: for most alchemists, spirits were not just the volatile particles of bodies, they were the powers hidden in objects – powers responsible for the generation and growth of a range of substances, including metals. This is apparent in two influential alchemical texts, i.e., pseudo-Lull's *Testamentum* (fourteenth century) and Pietro Bono's *Pretiosa margarita novella* (*New Pearl of Great Price*, fourteenth century). The author of *Testamentum* employs much of the terminology and concepts related to living organisms and calls the *multiplicativa virtus* "spirit," because it has the power of giving life to the dead body (*vivificare corpus mortuum*).[24]

Most alchemists maintained that the philosophers' stone transmuted metals by means of fermentation. According to the author of *Testamentum*, the stone needs to undergo a process of fermentation in order to become a perfect elixir. Such a process is a purification produced by digestion, requiring the action of the ferment. The latter has the power of converting matter into its own nature.[25] Pietro Bono paid special attention to fermentation, devoting a section of his work to ferment, to its properties and to its transmuting power. He started by saying that the *opus* is not complete if it does not include the work of ferments.[26] For Pietro, the stone is the same as the ferment, as it transforms the mass of the body it acts upon, but is not converted into the new substance.[27] These views are summed up in Martin Ruland's *Lexicon of Alchemy* (1612) where we read: "For even as a small modicum of ferment, or yeast, can leaven a large mass of flour, so does the chemical ferment assimilate itself to the thing to be fermented. [...] The Stone itself is the ferment; Gold and Mercury are also called Ferment."[28]

Alchemists often used theories based on the notions of *semina*, spirits and ferments to explain the generation and transmutation of metals.[29] The legacy of alchemy is apparent in Paracelsus and in his followers, notably in the medical works of Petrus Severinus and of Oswald Croll. They saw human physiology as the outcome of chemical reactions triggered by spiritual principles. Paracelsians saw spirits as the cause of metal generation and transmutation. Moreover, in his medical

[24] Pereira and Spaggiari 1999, 32–36. For the philosophical and medical sources of the *Testamentum* see Pereira 2003.

[25] Pereira and Spaggiari 1999, 136. "Et cum illis [e.g. gold and silver] debes facere fermentacionem tui lapidis cum naturali coniunctione, et deinde habebit perfectam ingressionem in omni alio metallo per medium fermenti, quod traxit in naturam propinquam vere medicine, que participat cum essencia perfeccionis perfectorum et cum [...] corruptione imperfectorum ad illam medicinam et fermentum."

[26] Bono da Ferrara 1976, 143. "Del fermento dunque, senza il quale l'arte dell'Alchimia non si può finire e fare perfetta."

[27] Bono da Ferrara 1976, 143. "...sì come il fermento della pasta vince la la pasta et a sé la converte sempre, così questa pietra converte a sé gli altri mettalli; e sì come una parte del fermento della pasta può convertire infinite parti della pasta a vicenda, e non esser convertito, e così questa pietra può convertire a sé la più parte de mettalli e non esser convertita."

[28] Ruland 1964.

[29] Cf. Hirai 2005 and Shackelford 1998.

tracts he explained digestion and growth as the outcome of fermentation.[30] In the macrocosm, Paracelsus argued, salt and sulfur produce fermentation in the bowels of the earth, whence minerals and rocks are generated. In addition, he saw the elixir as a medicine fermented from the seven metals.[31] Fermentation was often explored in connection with the alchemical transmutation and with the preparation of chemical medicines. Like pseudo-Lull and Pietro Bono, Andreas Libavius claimed that a small quantity of ferment brings about the transformation of a given substance into its nature. The ferment brings about the transmutation by means of spirits penetrating the bulk of the body ("totam penetrat massam").[32]

In the early decades of the seventeenth century, an increasing number of physicians and naturalists had recourse to fermentation to account for a variety of phenomena, as attested by Pietro Castelli, Daniel Sennert, Angelo Sala, Edward Jorden, among others. A physician and professor of medicine, first in Rome and subsequently in Messina, Pietro Castelli (c. 1575–1661) adopted chemical theories and experiments in his medical work. He dealt with digestion in his *Epistolae medicinales* (1626) and explained it as the action of an acid ferment found in the stomach. He ruled out the traditional medical views of digestion as *concoctio* produced by heat, since the latter, he claimed, played only an indirect role in digestion. Digestion was the outcome of fermentation, activated by an acid spirit in the stomach that he described by employing a typical Paracelsian parlance. Such a spirit—for Castelli—was "the alchemist of the stomach."[33] Fermentation occurs at various stages during digestion, and chyle is purged of its feces by fermentation. Daniel Sennert (1572–1637), an influential professor of medicine at the University of Wittenberg, referred to fermentation in his explanation of digestion. Sennert was a critic of the Paracelsians and did not rule out the action of heat in the digestion (notably in chylifaction). When he dealt with scurvy (*De scorbuto*), he argued that by means of a process analogous to fermentation the gross and fixed parts of food are made volatile and perfected, i.e., free of their excrement.[34]

A Calvinist apothecary from Vicenza, Angelo Sala (1576–1637) spent most of his life in Protestant countries (in German States and in the Low Countries). His chemical works are mostly practical, focusing on the preparation of chemical medicines, on vitriol, and on antimony.[35] A follower of Paracelsian iatrochemistry, Sala adopted a corpuscular matter theory and dealt with fermentation in his *Hydrelaeologia* (1633), a work devoted to distillation and to the extraction of spirits. Sala defined fermentation as the internal motion of the constituent particles of bodies.

[30] The links between alchemy and Paracelsus have been stressed by Pagel 1958 and Bianchi 1994. Kahn has pointed out that alchemy and Paracelsianism merged in the second half of the 16th century; see Kahn 2007, 597.

[31] Paracelsus 1929, vol. I, 30; and vol. VIII, 187–8.

[32] Libavius 1597, 74. For Libavius, see Hannaway 1975 and Moran 2007.

[33] Castelli 1626, 154: "Idem spiritus acidus ille est, quem ventriculi Alchimistam vocat Paracelsus." For Castelli, see Clericuzio 2010.

[34] Sennert 1641, 698. See Newman 2012, 119–121.

[35] On Angelo Sala see Gelman 1994.

Fermentation is generated by heat in the presence of moisture and brings about a new and nobler substance.[36] In *Saccharologia* (1637) he paid special attention to the fermentation of sugar, and described the resulting production of "phlogistic spirit."[37]

A physician from Kent, Edward Jorden (c. 1565–1633), studied in Padua and practiced in London and in Bath. He studied mineral waters and explained the generation of metals as fermentation. Following the Paracelsian outlook (he often referred to Paracelsus, Dorn and Croll), Jorden claimed that spirit was the plastic agent operating in the earth as a ferment: "There is a Seminarie Spirit of all minerals in the bowels of the earth, which meeting with convenient matter, and adiuvant causes, is not idle, but doth proceed to produce minerals, according to the nature of it, and the matter which it meets withall: which matter it workes upon like a ferment, and by his motion procures an actuall heate, as an instrument to further his work; which actuall heate is increased by the fermentation of the matter."[38]

Anthon Günther Billich (1598–1640), Sala's son in law and physician to the Count of Oldenburg, wrote an erudite work (*Anatome Fermentationis Platonicae*, 1639) containing a survey of philosophical and medical doctrines related to fermentation. He maintained that the origin of life and every physiological process were the outcome of fermentation.[39] A critic of the Paracelsian chemical principles, Billich, as Pagel put it, "allows the classical elements a prominent role in bringing about fermentation, although he regarded fire as the leader."[40] Billich maintained that chemists could not claim priority in the discovery of fermentation, which was well-known in the antiquity, as attested by Plato's *Timaeus*.

Jan Baptista van Helmont assigned a prominent role to fermentation both in natural philosophy and in medicine. In his influential *Ortus medicinae* (1648) he maintained that generation and growth, as well as transmutations were not the outcome of purely material change, but required also the action of ferments.[41] As Walter Pagel pointed out: "In according ferments a central position in his natural philosophy, van Helmont was indebted to alchemy in the first place and to Paracelsianism

---

[36] Sala 1647, 95. "Fermentatio igitur est motus quidam, seu alteratio, a calore interno, in humido agente inducta, qua diversae & inter se pugnantes, substantiae elementares, partim separantur, partim in unum nobiliorem mixtionis modum, ac unionem rediguntur, quod rerum fermentantium strepitu, pugna, & humidi turgescentia apparet, hac mediante res, ad subtiliores, spirituosas, & balsamicas, varieque operandi, & penetrandi virtutes exaltantur...."

[37] Sala 1647, 164

[38] Jorden 1631, 82. For Jorden, see Debus 1969.

[39] Billich 1646. "Omnis fere vita fermentum est. Nam quid aliud sit lactea illa spuma, ex qua nascimur? Nati sine fermento vitam non ducimus, sive valeamus, sive aegrotemus. [...]Fermento cor pulsat, ateriae saliunt, venae bulliunt, cibus coquitur, sanguis conficitur, corpus alitur. Fermentum est, vel certe fermenti particeps, quod expiratur, quod expuitur, quod excreatur, quod expectoratur, quod per alvum, per vescicam, per nares, quod per uterum excernitur."

[40] See Pagel 1982, 82. Billich 1646, 540 stated: "Fermentatio est motus terrae, vi ignis interni concitatae, ut beneficio aquae intermediae aerescat atque ignescat."

[41] Van Helmont 1648, § 1, 111. "Notitia fermenti, ut nulla in Scholis jejunior, ita nulla utilior. Fermenti nomen, ignotum hactenus, nisi in panificio: cum attamen nulla in rebus fiat vicissitudo, aut transmutatio, per somniatum appetitum hyles: sed duntaxat solius fermenti opera."

in the second. [...] The soul-like ferment vanquishes matter. Its luminous quality enables ferments to enter bodies which are thereby exalted and sublimated, reaching a level intermediate between the elemental and the spiritual, a mercurial existence."[42] For van Helmont, ferments are psycho-physical agents, principles of activity, God's gifts and the roots of activity in nature, prior to *semina*.[43] Ferments operate according to a kind of program established by God, they are endowed with the power of shaping the prime matter, i.e., water. Ferments dispose matter to receive the idea, or the first shape of individual objects. Van Helmont's ferments transform the mass into something akin to the image contained in the ferment. He compared this process with pregnancy, by stating that the "Image of the Ferment makes the mass pregnant with semen." Ferments operate by means of "odours" having a penetrative power that impregnates and disposes matter. In the human body, ferments are responsible for major physiological processes, notably, digestion and the origin and motion of blood. Food is assimilated by the action of a specific agent, a spiritual driving force immanent to the human body. Chemical reactions produced by acid juice in the stomach are directed by the ferment, since acids do not have any digestive power in themselves.[44] Digestion is a qualitative transformation of matter, not a change in bulk and size, a transmutation like the alchemical *opus*. So, he argues, digestion occurs by means of a true fermentative metamorphosis ('per veram fermentalem metamorphosin').[45]

12.3 Fermentation, Blood and Animal Heat

The young Robert Boyle's medical writings betray a strong Helmontian outlook, though they show no trace of the spiritual agents—that, as we have seen, played a central part in van Helmont's medicine. It is no surprise that Boyle stressed the importance of ferments and fermentation in *The Usefulnesse of Experimental Natural Philosophy* (1663), largely written in the 1640s and 1650s and extolling the importance of chemistry to improve the knowledge of the human body as well as to renew man's dominion over nature. Boyle wrote:

> And let me add, that he that thoroughly understands the nature of ferments and fermentation, shall probably be much better able then he that ignores them to give a fair account of divers phaenomena of severall diseases ... which will perhaps be never thoroughly understood without an insight into the doctrine of fermentation, in order to which, for that and other reasons, I design'd my historicall notes touching that subject.[46]

[42] Pagel 1982, 79–80. See also Giglioni 2000.

[43] van Helmont 1648, § 24–25, 36.

[44] van Helmont 1648, 218.

[45] van Helmont 1648, 220. For the Helmontian views of digestion see Multhauf 1955 and Clericuzio 2012.

[46] Boyle 1999–2000, vol. III, 321.

Boyle's work on fermentation does not survive, but there are fragments of his projected essay in his manuscripts. His investigation of fermentation started at an early stage of his career, under the influence of van Helmont, and was spurred by members of the Hartlib Circle, like George Starkey, Benjamin Worsley, and Frederick Clodius. Hartlib's associates explored fermentation as part of a wide range of chemical and medical investigations, dealing with nutrition, pharmacology and alchemy. Starkey wrote to Boyle about ferments in a letter of April-May 1651 dealing with transmutation. Echoing van Helmont, Starkey wrote: "we want a formal principle which is a *fermentum archeale*, which is the invisible seed...." On 3 February 1652 Starkey insisted on this point by claiming that "every transmutation presupposes a corruptive ferment."[47] As attested by Boyle's workdiaries of 1655, Starkey and Clodius sent Boyle recipes for the confection of liquors, the extraction of oils from vegetables and of spirit and fixed salt from urine. All these operations involved fermentation.[48]

Boyle followed with the utmost attention Worsley's experimental investigations of nitre.[49] In an undated paper on saltpeter, Worsley came to the conclusion that he had "found out by Experience a ferment, which mixt among fit Matter, will cause the whole at length to turne into the nature of nitrum."[50] Boyle explored the practical uses of fermentation, notably the new perspectives it opened to pharmacology and nutrition. For Boyle, fermentation can produce such change in vegetables as to make them fit to perform extraordinary cures, as attested by the preparation of medicines to cure diseases of the kidneys, and the confection of opiates.[51] It offers great potentiality to correct the taste and other qualities of food and drinks and to elaborate new techniques for the preservation of food. As he put it in *The usefulnesse*, "the distinct knowledge of the true nature and particular phaenomena of fermentation would enable men to prepare a great variety of drinks... to correct and meliorate both hard and liquid aliments...." This can be achieved by means of chemistry, since—he stated—"chemistry can enable us to confer a very grateful taste on very many of the things we eat, barely by a skillful and moderate untying the formerly clogged spirits, and other sapid parts contained in them." The fermentation of some aliments—he believed—could help keep them "long uncorrupted."[52] As one gathers from *The usefulnesse*, Boyle planned to write a natural history of fermentation, of which a list of headings is to be found among his papers, possibly written in the 1650s. It is apparent that in the 1650s Boyle focused on fermentation and on its medical uses, as attested by scattered notes on fermentation to be found among his

[47] Starkey 2004, 25; 80. For Starkey, see Newman 1994.

[48] Royal Society Boyle Papers (hereinafter BP), 8, 140v–146v.

[49] As Webster put it, Worsley's project of manufacturing saltpetre, which began in the mid-1640s, "was supported wholeheartedly by the Hartlib circle and it may well have provided one of the main incentives for Boyle's interest in experimental chemistry," Webster 1975, 379.

[50] Worsely 1653, 39/1/11A. See Newman and Principe, 2002, 240. According to Newman and Principe, Worsley's text was composed in 1653.

[51] See Boyle 1999–2000, vol. III, 355–6 and BP 26, fols 96–97 (workdiary April 1657).

[52] Boyle 1999, vol. 3, 350–7; 361.

manuscripts of mid-1650s and sections of *The usefulnesse*, written in the late 1650s – dealing with the fermentation of body fluids, notably urine and blood.[53] Boyle set out to explore the different stages of fermentation, the changes in color, odor and other properties of bodies fermented; which animals, vegetables and mineral bodies are capable of fermentation; "what things promote fermentation" and "what things oppose fermentation"; finally "A corollary of the difference betwixt fermentation and putrefaction."[54] Boyle scrutinized a number of fermentative processes paying special attention to the conspicuous alterations in the bulk of fermenting liquors and to the chemical change produced by fermentation: "ardent spirit of fermented sugar and highly rectified Spirit (sal ammoniac), being shaken together in the cold did presently begin and in a short time made a coagulum, that seem'd at first to take up the greatest part of the space possess'd by the two mingled Liquors, but afterwards subsided in the form of a white saline powder which I judg'd not to fill much above a quarter of that space."[55] As attested by a note to be found in his workdiary of late 1660s, Boyle tried to provide a corpuscular explanation of fermentation, framing the hypothesis that "energetic particles" activated the fermentative process.[56]

In the mid-seventeenth century there was a growth of interest in ferments and fermentation among physicians. This is shown in a wide range of medical works published in England and in Oxford university medical disputations too.[57] In propositions disputed in 1651, Ralph Bathurst rejected the Galenic view of digestion, arguing that digestion was performed by an acid ferment, which was secreted by the walls of the stomach.[58] Eminent English physiologists like Thomas Willis, Francis Glisson and Walter Charleton engaged in the study of fermentation in relation to digestion, respiration, the origin and the circulation of blood. In 1659, Thomas Willis published *Diatribae duae medico-philosophicae*, the first tract devoted to fermentation (*De fermentatione*), the second to fevers (*De febribus*). The former was meant to provide a theoretical introduction for the work on fevers. Willis' work on fermentation deals with the composition of mixed bodies by adopting the five chemical principles (e.g., spirit, salt, sulfur, water and earth). He defined fermentation in chemical and corpuscular terms, stressing that it brings about the transformation and perfection of bodies. For Willis, fermentation occurs in both natural and artificial bodies, "in all which is found an heterogeneity of parts or particles, to wit, there are in them some substances light, and alwaies endeavouring to fly away: and also there are others thick, earthy, and more fixed, which intangle the subtil particles, and detain them in their embraces, whilst they endeavour to fly away…."[59] Fermentation is a qualitative change, which is the outcome of "an intestine motion

[53] BP 25, p. 349; Boyle 1999–2000, vol. III, 321–322.

[54] BP, 28, fol. 403.

[55] BP, 26, fol. 138 (workdiary: early 1670s).

[56] BP 44, fol. 48.

[57] Webster 1975, 139.

[58] See Frank 1980, 107.

[59] Willis, 1659; English translation Willis 1684, 1. On Willis' doctrine of fevers see Bates 1981.

of particles, or the principles of every body, either tending to the perfection of the same body, or because of its change into another."[60] He identified the fermentative agent with spirits, "substances highly subtil, and aetherial particles of a more divine breathing, which our parent nature hath hid in this sublunary world, as it were the instruments of life and soul, of motion and sense…."[61] Because of their affinity with the corpuscles of sulfur, spirits produce with them a sweet, stable and lasting compound, which is the main component of both vital and animal spirits and the agent of fermentation.[62] Like van Helmont's *Archeus*, Willis' ferments are located in the main organs performing vital functions. Vital spirits originate from a small particle of spirit, which is activated in the heart by a ferment keeping blood in constant fermentation: "so by the fermentation, or accension which the blood suffers in the bosom of the heart, very many particles of spirit, salt and sulphur endeavour to break forth from its loosened frame: by which being much rarified, and like water boyling over a fire, the moved and boyling blood is carried through the vessels, not without great tumult and turgescency."[63] Willis adopted the iatrochemical outlook by claiming that an acid digestive ferment was found in the stomach. Through the action of the ferment, food is broken into small particles and chyle is fermented, acquiring a whitish colour, "by the reason that sulphureous particles are dissolved with the saline and mixed with the acid ferment."[64]

Willis explained fevers on the basis of the chemical composition and motion of blood. Blood is constant fermentation, which "depends on both on the heterogeneity of the parts of the blood itself, and on the various ferments, which are breathed into the mass of the blood from the bowels."[65] Willis' main argument was that fevers resulted from an abnormal fermentation in the blood, causing it to overheat. Abnormal fermentation might come from an alteration of the chemical composition of blood or by inordinate motions of its particles.[66] Hence, he advocated chemical therapies to heal distempers of blood, such as by introducing chemicals that can trigger fermentation. He maintained that, as ferments are required to make blood, "when they are wanting by nature, they are with good success supplied by art: for fixed salts, alkali salts, extracts, digestives, openers, and especially chalybeat remedies, help for this reason, that, as it were by a certain ferment, they restore anew the weak, or almost extinct ebullition, or boyling of blood."[67] It is apparent that Willis' inquiry into ferments and fermentation was not confined to medical theory. As

[60] Willis 1684, 8.

[61] Willis 1684, 3.

[62] Willis 1684, 4–7. For affinity in early17th-century chemistry see Newman 2012. I agree with Newman's argument that affinity had been firmly and explicitly espoused by at least one highly prominent chymist almost a century before the publication of Geoffroy's *Table des differents rapports* (100).

[63] Willis 1684, 11–12.

[64] Willis 1684, 12.

[65] Willis 1684, 53.

[66] Willis 1684, 50–52.

[67] Willis 1684, 53.

attested by his Oxford Casebook, Willis had recourse to fermentation in his medical practice as early as 1650. The case of Robert Wylde, a Royalist of Worcester reported in the entry bearing the date of September 10, 1650, provides evidence of Willis' chemically oriented approach to pathology. He wrote that in his patient "the blood mass emerges in an impure state and arrives at the kidneys full of tartarean feculencies. When this stone-bearing stuff reaches there… while the serum is being strained from the blood (or rather precipitated from a ferment) that tartarean and thicker matter, in the face of its Gorgon-like ferment, is straightway congealed into sand or little stones."[68]

Like Willis, Glisson and Charleton bracketed ferments and fermentation with spirits and investigated the chemical composition of blood, a process producing the exaltation and spiritualization of bodies. In his *Anatomia hepatis* (1654) Glisson explained animal heat as the outcome of fermentation in blood. Such a process results from the struggle of spirits trying to disentangle from coarser parts.[69] For Charleton, fermentation plays a central part in nature and in animal economy. He claims that ingested food undergoes fermentation so that the particles of spirits therein contained are activated and get into a state of volatility. A follower of Gassendi, Charleton states that blood, which is rich in vital spirits, is the product of chemical and mechanical processes.[70]

Like the iatrochemists, Boyle put special emphasis on fermentation as the key to understanding digestion, claiming that digestion and the generation of stones within the body are chemical processes brought about by active principles i.e., ferments. He spelled out his view in *The usefulnesse of natural philosophy*:

> … and it seems a mistake to imagine (how many soever do so) that heat must needs be the efficient of all the changes the matter of our aliments may happen to undergoe in a humane body: where there are streiners, and solvents, and new mixtions, and perhaps ferments, and diverse other powerfull agents, which by successively working upon the assumed matter, may so fashion and qualifie it, as, in some cases, to bring the more disposed part of it to be not unlike even fossile salts or other mineral substances.[71]

For Boyle, the assimilation of ingested food was the outcome of a fermentative process that in turn triggered a series of chemical reactions. Fermentation produces the breaking down of food and the assimilation of the nutrients into blood:

> For in fermentation, the sulphurous (as Chymists call them) active and the spirituous parts of vegetables, are much better loosened, and more intirely separated from the grosser and clogging parts, in most mixts, then they are by the vulgar ways of distillation, wherein the concrete is not opened by previous fermentation. And these nobler parts being incorporated with our aliments, are with them received freely, and without resistance carryed into the mass of blood, and therewith, by circulation, conveyed to the whole body where their operation is requisite.[72]

[68] Willis 1981, 69.

[69] Glisson 1654, 366. "Fermentatio autem est calor intus exoriens, ob luctam inter spiritus & partes crassiores; dum illi conantur sese expandere, atque avolare, hae vero illi nisui adversantur."

[70] Charleton 1659, 62, 65, 100.

[71] Boyle 1999–2000, vol. III, 319.

[72] Boyle 1999–2000, vol. III, 355.

12.4 Extraordinary Cures, Ferments and Effluvia

Ferments, fermentation and Helmontian medicine played a central part in the controversies over the extraordinary cures performed in England in 1665–1666 by the Irish healer Valentine Greatrakes, also known as "the Stroker." Boyle, like other natural philosophers, took an active part in the investigations of Greatrakes' supposed miraculous healing powers. Boyle's interest in Greatrakes' cures was not confined to the medical aspects of the case, as he dealt with the Stroker's supposed miraculous gift and his piety – a matter of contention among natural philosophers and divines. Boyle was a spectator of at least 60 performances by Greatrakes, who made his cures in Lady Ranelagh's house, where he also cured Boyle's brother-in-law. Boyle himself stroked a patient with the inside of Greatrakes' glove.[73] Furthermore, Boyle compiled a questionnaire aimed at acquiring information on the Stroker's cures and collected information on the Irish healer in his Diaries of 1666. In 1665 John Beale (a former member of the Hartlib Circle), sent Boyle via Oldenburg an early account of Greatrakes' cures, which he reported as being successful most times in treating "pox, scorbutus and withered limbs," yet he failed to cure the blind and to raise the dead.[74] Then Boyle asked the chemist Daniel Coxe to investigate the case. Coxe sent Boyle a detailed account of Greatrakes' religious beliefs and curative powers in a letter of 5 March 1666. He reassured Boyle about the Irish healer's piety and ruled out supernatural explanations, describing Greatrakes' gift of healing as "merely complexional."[75] Coxe went on by comparing Greatrakes' to Butler's cures as related by van Helmont. He added that active particles transpiring from Greatrakes' body had the power to "mitigate the furious Archeus" of the patient. On 9 March 1666 Boyle wrote a long letter to Henry Stubbe dealing with the Irish healer's powers and with a number of medical issues, notably fermentation and its role in diseases.[76] Stubbe was actively involved in the study of Greatrakes' cures. In the *Miraculous conformist*, dedicated to Willis and published in 1666 in the form of a letter to Boyle, Stubbe explained Greatrakes' cures along Helmontian lines, i.e., as an effect of fermentation. Yet, in the dedicatory epistle he added, "the gift of healing was bestowed on [Greatrakes] since the Restoration of His Sacred Majesty and the Restitution of the Doctrine and Discipline of the English Church." For Stubbe, Valentine Greatrakes' healing power was God's gift in the sense that he was given a peculiar temperament. He wrote that "his body being composed of some particular ferments, the effluvia whereof, being introduced sometimes by a light, sometimes by a violent Friction, should restore the temperaments of the debilitated parts, re-invigorate the bloud, and dissipate all heterogeneous ferments out of the Bodies of the diseased, by the eyes, nose, mouth,

[73] Greatrakes 1666, 71. On Greatrakes see Elmer 2013.

[74] Boyle 2001, vol. 2,506.

[75] Boyle 2001, vol. 3, 82–83.

[76] Boyle 2001, vol. 3, 93–107. For Stubbe, see Jacob 1983. On the relationship between Boyle and Stubbe see Jacob 1977.

hand and feet."[77] In his letter to Stubbe, Boyle endorsed the naturalistic explanation and recognized the role of ferments as specific agents of disease in the extraordinary cures performed by the Irish healer: "I doe not see why it may not be possible for the sanative, and perhaps Anodine steames of his Body to be of such a Texture that they may both reinvigorate the spirits & by appropriated Qualitys, oppose & subdue the morbifick Matter or Ferment."[78] In the same year (1666) John Beale wrote to Boyle saying that the study of fermentation disclosed most arcana of nature: "All motions in nature seem to follow the course of some Ferment."[79] Despite his commitment to explore fermentation as a cause of pathological conditions, Boyle was not so confident that all diseases originated from specific ferments.[80] In his view, fermentation was but one of the causes of abnormal heat of the blood, therefore suggesting that different causes could be discovered by means of more accurate chemical investigations.[81]

12.5 Respiration, Fermentation and Aerial Nitre

In the 1660s Boyle and the Oxonians took issue with van Helmont's (and Willis') ferments and spirits, and set out to investigate by means of experiments the chemical reactions occurring in the human body, notably the role of air in respiration and in blood heating. Their physiological research gave increasing importance to nitre and its role in respiration and animal heat. Several chemical philosophers, notably Michael Sendivogius, saw nitre as the vital substance (of celestial origin) contained in the air. Bathurst and his fellow physiologists in Oxford rejected the old view that the use of respiration was refrigerating blood and carrying off fumes, and maintained that the use of particles of aerial nitre was to transform blood. As documented by Frank, aerial nitre became the focus of Oxford physiological investigations for nearly two decades.[82] Yet, fermentation did not disappear from their medical agenda, it was rather reinterpreted as a series of chemical reactions. According to Bathurst, *spiritus nitrosus* i.e., a volatile salt contained in air, was necessary to the life of plants and animals – not for the refrigeration of blood, but because it was the food of vital and animal spirits. Bathurst describes the way corpuscles of aerial nitre

[77] Stubbe 1666, 10–11.

[78] Boyle 2001, vol.3, 103.

[79] Boyle 2001, vol. 3, 159.

[80] Boyle 2001, vol. 3, 104. "What you mention of Morbifick Ferments I know divers Ingenious Readers will approve and they seeme to be of good Use in the explication of Diseases. But whether all Diseases require Ferments, & whether your Doctrine about them be as well applicable to the rest, as to some, is a Disquisition that I shall willingly leave to those Learned Men of you Faculty that our Age & Country abounds with."

[81] Boyle 1999, vol. 3, 321.

[82] Frank 1980.

penetrate blood as a chemical process, similar to distillation.[83] Bathurst's lectures spurred John Mayow's research, which saw the light first in *Tractatus duo* (1668) and then in the *Tractatus Quinque* (1674). Mayow maintained that aerial nitre is necessary to both combustion and respiration and explored the composition of this salt. Nitre, Mayow stated, is composed of three parts: an extremely fiery and volatile acid salt (the aerial part that could be identified as nitric acid), an alkaline salt (potassium carbonate), and a saline volatile salt (ammonium carbonate). He reached the conclusion that part of the air is necessary for combustion, and that this part is present in nitre. The volatile component of nitre provides "food for fire," and also passes into the blood of animals by respiration. This brought about Mayow's rejection of spirits as *factotum* and their replacement with nitro-aerial spirit, again a sort of universal *explanans*.[84] On the grounds of his well-known experiments, he concluded that nitro-aerial particles contained in air mixing with the saline-sulphurous parts of blood brought about fermentation.[85]

In 1668 Willis issued *Pathologiae Cerebri*, adopting the aerial nitre theory. He explained the origin of muscular motion by means of aerial nitre, and maintained that muscular motion was produced by an explosion caused by the encounter of the spirito-saline particles of animal spirits (coming from the nerves) and the nitro-aerial ones contained in the blood.[86] In *De Sanguinis Incalescentia* (1670) Willis developed the notion of nitre and dropped the view of fermentation he had maintained in the *De Fermentatione*. In 1670 Willis denied that fermentation could produce heat in liquids. Having stated that fermentation was not the cause of the warming of blood, Willis suggested instead that heat was generated by the reaction of particles of nitre coming from air with those of sulphur contained in the blood.[87] Willis' main concern was to establish the chemical reactions occurring in blood and for this reason he wholeheartedly adopted the aerial niter as an *explanans*.

As attested by his well-known essay on nitre to be found in his 1661 *Certain physiological essays*, Boyle took special interest in spirit of nitre. He investigated its role in respiration too. He took active part in the Oxford physiologists' research on respiration, yet he did not subscribe to Mayow's view of nitre. For Boyle, the chemical properties of the aerial nitre were not understood properly, therefore he refrained from adopting the theory of aerial nitre as the vital component of air. In *General History of Air* (published posthumously in 1692), containing notes that he had been collecting for more than 20 years, Boyle criticized the theory of spirit of nitre as the vital part of the air. For Boyle, spirit of nitre was an "exceedingly corrosive" substance, which could scarcely be conceived as "refreshing to the nature of animals." In addition, he conceived of spirit of nitre as an acid spirit, whereas he classified

[83] Warton 1761, 208. Cf. Guerlac 1953 and Frank 1980, 115–139; 221–245.

[84] Mayow 1674, 48; Mayow 1926, 34. "With regard to the spirit of the chemists, which usually leads their band of elements, I am quite unable to understand what they mean by the very grand word spirit." Cf. Frank 1980, 266.

[85] Frank 1980, 268–269.

[86] Willis 1684, 3.

[87] Willis 1684, 21–23.

spirit of blood as an alkaline volatile salt. These two substances, he concluded, were of "opposing natures."[88]

Throughout his career, Boyle kept investigating the composition and the properties of blood by means of a series of chemical experiments, as shown by his *Memoirs for the History of Human Blood, Especially the Spirit of that Liquor* (published in 1684, but largely written in the late 1660s). The encouragement to write a natural history of blood came from Locke, to whom Boyle addressed the Preface to the work. At the outset of his work on blood, Boyle adopted the Paracelsian stance, stressing that chemistry disclosed invisible, hidden active substances contained in blood. He insisted on the importance of chemical investigation of blood, saying that most of the extant studies on blood "consisted much more of observations than experiments; being suggested far more by the phenomena that nature her self has afforded physicians, than by trials industriously made, to find what she will not, unsolicited by art, discover."[89] Boyle's main goal in this work was to establish by chemical experiments the composition and properties of blood. As part of his chemical exploration of blood composition, he thoroughly investigated the chemical change air produced in blood.

Boyle dealt with blood fermentation as the cause of fevers in *Specifick medicines*, published in 1685, but conceivably written in 1680. He explicitly referred to the research on fermentation carried out by Willis and other physiologists: "many modern physitians, especially since the learned Dr Willis's notions came to be in request, have looked upon fevers and agues to consist in, or be produced by vitious fermentation of the blood."[90] He argued that blood fermentation might be altered in different ways, bringing about various distempers:

> as there is oftentimes a vitious fermentation of the blood, so there may be sometimes a want of fermentation, or a certain sluggishness, upon whose account, either the brisk intestine agitation, that it ought to have as a warm fluid of such a nature as 'tis wont to be in sound persons, or a due quickness of circulation through the heart is wanting: to which sluggish state of the blood, if it be obstinate and lasting, several distempers are wont to be consequent.[91]

Boyle was too careful an experimental philosopher to commit himself to the view that fermentation accounted for all kinds of diseases. He maintained that a number of diseases could not by explained as the outcome of imperfect fermentation occurring in blood. In addition, he expressed doubts about the analogies between blood fermentation and the fermentation of other liquors such as wine and vinegar, an analogy that was widely employed by seventeenth-century iatrochem-

[88] Boyle 1999–2000, vol. XII, 32.

[89] Boyle 1999–2000, vol. X, 5. For the composition of, and the material related to, Boyle's *History of blood*, see Knight and Hunter 2007. For the manuscripts, see Boyle 2013, http://www.ncbi.nlm.nih.gov/pmc/articles/PMC1871712/.

[90] Boyle 1999–2000, vol. X.

[91] Boyle 1999–2000, vol. X, 387.

ists. Boyle's aim was to investigate the specific chemical substances responsible for blood fermentation.[92]

12.6 Epilogue

Thanks to the Paracelsians' and van Helmont's contributions, the study of fermentation flourished in Europe in the second half of the seventeenth century. In England, Boyle and the Oxford physiologists gave a substantial impetus to research on fermentation and to its uses in medicine. They reinterpreted the Paracelsian and Helmontian ferments as units of matter and tried to establish the chemical rationale for the fermentation. In addition, they experimentally explored the composition of blood and the chemical reactions responsible for its fermentation.

A major impulse to explore ferments and fermentation came from the Great Plague. A number of physicians endorsed Helmontian doctrines, notably the view that pestilential seeds and ferments played a central part in the origin of plague. Two champions of the Helmontian revival of the mid-1660s, George Thomson and William Simpson stood out for their insistence upon the importance of fermentation as a key to the understanding of physiology and pathology.[93] In a work containing a head-on attack on Galenism and the medical establishment, Thomson extolled van Helmont because he disclosed "the seminal virtues and properties of things... bringing to light the doctrine of fermentation, the original and activity of spirits."[94] Thomson's works on plague are a synthesis of empirical observations and Helmontian notions such as *Archeus*, spirits and ferments. This is apparent in his understanding of plague symptoms: "when the natural ferment of the stomack in the Pest is so far lost, that instead of white a black juice is engendered, it is a certain sign of the abolition of spirit and consequently of approaching death."[95] In 1670 Thomson published a tract on blood, where he opposed bloodletting and insisted on the spiritual nature of active principle, stressing that ferments are the source of life as they are responsible for digestion and the origin of vital spirit in blood.[96]

Like Thomson, Simpson was engaged in providing chemical foundation to medicine. In *Zenexton anti-pestilentiale* he argued that: "Ferments [are] ... certain powers in Nature whereby all things are put into a way of change... Ferments are Parents of transmutation out of one form into another, or from one degree to another, whereby things are brought to their highest energy."[97] Simpson continued his

[92] Boyle 1999–2000,vol.X, 42.

[93] For Thomson and the Helmontians' critique of traditional medicine, see Cook 1986, 145–162. Clericuzio 1993, 319–26; Wear 2000, 368–72; 416–422.

[94] Thomson 1665, 4–6.

[95] Thomson 1666, 112–113.

[96] Thomson 1670, 6.

[97] Simpson 1665, 9–10.

research on fermentation and in *Zymologia physica* (1675) identified the particles of acid and those of sulphur as the agents of any fermentative process.[98]

Approaches to the study of fermentation diverged both in methods and aims. Whereas the Helmontians explained it as the outcome of ferments, i.e, semi-material principles, Boyle and the Oxford physiologists focused on the chemicals responsible for the fermentative process. There is little doubt that fermentation occupied a central part in English science and it comes as no surprise that Newton paid special attention to it. In *Hypothesis explaining the properties of light*, sent to Oldenburg on 7 December 1675, Newton maintained that "The whole frame of nature may be nothing but aether condensed by a fermental principle." [99] Evidence of his continuing research on fermentation is the famous "Query" 31 of 1717 *Opticks*, where Newton deals with active principles. He states that forces acting at very small distance are "…the cause of fermentation, by which the heart and blood of animals are kept in perpetual motion and heat." According to Newton, air "abounds with acid vapours fit to promote fermentation, as appears by the rusting of iron and copper in it, the kindling of fire by blowing, and the beating of heart by respiration."[100]

References

Anstey, Peter R. 2000. *The philosophy of Robert Boyle*. London: Routledge.

Anstey, Peter R. 2014. Philosophy of experiment in early modern England: The case of Bacon, Boyle and Hooke. *Early Science and Medicine* 19: 103–132.

Bates, Don G. 1981. Thomas Willis and the fever literature of the seventeenth century. In *Theories of fever from antiquity to the enlightenment*, ed. W.F. Bynum and V. Nutton, 45–70. London: Wellcome Institute for the History of Medicine.

Bertoloni Meli, Domenico. 2011. *Mechanism, experiment, disease: Marcello Malpighi and seventeenth-century anatomy*. Baltimore: Johns Hopkins University Press.

Bianchi, Massimo L. 1994. The visible and the invisible. From alchemy to paracelsus. In *Alchemy and chemistry in the 16th and 17th centuries*, ed. P. Rattansi and A. Clericuzio, 17–50. Dordrecht: Kluwer Academic Publishers.

Billich, Anton Gunther. 1646. *Anatome Fermentationis Platonicae*. In *De Sanguinis Generatione et Motu Naturali*, Hermann Conring. Leyden, apud Ludovicum Elzevirium.

Bono da Ferrara Pietro. 1976. *Preziosa Margarita Novella*. Trans. Chiara Crisciani. Firenze: La Nuova Italia.

Boyle, Robert. 1999–2000. *The works of Robert Boyle*, eds. Michael Hunter and E. B. Davis. 14 volumes. London: Pickering & Chatto.

Boyle, Robert. 2001. In *The correspondence of Robert Boyle*, vol. 6, ed. M. Hunter, A. Clericuzio, and L.M. Principe. London: Pickering & Chatto.

Boyle, Robert. 2013. *Unpublished material relating to Robert Boyle's memoirs for the natural history of human blood*, eds. M. Hunter and H. Knight. Robert Boyle Project Occasional Papers

[98] Simpson 1675.

[99] Newton sent "An Hypothesis explaining the Properties of Light" to Oldenburg on 7 December 1675. His paper was read at a meeting of the Society on 9 December 1675. Newton 1959, 362–386.

[100] Newton 1952, 399.

No. 2 online. http://www.ncbi.nlm.nih.gov/pmc/articles/PMC1871712/. Accessed 29 Oct 2013.

Castelli, Pietro. 1626. *Epistolae Medicinales*. Roma: Mascardi.

Chalmers, Alan. 2012. Intermediate causes and explanations: The key to understanding the scientific revolution, *Studies in History and Philosophy of Science*, Part A 43(4): 551–562

Charleton, Walter. 1659. *Natural history of nutrition, life, and voluntary motion*. London: Henry Herringman.

Clericuzio, Antonio. 1993. From van Helmont to Robert Boyle: A study of the transmission of chemical and medical theories in seventeenth-century England. *The British Journal for the History of Science* 26: 303–334.

Clericuzio, Antonio. 1998. The mechanical philosophy and the spring of air. New light on Robert Boyle and Robert Hooke. *Nuncius* 13: 69–75.

Clericuzio, Antonio. 2000. *Elements, principles and corpuscles: A study of atomism and chemistry in the seventeenth century*. Dordrecht: Kluwer.

Clericuzio, Antonio. 2001. Gassendi, Boyle and Charleton on matter and motion. In *Late medieval and early modern corpuscular matter theories*, ed. C. Lüthy, J. Murdoch, and W.R. Newman, 467–482. Leiden: Brill.

Clericuzio, Antonio. 2010. Chemical medicines in Rome: Pietro Castelli and the Vitriol Debate (1616–1626). In *Conflicting duties: Science, medicine and religion in Rome, 1550–1750*, ed. M.P. Donato and J. Kraye, 281–302. London: The Warburg Institute.

Clericuzio, Antonio. 2012. Chemical and mechanical theories of digestion in early modern medicine. *Studies in the History and Philosophy of Biology and Biomedical Sciences* 43: 329–337.

Cook, Harold J. 1986. *The decline of the old medical regime in Stuart London*. Ithaca: Cornell University Press.

Davis, Audrey B. 1973. *Circulation physiology and medical chemistry in England, 1650–1680*. Lawrence: Coronado Press.

Debus, Allen G. 1969. Edward Jorden and the fermentation of the metals: An iatrochemical study of terrestrial phenomena. In *Towards a history of geology*, ed. C.J. Schneer, 100–121. Cambridge, MA: MIT Press.

Elmer, Peter. 2013. *The miraculous conformist: Valentine greatrakes, the body politic, and the politics of healing in restoration Britain*. Oxford: Oxford University Press.

Frank, Robert G. 1980. *Harvey and the Oxford physiologists*. Los Angeles: University of California Press.

Gelman, Zahkar E. 1994. Angelo Sala, an iatrochemist of the late Renaissance. *Ambix* 41: 142–160.

Giglioni, Guido. 2000. *Immaginazione e malattia. Saggio su Jan Baptiste van Helmont*. Milano: Franco Angeli.

Glisson, Francis. 1654. *Anatomia hepatis*. London: O. Pullein.

Greatrakes, Valentine. 1666. *A brief account of Mr Valentine Greatrak's*. London: J. Starkey.

Guerlac, Henry. 1953. John Mayow and the Aerial Niter. In *Actes du VIIe Congrès International d'Histoire des Sciences*: 332–349. Paris: Académie Internationale d'Histoire des Sciences.

Hannaway, Owen. 1975. *The chemist and the word. The didactic origins of chemistry*. Baltimore: Johns Hopkins University Press.

Hirai, Hiro. 2005. *Le Concept de Semence dans les Théories de la Matière à la Renaissance, de Marsile Ficin à Pierre Gassendi*. Turnhout: Brepols.

Hunter, Michael. 1997. Boyle versus the Galenists: A suppressed critique of seventeenth-century medical practice and its significance. *Medical History* 41(3): 322–361.

Hunter, Michael, and Edward B. Davis. 1996. The making of Robert Boyle's *Free enquiry into the vulgarly receiv'd notion of nature* (1686). *Early Science and Medicine* 1(2): 204–271.

Jacob, James R. 1977. *Robert Boyle and the English revolution*. New York: Lenox Hill Publishers.

Jacob, James R. 1983. *Henry Stubbe, radical protestantism and the early enlightenment*. Cambridge: Cambridge University Press.

Jorden, Edward. 1631. *A discourse of naturale bathes, and minerale waters*. London: Thomas Harper.

Kahn, Didier. 2007. *Alchimie et Paracelsisme en France à la fin de la Renaissance (1567–1625)*. Geneva: Droz.

Kaplan, Barbara B. 1993. *"Divulging of useful truths in physick": The medical agenda of Robert Boyle*. Baltimore: The Johns Hopkins University Press.

Knight, Harriet. and Michael Hunter. 2007. Robert Boyle's *Memoirs for the natural history of human blood* (1684): Print, manuscript and the impact of Baconianism in seventeenth-century medical science. *Medical History* 51(2): 145–164.

Libavius, Andreas. 1597. *Alchemia*. Frankfurt: impensis Petri Kopffij.

MacIntosh, John J. 2005. *Boyle on atheism*. Toronto: University of Toronto Press.

Mayow, John. 1674. *Tractatus quinque*. Oxford: Sheldonian Theatre.

Mayow, John. 1926. *Medico-physical works*. Oxford: The Alembic club.

Meinel, Christoph. 1988. Early seventeenth-century atomism. Theory, epistemology and the insufficiency of experiments. *Isis* 79: 68–103.

Moran, Bruce. 2005. *Distilling knowledge. Alchemy, chemistry, and the scientific revolution*. Cambridge: Harvard University Press.

Moran, Bruce. 2007. *Andreas Libavius and the transformation of alchemy: Separating chemical cultures with polemical fire*. Sagamore Beach: Science History Publications/Watson Pub. International.

Multhauf, Robert P. 1955. J. B. van Helmont's reformation of the Galenic doctrine of digestion. *Bulletin of the History of Medicine* 29: 154–163.

Newman, William. 1994. *Gehennical fire: The lives of George Starkey, an American alchemist in the scientific revolution*. Cambridge: Harvard University Press.

Newman, William. 2006. *Atoms and alchemy. Chymistry and the experimental origins of the scientific revolution*. Chicago: University of Chicago Press.

Newman, William. 2009. Brian Vickers on alchemy and the occult: a response. *Perspectives on Science* 17: 482–506.

Newman, William. 2012. Elective affinity before Geoffroy: Daniel Sennert's atomistic explanation of vinous and acetous fermentation. In *Matter and form in early modern science and philosophy*, ed. G. Manning, 99–124. Leiden: Brill.

Newman, William, and Lawrence M. Principe. 1998. Alchemy vs. chemistry: The etymological origins of a historiographic mistake. *Early Science and Medicine* 3: 32–65.

Newman, William, and Lawrence M. Principe. 2001. Some problems with the historiography of alchemy. In *Secrets of nature: Astrology and alchemy in early modern Europe*, ed. Wiiliam R. Newman and Grafton Anthony, 385–431. Cambridge, MA: MIT Press.

Newman, William, and Lawrence M. Principe. 2002. *Alchemy tried in the fire: Starkey, Boyle, and the fate of Helmontian chymistry*. Chicago: University of Chicago Press.

Newton, Isaac. 1952. *Optics*. New York: Dover.

Newton, Isaac. 1959. In *The correspondence of Isaac Newton*, vol. 7, ed. H.W. Turnbull, J.F. Scott, and L. Tilling. Cambridge: Cambridge University Press.

Pagel, Walter. 1958. *Paracelsus: An introduction to philosophical medicine in the era of the renaissance*. Basel: Karger.

Pagel, Walter. 1982. *Joan Baptista van Helmont. Reformer of science and medicine*. Cambridge: Cambridge University Press.

Pagel, Walter, and P. Rattansi. 1964. Vesalius and Paracelsus. *Medical History* 8: 309–328.

Paracelsus. 1929. *Paracelsus: Sämtliche Werke*, ed. Karl Sudhoff. 14 volumes. Berlin: Oldenbourg München.

Pereira, Michela. 2003. L'Alchimista come medico perfetto nel *Testamentum* pseudolulliano. In *Alchimia e medicina nel Medioevo*, ed. C. Crisciani and A. Paravicini Bagliani, 77–108. Tavernuzze-Firenze, Sismel: Edizioni del Galluzzo.

Pereira, Michela, and Barbara Spaggiari (eds.). 1999. *Il «Testamentum» alchemico attribuito a Raimondo Lullo*. Tavernuzze-Firenze, Sismel: Edizioni del Galluzzo.

Principe, Lawrence M. 2012. *The secrets of alchemy*. Chicago: University of Chicago Press.
Royal Society Boyle Papers. London: Royal Society.
Ruland, Martin. 1964. *Lexicon of Alchemy*. Trans. by E.A. Waite. London: Watkin.
Sala, Angelo. 1647. *Hydrelaeologia*. In *Opera Medico-Chymica*. Frankfurt: Beyer.
Sennert, Daniel. 1641. *Medicina Practica*. Venice: Francesco Baba.
Shackelford, Jole. 1998. Seeds with a mechanical purpose: Severinus' Semina and seventeenth-century matter theory. In *Reading the book of nature*, ed. A.G. Debus and M.T. Walton, 15–44. Kirksville: Thomas Jefferson University Press.
Simpson, William. 1665. *Zenexton anti-pestilentiale*. London: George Sawbridge.
Simpson, William. 1675. *Zymologia physica, or a brief discourse of fermentation, from a new hypothesis of Acidum and Sulphur*. London: W. Cooper.
Starkey, George. 2004. In *Alchemical laboratory notebooks and correspondence*, ed. Newman William and Lawrence M. Principe. Chicago: Chicago University Press.
Stubbe, Henry. 1666. *Miraculous conformist*. Oxford: R. Davis.
Thomson, George. 1665. *Galeno-pale, or a chymical trial of the Galenists, that their dross in physick may be discovered*. London: Edward Thomas.
Thomson, George. 1666. *Loimotomia, or, The pest anatomized*. London: N. Crouch.
Thomson, George. 1670. *Aimatiasis*. London: N. Crouch.
van Helmont, Jan Baptista. 1648. *Ortus medicinae*. Amsterdam: Apud Ludovicum Elsevirium.
Vickers, Brian. 2008. The 'new historiography' and the limits of alchemy. *Annals of Science* 65: 127–156.
Warton, Thomas. 1761. *The life and literary remains of Ralph Bathurst*. London: printed for R. and J. Dodsley, C. Bathurst; and J. Fletcher.
Wear, Andrew. 2000. *Knowledge and practice in English medicine, 1550–1680*. Cambridge: Cambridge University Press.
Webster, Charles. 1975. *The great instauration*. London: Duckworth.
Willis, Thomas. 1659. *Diatribae Duae Medico-Philosophicae Quarum Prior Agit de Fermentatione, Sive de Motu Intestino Particularum in Quovis Corpore: Altera de Febribus*. London: J. Martin, J. Allestry & T. Dicas.
Willis, Thomas. 1684. *Practice of Physick*. London.
Willis, Thomas. 1981. In *Willis's Oxford casebook*, ed. K. Dewhurst. Oxford: Sandford Publications.
Worsely, Benjamin. 1653. Observations about Saltpetre. In *The Hartlib Papers*. 39/1/11A.

Chapter 13
Boyle, Malpighi, and the Problem of Plastic Powers

Ashley J. Inglehart

Abstract In this paper, I examine Boyle's views on generation and his appeal to plastic powers as an explanatory agent following a brief overview of the secondary literature. Further, I also look at the relationship between Boyle and the Italian anatomist Marcello Malpighi, who was one of the most prolific medical writers in the second half of the seventeenth century. This paper looks at their correspondence, which included not only ideas about experiments and the mechanical philosophy, but mineral samples as well. I argue that Malpighi appropriated Boyle's notion of plastic powers in his own writings as part of a mechanical account of generation. Thus, Boyle's description of a plastic power was perceived as being mechanical by one of his own contemporaries similarly committed to the mechanical philosophy. In both Boyle and Malpighi, moreover, we see a shared philosophy regarding mechanical explanation. For each of them, mechanical explanations involve physical agents acting on matter. Their shared view implies an important shift in which the power of an explanatory agent like a plastic power is in its mechanical mode of operation.

Keywords Robert Boyle • Marcello Malpighi • Plastic power • Generation • Mechanical explanation

I would like to thank Domenico Bertoloni Meli, William Newman, the editors of this volume, and two anonymous reviewers for many helpful suggestions and constructive criticisms on earlier drafts of this paper. Parts of this paper were presented at both the *Early Modern Medicine and Natural Philosophy* workshop held at the University of Pittsburgh and the 2012 *Workshop on the History of Biology in Honor of Fred Churchill* held at Indiana University. I wish to thank the audience members at each for insightful questions.

A.J. Inglehart (✉)
Department of History and Philosophy of Science, Indiana University,
Bloomington, IN, USA
e-mail: asjingle@umail.iu.edu

P. Distelzweig et al. (eds.), *Early Modern Medicine and Natural Philosophy*,
History, Philosophy and Theory of the Life Sciences 14,
DOI 10.1007/978-94-017-7353-9_13

13.1 Introduction

This paper begins with the famed English chymist[1] Robert Boyle and the attempts throughout his corpus to explain the process of generation. The problem of generation is, essentially, one of how things in nature *come to be*. Thus, it is a much broader concept than that of reproduction, as the latter is restricted to how animals replicate themselves. Generation, in contrast, is the process by which a thing is formed and thus is not restricted to animals, but includes also the originative processes of both plants and minerals. Moreover, any kind of formative process requires something to organize the matter whence it is made into that which it will become. The agent responsible for this organization of matter has been conceived throughout history in a vast number of ways, from the *λόγοι σπερματικοὶ* of Plotinus to the formative faculties of Aristotle and later of Galen.[2] Yet, the distinctive feature throughout the history of these concepts prior to Boyle was that the explanatory power of these organizing principles came from the soul or faculty with which each was associated. The dramatic shift for which Boyle is responsible is that plastic or seminal principles are causes in virtue of their effects upon matter.

By the seventeenth century, 'plastic power' or 'plastic force' had become blanket terms for the formative power of generation, and they were nearly interchangeable with 'seminal principle'.[3] Seminal principles and their related plastic powers are potentially at odds with the mechanical philosophy. The tension arises in there being particles of matter that have powers which are not reducible to either matter or motion. As one of the most influential scholars of his time, Boyle was largely responsible for popularizing the so-called Mechanical Philosophy. Consequently, his appeal to any formative powers warrants further discussion.

In this paper, I examine Boyle's views on generation and his appeal to plastic powers as an explanatory agent following a brief overview of the secondary literature. I aim to show that despite whatever tensions seminal and plastic powers might hold within a reductionist ontology, Boyle's account of plastic powers nonetheless constitutes a mechanical explanation. Moreover, I show how Malpighi appropriated this concept and was influenced specifically by Boyle's treatises on minerals. In both Boyle and Malpighi, we see a shared philosophy regarding mechanical explanation. For each of them, mechanical explanations involve physical agents acting on

[1] For an explanation of this spelling, see Newman and Principe 1998.

[2] See Hirai 2005, 24 ff. and Preus 2002, 46 ff. for excellent discussions on the *λόγοι σπερματικοὶ*. For more on the formative faculties of Aristotle and Galen, see Aristotle 2008, GA, 730 b: 10 ff. and Galen 1916, II: 5 respectively. In the case of Aristotle, this formative power is linked to the formal and efficient cause contributed by the male semen. Epicurean atomism does provide an alternative explanation for generation. However, this view is characterized by its lack of such organizing principles, appealing instead to the random collision of atoms. This random chance made the view (in conjunction with its association with atheism) unappealing to many seventeenth-century authors.

[3] Hunter 1950, 199–200. See also Hirai 2007.

matter and thus producing noticeable, qualitative changes. Their shared view implies a kind of methodological shift in which they emphasize the production of those effects as the source of explanatory power.

13.2 Historiography

Earlier works on Boyle's mechanical philosophy, exemplified by Marie Boas Hall's scholarship, depict him as the defender of a rational, mathematically-based mechanical science that would ultimately replace the unscientific beliefs of alchemists.[4] It has since been shown, however, that Boyle's chymistry and his more general philosophy is very much informed by his alchemical pursuits, and moreover that the distinction between chemistry and alchemy simply cannot be made within a seventeenth-century context.[5] More recently, the secondary literature on Boyle's mechanical philosophy can be characterized into two groups: those who argue that Boyle's experimental chemistry is ultimately incompatible with and independent of his mechanical program, and those who argue that his experimental chymistry is central to and perfectly consistent with his mechanical philosophy. In this long-standing debate that spans more than twenty years, Alan Chalmers and Antonio Clericuzio belong to the former, whereas Andrew Pyle, William Newman, and (to a lesser extent) Peter Anstey belong to the latter.

A re-examination of Boyle's mechanical approach is much indebted to Antonio Clericuzio's groundbreaking 1990 paper, "A redefinition of Boyle's chemistry and corpuscular philosophy".[6] Within the paper, Clericuzio challenges the traditional picture that Boyle's corpuscular theory was thoroughly mechanistic, looking both at Boyle's chymical influences and his employment of non-mechanical explanations such as alchemical spirits. Clericuzio very sharply distinguishes between theories of matter which are subtly, yet importantly, different. His work, in that regard, is important because the applications of 'mechanical' are often murky and in need of disentangling. Further, it's especially significant in the attention that it brought to Boyle's corpuscular theory of matter. Clericuzio argues that Boyle's work could be described as corpuscular, but not mechanical. Clericuzio has since provided a more detailed analysis of Boyle's corpuscular theory in his 2000 book, *Elements, Principles and Corpuscles*. Within it, he argues that several of Boyle's agents—including seminal principles—are not in fact mechanical, despite Boyle's repeatedly asserting them to be.

[4] Cf. Hall 1958.

[5] Cf. Newman 2006, 9.

[6] Clericuzio 1990.

Following Clericuzio's original paper, Alan Chalmers published in 1993 a rather controversial paper entitled, "The Lack of Excellency of Boyle's Mechanical Philosophy." In the paper, he argues that, "far from there being an intimate and productive link between Boyle's mechanical philosophy and his science, his scientific successes were achieved *in spite*, rather than because, of his allegiance to that philosophy."[7] He goes to argue as well that, "the case that Boyle makes for the mechanical philosophy can be seen to be very weak indeed."[8] At the crux of Chalmers' argument is Boyle's reliance in his chemistry on properties of matter (like plastic powers) which are not reducible to shape, size, and motion. In response to "The Lack of Excellency of Boyle's Mechanical Philosophy," Andrew Pyle and Peter Anstey engaged Chalmers in a three-way discussion published in *Studies for History of Philosophy* in 2002.[9] Andrew Pyle wrote "Boyle on science and the mechanical philosophy: a reply to Chalmers," which makes two counter arguments. The first, to which I am sympathetic, is that Chalmers assumes on the part of Boyle a strictly Cartesian program.[10] In the second half of his response, Pyle argues that even if we were to restrict Boyle to a strictly Cartesian view, this more reductive program does actually play a positive role in Boyle's empirical research, namely by emphasizing, "the centrality of local motion in all physical explanation, and invit[ing] us to posit, as explanatory properties in our intermediate-level explanations, only properties that look mechanizable."[11]

Anstey's response to Chalmers, "Robert Boyle and the heuristic value of mechanism," involves a somewhat weaker connection between Boyle's experimental chemistry and his mechanical philosophy. He argues that the latter serves as, "a heuristic structure that motivates and drives Boyle's experimental programme."[12] Further, Anstey argues that Boyle's most important discoveries were both informed and confirmed by his mechanical philosophy, in much the same way that theory informs and confirms experiment. The same year (2002a), Anstey provided the first comprehensive analysis of the treatment of seminal principles within Boyle's corpus in "Boyle on Seminal Principles." Anstey's paper categorizes Boyle's works on seminal principles by topic so that studying them in his corpus becomes feasible. The amount of Boyle's work addressed by this piece and the manner in which it is organized are quite impressive. In the article, we are presented with a Boyle in tension who thus posits seminal principles only with reticence. In Boyle's account, Anstey argues, appeals to a seminal or plastic power are made in order to explain those features of generation that are beyond the explanatory means of the

[7] Chalmers 2002.

[8] Ibid.

[9] Anstey's and Pyle's responses originated from a 1997 symposium on Chalmers's work, "What is this thing called Science?"

[10] Pyle 2002, 176.

[11] Ibid. 189.

[12] Anstey 2002c, 164.

corpuscular hypothesis. In short, Anstey presents Boyle as caught in a catch-22. On the one hand, if seeds operate mechanically, then Boyle has no reason to appeal to them as an explanatory agent in the first place. On the other, in admitting nescience, Boyle has to allow for the possibility of non-mechanical causes.[13]

Anstey maintained that same position shortly after in his subsequent debate with Jonathan Walmsley in, "Robert Boyle and Locke's 'Morbus' Entry: A Reply to J.C. Walmsley."[14] Their dispute over Boyle was within the context of an entry entitled "Morbus"—composed in John Locke's handwriting—which Walmsley understood to be "vociferous in criticism" of Boyle's corpscularianism and accepting instead of a distinctly non-mechanical program.[15] While Locke's views on disease and possible authorship of the Morbus entry remain outside the scope of this paper, the episode is nonetheless significant because Walmsely's "analysis and conclusions are predicated upon the view that Boyle adhered to a strictly mechanical explanation of the operation of seeds."[16] Anstey explicitly denies this claim, citing the same two-horned dilemma described above.[17]

Walmsley replied in the same volume of *Early Science and Medicine* with "'Morbus,' Locke and Boyle- A Response to Peter Anstey".[18] He responds to Anstey's denial that Boyle has an explicitly mechanical account of seminal and plastic powers, stating that

> But the cause [of generation] itself need not act in an extra-mechanical way. I see no examples of Boyle "declaring his nescience," nor do I believe that there was an "unresolved tension" in his thought on this point – whenever he addressed himself to the question to the action of seminal principles, Boyle was clear that they were complex mechanical entities acting in a mechanical way to produce complex mechanical effects. There is nothing in the remarks that Anstey makes that shows Boyle to have had anything less than full confidence in mechanical explanations for the action of seminal principles.[19]

Walmsley goes on to suggest that the source of Anstey's trouble lies in his affinity with Clericuzio, as the latter rejects that the plastic power of seeds are mechanical. The problem, however, is that Anstey does not claim—as Walmsley would have us believe—that "Boyle maintained a 'nescience' about the *modus operandi* of seminal principles."[20] But, herein lays a key difference that I hope to elucidate. Anstey's claim is that Boyle is committed to an ontological nescience about the nature of any seed's plastic power. At the heart of the matter is this: Must the plastic power (or any chymical power, for that matter) be necessarily reducible to the

[13] Anstey 2002a.

[14] Anstey 2002b, 358–377.

[15] Walmsley 2000, 367–393. Cf. Anstey (2002b), 358.

[16] Anstey (2002b), 362.

[17] Ibid. 359–362.

[18] Walmsley 2002, 378–397.

[19] Ibid. 394.

[20] Ibid. 393.

mechanical qualities of matter *per se*, or is it sufficient that its mode of operation is reducible to shape, size, and motion?

The controversy surrounding Boyle's mechanical philosophy took on a new dimension with the publication of William Newman's 2006 book, *Atoms and Alchemy: Chymistry and the Experimental Origins of the Scientific Revolution.* Part of Newman's argument involves alchemy's contribution to corpuscular theories of matter and the mechanization of nature, the latter of which is frequently considered to be an essential part of the so-called Scientific Revolution. The last third of the book focuses on Robert Boyle's theory of matter and influences, most especially the Aristotelian chymist Daniel Sennert. In dealing explicitly with Boyle's mechanical philosophy, Newman criticizes previous authors such as Clericuzio and Chalmers for their "implicit reliance on Cartesianism in framing their definitions of the mechanical philosophy."[21] Newman goes on to argue that, "[t]he fact that Boyle does not attempt to reduce all phenomenal change to the level of the *prima naturalia* or initial particles does not mean that his chymical explanations are not mechanical, since the aggregate corpuscles are also endowed with mechanical affections having explanatory force."[22]

Malpighi's discussion of plastic powers has not elicited quite the same response as has Boyle's, but it is nonetheless not without some controversy. In the monumental work, *Marcello Malpighi and the Evolution of Embryology*, Howard Adelmann writes that the plastic virtue is essentially, "a combination of the formal and efficient causes of Aristotle and the plastic or formative faculty of Galen."[23] This reading of Malpighi is problematic because Malpighi's career is marked by his attempt to explain anatomy in terms of the parts of a machine and with explanations totally devoid of reliance on the soul. Adelman's influence can be seen more recently in Catherine Wilson's 1995 book, *The Invisible World*, where she criticizes Malpighi as "thoroughly opportunistic and philosophically inconsistent in his explanatory apparatus, employing now the terms of Cartesian mechanism, which would explain growth as the process of fluid and particle accretion, now the language of plastic powers and unfolding, as each seemed suitable."[24] Domenico Bertoloni Meli has recently argued contra Wilson that Malpighi likely borrowed the expression "plastic virtue" from Boyle. He bases that argument on Malpighi's claim that any agent – even something like a soul – must act physically upon matter as a machine, regardless of its nature. Here, Malpighi presents a virtually identical philosophical stance as taken by Boyle in the *Origin of Forms and Qualities.*[25]

I hope to expand up on this view by investigating more deeply the relationship between Boyle and Malpighi. The correspondence between Boyle and Malpighi

---

[21] Newman 2006, 178.

[22] Ibid. 182.

[23] Adelmann 1966, vol. II: 866, n. 12.

[24] Wilson 1995, 128.

[25] Bertoloni Meli 2011, 232–3.

shows that they were familiar with each other's works and that each scholar was highly regarded by the other. This exchange included not just ideas about the new experimental or corpuscular philosophy, but material objects such as mineral samples. Their correspondence and mutual influence has not yet received much attention in the secondary literature. What is especially telling is that Malpighi makes no references to a plastic power until after Boyle published *Origin and Virtues of Gems*, despite the fact that *Origin of Forms and Qualities* was in print several years before Malpighi's treatises on the chick. This fact, along with their dialogue on minerals, indicates that Boyle and Malpighi not only shared a similar philosophical treatment of plastic powers, but also had in mind the same specific, corpuscular processes when describing agents involved in generation. The take-away is that Malpighi's appeal to plastic powers as an "explanatory apparatus" *just is* a means of describing generation as "the process of fluid and particle accretion."

13.3 Boyle on Generation

In order to understand Boyle's views on generation, one must first distinguish between what he calls a seminal principle, a petrifick sprit, and a plastic power. These agents lack sharp distinctions, and there is much overlap between them. Boyle tells us, for example, that the source of the petrifick spirit *could* be a seminal principle (though he ultimately will deny this). Seminal principles are most often associated with plastic powers for Boyle. Further still, Boyle claims that the petrifick spirit is *almost like* a plastic power. That said, Boyle clearly has different processes associated with the production of minerals and the generation of living organisms. Thus, it is only fitting that he would conceive of their respective agents differently.

A plastic power—which is the focus of this paper—just is a molding agent that is responsible for organizing matter in generation. When Boyle uses the term, he seems to have one of two ideas in mind.[26] The first is the more general notion of a formative power. The source of that power could come from a number of agents, including the Galenic formative faculty, the soul or even on occasion the Neoplatonic world soul. It becomes clear from Boyle's writings, however, that the only agent which could have that kind of formative power for the purposes of generation within his own framework is a seed, or a seminal principle. Boyle more frequently uses the term in a restricted sense, which is the power of a seed to generate a body.[27]

[26] For a list of Boyle's use of the term 'plastic', see Table 13.1.

[27] It is in this more restricted sense that Peter Anstey is right to say that Boyle, "uses the term 'plastick power' to describe the ability of seminal principles to generate bodies." See Anstey (2002a), 602.

Table 13.1 List of Boyle's use of the term 'plastic'

	Reference to 'Plastick'			Explaining			In the context of					
	Title	Year Pub	Page Reference	Minerals	Animals	Plants	Seminal principles	Power of the Soul	Universal World Soul	Chymists	God	Other
1	History of Fluidity	1661	v. 2, p. 192	x								
2	Sceptical Chymist	1661	v. 2, p. 258	x	x	x	x			x		Formative Power
3	Sceptical Chymist	1661	v.2, p. 276		x	x	x					
4	Sceptical Chymist	1661	v.2, p. 348	x						x		
5	Sceptical Chymist	1661	v.2, p.351	x		x	x					
6	Usefulnesse of Natural Philosophy	1663	v. 3, p. 225								x	Living engines
7	Origin of Forms and Qualities	1666	v. 5, p. 362–3	x			x					Seminal form; denies
8	Origin of Forms and Qualities	1666	v. 5, p. 383–4		x			x				Contra Harvey
9	Origin of Forms and Qualities	1666	v. 5, 389			x						Plant buds
10	Origin of Forms and Qualities	1666	v. 5, p. 433	x			x					Transmutation
11	Origin of Forms and Qualities	1666	v. 5, p. 436				x					Transmutation
12	Usefulnesse of Experimental Natural Philosophy, 2nd Tome	1671	v. 6, p. 498									Manual skill, plastic art; plaster
13	Origin and Virtues of Gems	1672	v. 7, 28–29	x	x	x	x					
14	Origin and Virtues of Gems	1672	v. 7, p. 30	x								Plastic form

15	Excellency of Mechanical Hypothesis	1674	v. 8, p. 108						x			
16	Considerations About the Possibility of the Resurrection	1675	v. 8, p. 302									Deceased matter
17	Considerations About the Possibility of the Resurrection	1675	v. 8, p. 303			x						Ashes and plants
18	About the Mechanical Production of Tasts	1675	v. 8, p. 371			x		x				Tastes of plants
19	High Veneration to God	1684	v. 10, p. 172		x						x	Semen Animatum and preformation
20	Notion of Nature	1685	v. 10, p. 459		x	x			x			
21	Notion of Nature	1685	v. 10, p. 467								x	
22	Notion of Nature	1685	v. 10, p. 500		x							Aristotelian account of monsters; denied
23	Notion of Nature	1685	v. 10, p. 533		x							Aristotelian account of monsters; denied
24	Notion of Nature	1685	v. 10, p. 558		x	x		x				Peripatetics
25	Christian Virtuoso	1744	v. 12, p. 446		x							Eggs
26	Fragments of Boyle's Essay on Spontaneous Generation	2000	v. 13, p. 286									Eggs and silkworms
27	Generation of Minerals	2000	v. 13, p. 373–6	x			x					Petrifick Spirit

To take a closer look at Boyle's treatment of seminal principles, his discussion in *Origin of Forms and Qualities* (1665) makes it clear that Boyle conceives of their operations in terms of homogeneity of matter and transmutation. The discussion of homogeneity of matter fits with the generation project of the *Origin of Forms and Qualities*, since Boyle is attempting to explain how qualitative changes can mechanically derive from some universal, catholic matter. In the second half of the essay, Boyle cites experiments involving substantial changes, which he attempts to describe in terms of corpuscular philosophy. The example involves the development of a chick from diaphanous fluid within the egg.

Boyle explains that the substance of the egg undergoes a great change because of incubation and then is turned into a chick. That change, however, had not "been taken notice of, for the same purpose" that Boyle had, which is to explain the changes of a substance.

He states that any part of the diaphanous white of the egg is like the other, emphasizing a "similarity of substance." Appealing then to observations, he writes about the "Rudiments of the Chick, lodg'd in the Cicatricula." Given his views on generation, those rudiments are likely seminal rudiments. The cicatricula, which he describes as a "white Speck up on the Coat of the Yolk," is nourished by the white of the egg until it becomes a great chick. On this view, the cicatricula belongs neither to the white nor the yellow of the egg. Rather, its contents act upon the fluids of the egg, both of which later act as nutriment.[28] He emphasizes the many different qualities that come from the uniform, diaphanous substance of the white of the egg, such as the various colors of the bird's parts, fluids and solid parts such as bones. The speck will become a chick with a head, beak and claws before turning to the yolk for digestion, which is "reserv'd as a more strong and solid Aliment".[29] Boyle's focus, however, is on the nutritive white of the egg, which he describes as being so soft

> … that by a little Agitation it may be made Fluid, and is readily enough dissolvable in cold water, this very Substance, I say, being brooded on by the Hen, will within two or three weeks be transmutated into a chick furnished with Organical parts…[30] so that here we have out of the White of an Egg, which is a Substance Similar, Insipid, Soft, (not to call it Fluid), Diaphanous, Colourlesse, and readily dissoluble in cold water, out of this substance, I say, we have by the new and curious Contrivement of the small parts consisted of, an Animal…[31]

Hence, Boyle describes the change of a quality-less, homogeneous substance into the parts of the chick in terms of transmutation. He extends that reasoning to the buds of plants also, describing how the buds transmute the sap, a "flegmatick Liquor,

[28] Oddly, Harvey doesn't seem to hold the latter view. Cf. Harvey 1965, 307.

[29] Boyle et al. 2000, vol. V: 382, *Origin of Forms and Qualities*.

[30] Harvey also describes the chick's formation this way. Cf. Harvey 1965, 370.

[31] Boyle et al. 2000, vol. V: 383, *Origin of Forms and Qualities*.

that seems Homogeneous enough," into the bodies of plants which are endowed with various colors, medical virtues, and diverse qualities.[32]

A final example demonstrating the relationship between transmutation and seminal principles in the *Origin of Forms and Qualities* can be found much later in his section on experiments when he is addressing transmutation specifically.[33] Here, Boyle complains of learned individuals who accept that elements can be transmuted into one another, and that they may be, "artificially destroy'd, and (without the intervention of a Seminal and Plastick power) generated or produc'd…"[34] He then goes on to consider the transmutation of rain water into earth. Of this process, Boyle writes that it would be a "very great difficulty to conceive, how a perfectly and exquisitely Homogeneous Matter should, without any Addition, or any Seminal and Plastick. Principle, be brought to afford great store of a Matter of much more Specifick Gravity then it self…"[35] Worth noting is that the seminal principle differs from something like rennet in that it not only is the agent of active change, but it acts as an organizing principle upon that matter. That organizing, plastic principle is unique to seeds.

A petrifick spirit, contrarily, is the result of a mixture. As such, it acts less as an organizing principle and is instead responsible for petrification of minerals by causing coagulation and reorganization of ingredients.[36] In his later work, Boyle seems hesitant even to discuss the formation of minerals in terms of generation. Boyle's first description of the coagulation of minerals occurs in *History of Fluidity*.[37] Within the *History of Fluidity*, Boyle discusses the coagulation of bodies, like the curdling of milk by saline liquors, and he combats the chymical theory that salts are the agents responsible for hardening and coagulation.[38] One such counter-example he provides involves ice, which becomes hard by the cold rather than by salt. Boyle also considers egg shells, which are soft immediately after being laid, but soon harden from their interaction with air. He then continues with what he takes to be the more convincing counter-example: the internal fluid of the egg. He writes that the white of the egg coagulates and grows to create the various parts of the chick,

[32] Ibid. 389.

[33] Boyle's comments here are in reference to van Helmont's experiments, which involve a tree which grew over several years with soil and water. Because the amount of the soil weighed the same, van Helmont concluded that the tree's additional mass was derived from the addition of water. Thus, water was the element from which a number of living beings were produced, and that process reflected a divine, formative power. For more on this, see Pagel 1982, 56; and Newman and Principe 2002, 77–80.

[34] Boyle et al. 2000, vol. V: 433 *Origin of Forms and Qualities*.

[35] Ibid. 436.

[36] This spirit could take the form of vapor, steam, or a liquor. See Boyle et al. 2000, vol. XII: 371.

[37] This essay was later published along with several others in 1661 as part of Boyle's *Certain Physiological Essays*.

[38] Boyle et al. 2000, vol. II: 187, *Certain Physiological Essays*.

such as its beak and bones. All of that hardening from fluids occurs, argues Boyle, without salts.

Boyle first gives a detailed account of his petrifick spirit within the essay, "Thoughts and Observations about the Generation of Minerals," which was originally intended to be an addendum to the *History of Fluids.*[39] Throughout the work, Boyle defends the thesis that minerals develop from a liquid state. He comes to this conclusion after rejecting other possibilities, which include the creation of minerals simultaneously to that of the Earth, leaving their *production* per se unexamined; as well as the Epicurean thesis that minerals are formed by the "casuall Coalition of congruous particles."[40] Though Boyle acknowledges that minerals may have been developed by seminal principles via an internal heat, he ultimately thinks that mineral formation occurs by coagulated fluids resulting from a "petrifick spirit," which he describes as "*an Almost plastick Agent.*"[41]

He goes on to describe the petrifick spirit as either an almost plastic agent, a sensible ingredient, or something like a ferment. In truth, Boyle sees each of these as different ways of describing the same properties of his Petrifick Spirit. The agent is plastic in the sense that it is a formative force capable of producing and solidifying new bodies. That the spirit acts like a sensible ingredient is on account of its being distinctly physical. Finally, it works like a ferment in the sense that it acts upon and coagulates passive matter. Each of these analogies describes the process which Boyle has in mind, namely that of matter acting upon matter and their resulting interaction. Further inspection of how the Petrifick Spirit is said to work helps to clarify the sort of agent that Boyle had in mind. Though it results in coagulation, it does so "less like that of the Cicatricula of an Egg" and more like the coagulation of milk curds with rennet. Recall that in the case of the egg, the coagulation that takes place is from homogenous material. With milk and rennet, however, a mixture takes place. Though Boyle is hesitant to characterize the process, he goes on to write of it that,[42]

> … to say little rather than Nothing at All to this difficulty,[43] I must put you in Mind that when I lately ‹spoke of› the Action of this Petrifick Spirit. I ‹made› express mention of a duely dispos'd matter. For in these Coagulations I take the Effect to be a thing that results as well from the Disposition of the Patient, as the Efficacy of the Agent; as wee see that Rennet of which ‹very› little will coagulate ‹a› great quantity of Milk, has no such operation at all upon water or Wine ‹&c.› Nor have I know [sic] the vapour of Lead to coagulate any other Liquors then Quicksilver.[44]

---

[39] This treatise was published along with several other unpublished treatises of Boyle's as part of the collection of Michael Hunter and E.B. Davis. For the remainder of the paper, its title will be abbreviated as "Generation of Minerals."

[40] Boyle et al. 2000, vol. XIII: 366, *Generation of Minerals*.

[41] Ibid. 372, *Of the petrific spirit*.

[42] His apprehension likely stems from the fact that the petrifick spirit might have, "sunder ways operating unknown." Boyle et al. 2000, vol. XIII: 372, *Generations of Minerals*.

[43] The "difficulty" he mentions here is that of explaining how the petrifick spirit works.

[44] Ibid. 373.

Thus, matter involved in creating minerals is "duely dispos'd" in the sense that it is both the active agent and the passive matter that is acted upon.

Boyle's "duely dispos'd" matter is clarified in his *Origin of Virtues of Gems*, published in 1673, 12 years after he wrote the manuscript on the generation of minerals. Boyle continues to defend his thesis that minerals—and specifically gems—are formed originally by hardening a liquid. In arguing thus, he presents several arguments that demonstrate that the heterogeneity of matter was, for Boyle, a key factor in the way in which minerals are formed. These include gems with heterogeneous matter or gems with extraneous mineral bodies within them.[45] In such examples, Boyle argues that this mixture would have occurred whilst they were still in liquid form. He writes, for example, that

> … some (at least) of the Real Virtues of divers Gems may be derived from this [Petrifick Spirit], That whilst they were in a fluid form, (or at least not yet Hard'ned,) the Petresecent substance was mingled with some mineral solution or tincture, or with some other impregnated liquor, and that these were afterwards Concoagulated, or united and hardened, into one Gem…[46]

Boyle describes the petrifick spirit working upon the corpuscles of varying mineral solutions or liquors in a fluid form— both an agent and a patient—as a result of these fluids mixing, whereby the corpuscles of each are mechanically affected in such a way that they coagulate together and their texture hardens. As the physical result of mixture, the petrifick spirit does not present the same challenge for Boyle's thesis as does his appeal to seminal or plastic powers, which are responsible for the organization of catholic matter. Consequently, I wish to make plastic powers—those powers in generation not reducible to matter or motion—the focus of this chapter.

13.4 Plastic Powers as Explanation

Are the formative powers of seminal principles—plastic powers—introduced in just those cases where Boyle lacked a mechanical explanation, or is Boyle providing a kind of mechanical explanation despite his appeal to agents that go against the sparse ontology of someone like Descartes? Put another way, could Boyle's account of seminal principles and their corresponding plastic powers be rightly called a mechanical explanation? Herein lays the heart of Anstey's two-horned dilemma. Anstey is absolutely right to claim that Boyle's seminal or plastic principles commit him to nescience about the nature of the agents responsible for generation. But, the question at hand is whether that nescience precludes Boyle from providing for them a distinctly mechanical explanation, nonetheless. I would argue that it does not,

[45] Ibid. 25 ff.

[46] Ibid. 45.

insofar as the explanatory power of those agents comes from the structural or textural changes which those agents cause upon passive matter. Seeds can operate mechanically—as Boyle understands the term—regardless of the nature of their plastic powers.

Boyle deals explicitly with what he takes to be a mechanical explanation in his 1674 treatise, *On the Excellency and Ground of the Corpuscular or Mechanical Philosophy*. Beginning with his mechanical philosophy, Boyle recognizes that there are different sects of Mechanical philosophers. Of other Mechanical philosophers, he complains, "that they think [the Mechanical Philosophy] pretends to have Principles so Universal and so Mathematical, that no other Physical Hypothesis can comport with [the Mechanical Philosophy], or be tolerated by it."[47] That kind of presumption Boyle inevitably describes as a mistake because mechanical principles are universal. They should, consequently, be inclusive and applicable to many explanations, rather than be exclusive and rejecting of certain explanations.[48] This claim is not insignificant either because it entails that on Boyle's view a mechanical explanation must be able to account for those things, like plastic powers, that would be otherwise incompatible with the ontological commitments of the mechanical philosophy.

Boyle elaborates upon his main complaint against those who would appeal to agents like a plastic power or Platonic world-soul, which is not their postulation of the existence on such powers. Rather, his criticism is that they fail to explain *how* these agents work. The main topic which naturalists should seek to explain, he argues, is not so much what the agent *is*. Rather, the naturalist should focus on what changes the agent makes, and "after what manner, those changes are effected." These changes, according to Boyle, can inevitably be explained in terms of and are reducible to matter and motion:

> So that the *Mechanical* Philosopher being satisfied, that one part of Matter can act upon another but by vertue of Local Motion, or the effects and consequences of Local Motion, he considers, that *as*, if the propos'd Agent be not Intelligible and Physical, it can never Physically *explain* the *Phænomena*; *so*, if it be Intelligible and Physical, 'twill be reducible to *Matter*, and some or other of those onely Catholick affections of Matter, already often mentioned.[49]

Here, it might be helpful to distinguish between a mechanical explanation and mechanical ontology. Dennis Des Chene has recently provided an excellent discussion of this distinction, arguing that "One could, and many philosophers did adopt mechanism as a method of explanation without adopting a mechanist ontology."[50] An organizing principle such as a plastic power undoubtedly causes rather obvious conflict with an ontology that only includes matter devoid of such powers. Yet, we

---

[47] Boyle et al. 2000, vol. VIII: 109.

[48] Ibid.

[49] Ibid.

[50] Des Chene 2005, 245.

ought not to assume that Boyle has such a strictly bare ontology. Indeed, as touched upon in the historiography section, Boyle is somewhat notorious for positing corpuscles with various chymical powers, and yet he was still largely viewed by contemporaries as one of the most influential proponents of the mechanical philosophy. Indeed, he is largely responsible for the term.[51]

Moreover, Boyle is especially keen to maintain a methodological agnosticism about those things in nature that he has no experimental means to observe. Given that methodological agnosticism, it's not surprising that he would accept nescience about the nature of agents involved in generation so long as he can explain their effects as reducible to the mechanical affections of matter. This black-boxing of the agent and insistence on its working as a physical agent in accordance to the mechanical laws of nature became a common application of the mechanical philosophy. Thus, I wish to put aside questions of Boyle's ontology and focus instead on his own understanding of mechanical explanation.

That Boyle's own treatment of plastic powers follows that prescription of a mechanical explanation is especially evident from his retorts to possible objections. In his *Origin of Forms and Qualities* Boyle considers the possible objection that the chick is fashioned by the soul, "lodg'd chiefly in the Cicatricula, which by its Plastick power fashions the obsequious Matter…" Boyle very likely had William Harvey in mind when presenting that objection. First, the passage bears resemblance to Harvey's own view. Moreover, Boyle's explicit reference to the cicatricula strongly suggests that he was thinking of William Harvey.[52]

To this complaint, Boyle answers that this objection does not invalidate his claim that the chick is a mechanically contrived engine. For, as he writes,

> let the Plastick Principle be what it will, yet still, being a Physical Agent, it must act after a Physical manner, and having no other matter to work upon but the White of the Egg, it can work up that Matter but as Physical Agents, and consequently can but divide the Matter into minute parts of several Sizes and shapes, and by Local Motion variously context them, according to the Exigency of the Animal produc'd, though from so many various Textures of the produc'd parts there must naturally emerge such differences of Colours, Tasts, and Consistencies, and other Qualities as we have been taking notice of. That which we are here to consider, is not what is the Agent or Efficient in these Productions, but what is done to the Matter to effect them.[53]

Note how Boyle ends this discussion by emphasizing how the explanatory power of the plastic power comes from its physical effects upon matter. Boyle elaborates on that point, stating that,

---

[51] Newman 2006, 178.

[52] In a previous work, Boyle claims that Harvey made evident both that the cicatricula is the source of the chick, and that it belongs to neither the white nor the yolk. See Boyle et al. 2000, vol. VI: 511.

[53] Boyle et al. 2000, vol. V: 383–384.

> And when Man himself, who is undoubtedly an Intelligent Agent, is to frame a Building or an Engine, he may indeed by the help of Reason and Art, contrive his Materials curiously and skillfully, but still all he can do, is but to move, divide, transpose and context the several parts, into which he is able to reduce the Matter assign'd him.[54]

Boyle goes on to explain how the external heat of incubation puts the parts of the substance into motion so that, "the Formative Power (whatever that be) doth any more then guide these Motions, and thereby associate the fitted Particles of Matter after the manner requisite to constitute a Chick…"[55] Thus, throughout his response to what is most likely Harvey's view, Boyle places the explanatory focus upon the material effects and modes of operations.

Recall that in the *Origin of Forms and Qualities*, Boyle extended his reasoning of the chick's formation to that of the plant. In the same vein, he considers a similar objection to his theory of the chick, namely that the diverse qualities of the resulting plant are, "the productions of the Plastick Power residing in prolific Buds…" To this objection, Boyle replies that he, "shall return the same Answer that I did to the like Objection, when 'twas propos'd in the First Observation."[56] In other words, whatever formative agent that plastic power might be, it must act in a physical manner.

In the *Origins and Virtues of Gems*, Boyle also considers the objection to his thesis of minerals developing fluids that a seminal principle must be involved. He replies by stating that there is "no absurdity" in seminal or plastic powers being harbored in liquid principles, citing the parts of animals which are turned to bones and the juices in trees which become bark.[57] His main point is that regardless of the nature of the agent, it must act upon the matter in exactly the same fashion, in this case by altering the mixed liquids mechanically. Thus, it is enough for Boyle to admit the possibility of his opponent's agents or principles so long as the process by which they work is the same.[58] In this manner, Boyle is able to provide an inclusive application of his mechanical philosophy as described earlier.

Avoidance of anachronism requires that we look at Boyle both as he understood his own work and as he was understood by his contemporaries. I hope to have demonstrated that Boyle took a mechanical account to be one in which the force of an explanation comes from the agent's mode of operation upon matter, the effects of which are reducible to the more catholic affections of matter. Those agents, then, may very well include matter that is endowed with certain powers. Seminal or plastic powers rightly meet this specification because- regardless of their nature- they

---

[54] Ibid. 384.

[55] Ibid. 384.

[56] Ibid. 389.

[57] Boyle et al. 2000, vol. VII: 28–29, *Origins and Virtues of Gems*.

[58] Boyle does not restrict that line of reasoning to agents of generation. He gives a similar response when addressing Aristotelian claims of the supposed gravity and levity of the elements, as well as the possibility of incorporeal agents such as angels. See Boyle et al. 2000, vol. X: 205, *Notion of Nature* and Principe and Boyle 1998, 208 respectively.

must and do act as physical agents on matter. For the latter purpose of understanding Boyle's reception by his own contemporaries, Malpighi proves to be especially helpful.

13.5 Malpighi and Boyle

Malpighi appropriated Boyle's notion of plastic powers in his own writings. This usage by Malpighi is part of a mechanical account of the generation of bodies from heterogeneous fluids. Moreover, and perhaps more importantly, it shows that Boyle's description of a plastic power was perceived as being mechanical by one of his own contemporaries similarly committed to the Mechanical Philosophy. That appropriation, however, was based on the works of Boyle's that Malpighi had available to him. In short, Malpighi understands a plastic power to work very much like and to be related to Boyle's petrifick spirit. Given what resources Malpighi had available to him, that Boyle had such a concept for the generation of animals in mind was a reasonable assumption on the part of Malpighi.

The essay in which Boyle is most explicit in print about the petrifick spirit remains his *Origin and Virtues of Gems.* This treatise was especially influential upon Malpighi. In a letter from Henry Oldenburg, dated 1672, Oldenburg explains that he is sending to Malpighi, along with the copper engravings for the illustrations of his embryological treatises of the chick and their resulting figures, a Latin edition of Boyle's *Origin and Virtues of Gems* prior to its being published in Latin.[59] Boyle's essay later came out in 1673 as *Exercitatio de origine et viribus de gemmarum*, and that letter of Oldenburg's was published with Malpighi's first embryological treatise, *Pulli in Ovo.* That latter fact is of interest, too, because it means that any of Malpighi's contemporaries reading his earlier works on the chick would have known of his receiving and likely reading Boyle's work.

Pulli in Ovo was the first of two treatises on the development of the chick egg. Both of these were written by Malpighi the previous year, 1671. They were published by the Royal Society in 1673 and 1675, respectively. From his observations on the egg, Malpighi concludes in *Pulli in Ovo* that

> it is consequently right and proper to surmise [what] we are considering the primeval and simultaneous genesis of the [parts] of animals. For we may surmise that the chick together with the bounding saccules of almost all its parts lies concealed in the egg, floating in the colliquament, and that the nature of the latter results from the integration of the mingled nutritive and fermentative juices, through the joint action of which, when aroused, the blood is produced in successive steps and the parts formerly outlined erupt and swell out.[60]

---

[59] Oldenburg et al. 1969, IX: 229–230.

[60] I have changed the translation slightly. Cf. Adelmann 1966, vol II: 957, *Pulli in Ovo*.

Malpighi's second treatise on the chick egg, *De Ovo Incubato*, supports his previous claims. He maintains the "conjecture" that the parts of the chick are pre-formed within the egg after the simultaneous genesis of parts. At various stages Malpighi provides a chymical analysis by heating the substance of the *chorion* and amniotic fluids of the egg to see if they coagulate, in an attempt to understand its means of production.[61]

Previous commentators have pointed out that the "simultaneous genesis of parts" within the saccules puts him in stark contrast with proponents of pre-existence like Malebranche or Swammerdam.[62] The latter assume that the rudimentary parts from which the embryo developed existed before fertilization. Development of the embryo before fertilization entails that the generation of the embryo is subsumed under growth, bypassing the problem of generation all together. For Malpighi, however, the pre-formed embryo does not exist before fertilization. Rather, the development of animal-embryos begins at fertilization from heterogeneous male and female components, and this would entail that some kind of *production* occurs.[63] That production happens via the simultaneous genesis of parts.[64] But explaining just *how* those rudimentary parts are produced remains a difficult task. One could imagine, then, how a plastic and organizing force that could explain how the rudiments form from fluids mechanically mechanically would be an especially helpful explanatory apparatus. Such an apparatus Malpighi gets from Robert Boyle.

The peak of the exchanges between Boyle and Malpighi happened after the death of Oldenburg primarily through Carlo Ronchi, a resident in London from Italy. At the beginning of 1686, Malpighi wrote to Ronchi expressing his deep admiration of Boyle and posed questions for Boyle about the latter's work on blood.[65] Moreover, accompanying this letter was a package of books and Porretta crystals.[66] Boyle

[61] See, for example, Adelmann 1966, vol II: 1006 and 1009, *Pulli in Ovo*; Also, Cf. Bertoloni Meli 2011, 231.

[62] Cf. Roe 1981, 6; Roger 1997: 259 ff.

[63] Adelmann 1966, vol. II: p. 885.

[64] It might be easy to confuse Malpighi's views with those of Harvey's metamorphosis. There are some similarities, as each consists of parts which form simultaneously and exist before the generation of the animal. It would be a mistake, however, to do so. First, Harvey's conception of metamorphosis is related to his ideas about spontaneous generation. (Cf. Harvey 1965, 335.) Malpighi rejected Harvey's views on spontaneous generation. More importantly, however, Malpighi does not deny that growth and gradual development occur within the egg. Metamorphosis entails the breaking-up of homogenized material into different parts. Malpighi's view, however, simply entails the enlargement of parts which were produced simultaneously.

[65] February 6th 1686, NS; January 26th, OS. Boyle's *Memoirs for the Natural History of Human Blood* was published in English shortly before this correspondence in 1684. See Boyle 2000, V. 10:3 ff.

[66] Malpighi and Adelmann 1975, 1135. As late as 1911, Porretta, located near Bologna, was noted for having "remarkably hallowed crystals." See Chisholm 1911, 433.

responded through Ronchi shortly after expressing his gratitude for the crystals, his high regards for Malpighi, and cheers for Malpighi's good health.[67]

There was much interest in the rest of the community of Italian anatomists in Malpighi's correspondence with Boyle, and they wrote frequently to him inquiring about it. Most note-worthy for our purposes was Malpighi's correspondence with Lorenzo Bellini. A letter from Malpighi to Bellini, who was also interested in the generation of minerals, shows the relevance which the crystals sent to Boyle by Malpighi had to the theory of minerals. In the letter, Malpighi explains that one particular crystal confirms Boyle's opinion that minerals originate from fluid material. The crystal had an air bubble that would move, as if the internal contents of the stone had not entirely solidified.[68]

Incidentally, Malpighi also sent a second package to Boyle including more crystals later that year. It seems to have been lost, however.[69] Nonetheless, Malpighi's correspondence is very telling about the relationship between him and Boyle. Given that fact, Boyle's description of coagulation in those works published in Latin can help to contextualize Malpighi's own references to plastic virtues or powers in his own writings.

13.6 Malpighi's Plastic Power

Malpighi's first and most noted reference to plastic powers is found in his epistle to the Lyon physician, Jacob Spon. Malpighi sent a copy of the letter, dated November 1681, to the Royal Society in August of 1683. It was subsequently published in 1684. In the epistle, Malpighi describes the fecundation of several animals such as butterflies and a cow, as well as the fecundation of plants. It is not unlikely that Malpighi had Boyle specifically in mind when writing to Spon, either. Accompanying the copy of the epistle that Malpighi sent to the Royal Society was a letter to Dr. Francis Aston.[70] In that letter, Malpighi explained that he had already sent a copy of the epistle to the "Illustrious Robert Boyle".[71] In fact, a copy of the epistle to Spon is still present in the archive of Boyle's manuscripts and papers.[72]

[67] Malpighi and Adelmann 1975, 1156. The sentiments described by Ronchi on behalf of Boyle were likely sincere too, as evident by Boyle's later requesting Malpighi to host his nephew. Cf. Ibid. 1258.

[68] Ibid. 1212.

[69] Ibid. 1250, 1258.

[70] Aston replaced Oldenburg as the Royal Society's secretary after the latter's death in 1677.

[71] Malpighi and Adelmann 1975, 910–911.

[72] Boyle, Robert 2014, BP 17, fol. 116v–125v.

In the epistle, Malpighi makes only three references to plastic powers, the first of which is in passing.[73] The latter two references are in a later passage. He begins by describing that in butterflies, a sticky ichor drips from a structure attached at the end ovary through the vagina. Malpighi writes (boldface mine),

> By this ichor the semen of the male and also another humor emited by a lateral chamber are received and maintained, and by all three of them eggs passing through the vagina are moistened and fecundated; and thus that **plastic power** is preserved for many days and communicated to eggs emerging at subsequent times. This, we may infer, is also true of the hen, where the energy of the semen received in a single mating is preserved to no small degree, with the result that eggs are fecundated even for some time thereafter. And since in the hen's egg Nature does not scatter and sprinkle the semen of the cock or another menstruum fecundated by the semen upon the cicatricula alone, in which the rudiments of the parts lie concealed, but also moistens with **plastic force** the entire egg (that is to say, the aliment in the form of albumen and yolk), so that the whole is fecundated, and residue, too, the uterus is analogous to the hen's egg, because it is swollen with humor and surrounds the ovule, it is therefore probable that the uterus and the humors it contains are also fecundated.[74]

Given Malpighi's general mechanical program, that he had in mind a more traditional, Galenic notion of the term 'plastic' in the passage is unlikely. Moreover, such a reading would be an awkward one, too. This part of Malpighi's discussion is immediately preceded by a description of fluid particles in motion in which he explains that the male semen is able to fecundate eggs by mixing with an ichor from the vagina. In other words, fecundation occurs as the result of mixing two fluids. Further, that process is explained entirely in terms of particles of matter and motion. From all of this, Malpighi concludes that the compendium, the outermost boundaries of the chick, lie concealed within the cicatricula before incubation. Those first rudiments become visible only through motion and are formed separately from fluid.

Malpighi's *Vita*, from his *Opera Posthuma*, explains more explicitly just how those rudiments become visible through motion. He concludes from several observations on incubated eggs that the fecundated egg contains the rudimentary parts.[75] Those rudimentary parts, Malpighi explains, are nothing more than a collection of fluid confined by an embankment. After incubation, the fluid becomes thinner as the solid parts are dissolved into furrows so that the humors flow and are confined to the cicatricula. Within the cicatricula, in the same manner described in his embryological treatises, the parts swell and grow making the first filaments visible. That happens, he explains, when the outermost parts are constructed, giving

[73] The first mention of plastic virtues can be found in the very first sentence of the second half of the epistle to Spon. Malpighi refers to it in passing in the context of the uterus, which he calls the "workshop" or "office" of the plastic virtue. "…plasticae virtutis officinam contemplemur exarando ea…" See Malpighi 1684, 630.

[74] Malpighi 1684. Translated in Adelmann 1966, vol. II: 861–2.

[75] Malpighi presents the same view as in his embryological treatises, but he is much more explicit about it.

rise to cavities which are then filled by fluid, causing the early formation of the spine to emerge.[76]

In the context of this discussion, that is, the initial simultaneous development of the parts within the compendium, there is another reference to these plastic powers. Malpighi writes,

> [Nature] begins to form the rudiments of the parts to be delineated… by means of whose pores, as by so many glandular sieves, she separates the infenced fluid from the fluid in which it is immersed; and the fluid thus confined is pervaded by the plastic spirit and organized, when unsuitable substances have been transpired and its parts have been properly adapted.[77]

Here, too, Malpighi appeals to the example of butterflies, whose wings and antennae are first sketched out in their outer parts, and then filled with fluid. Recall how, in *Pulli in Ovo* Malpighi describes the saccule as floating in a liquor of colliquament. The fluid which fills out the cavities formed by the outer parts later solidifies, and the parts are drawn out from their saccules. The *carina* (rudimentary parts of the spine) is formed similarly. Boyle and Malpighi are strikingly similar in this regard: Whereas Malpighi describes the organization and hardening of the fluid from the saccules to create the rudimentary parts of the chick by the plastic power, so too does Boyle describes the production of complex crystal formations. Both scholars describe the formation of a new being from the organized, coagulation of fluids resulting from mixture.

Another passage from Malpighi's posthumous work confirms that he wrote on plastic powers within the context of Boyle's work on minerals. In this particular essay, Malpighi is describing how fecundation of fruit occurs, especially those cases in which multiples occur. To explain, Malpighi appeals to juices and the power of the colliquament. In the case of unfertilized eggs Malpighi explains that because "the plastic virtue misses by chance", there is merely an accumulation of particles. Malpighi compares the fruit to flowers, stating that multiples arise from an abundance of mixed juices and of floating particles.

Malpighi continues by comparing this process to the formation of stones. "Similar phenomena," he writes, "also occur not infrequently in the concretion of stones." For an example, he looks at jet, which forms a kind of "egg" from fluid materials in which diverse tinctures and particles are present due to the different densities (*gravitas*) of the particles. That process causes a kind of resistance which creates layers, like those of onions. The onion described in this passage is reminiscent of the layering displayed in his sketches from the embryological treatises. Indeed, Malpighi goes on to compare the stones to his observations in the incubated egg. Here, the center circle is the cicatricula, which contains the chick's rudiments (the *carina*). He explains that the concentric circles of the stone exhibit a similar

---

[76] Malpighi, *Opera Posthuma, Vita.* Translated in Adelmann 1966, vol. II: 866.

[77] Ibid. 866.

appearance (*species*) to the incubated egg, in which the many circles form around the *carina* by means of fermentation and get larger.[78]

As with previous works, Malpighi continues to analyze their continued development in terms of the concretion of juices and the movement of fluid particles. Later, he mentions Boyle by name, stating that stones and gems are thought to derive their origins from fluids by "the most learned Boyle and Steno" (boldface mine):

> Succedunt autem in prima gagatis productione tot ova, non quia lapides ab ovo viventium more ortum necessariò trahant, sed materiæ necessitate. Constat namque ex his, quæ à Doctissimis D.D. *Boyle*, & *Stenone*, habentur lapides & **gemmas** suam originem à **fluidis** trahere. Et quoniam primum fluidum salibus, & particulis fossilium, & **mineralium** ad minima **solutis impregnatum** turget, ideo intestino suscitato motu exagitatur, & ambientis pondere premitur, unde à centro extrusæ graviores medio **fluido** aequali vi circulum efformant, cui aliæ succedentes consimili compressione extrusæ alium addunt, quod pariter repetitur secundum copiam primi **fluidi**, & ejusdem heterogeneitatem.[79]

What is more striking than Malpighi's explicit mention of Boyle is the extent to which Malpighi's language is parallel to that of Boyle's own in the passage of the *Exercitatio De Origine Et Viribus Gemmarum* (*The Origin of Gems and Virtues)*, in which Boyle explains specifically how he thinks that the petrific spirit acts upon matter:

> Atque ut hactenus dicta in rem meam vertam, existimem, quasdam (saltem) ex veris **Gemmarum** quarundam Viribus posse ex eo derivari, Quod dum **fluidam** illae formam obtinebant (vel saltem necdum durantae errant) substantia Petrifica **Minerali** cuidam **Solutioni tincturaeve**, vel alii cuidam **liquori impraegntato**, comixta, haeque postmodum substantiae coagulatae, vel unitae durataeve fuerint in unam **Gemmam**...[80]

That Malpighi's vocabulary mirrors Boyle's is especially striking given that Malpighi references such phenomena as impregnated liquors and stones all within a discussion of the fecundation of fruit. Outside of the context of Boyle's own works, such references seem out of place.

Moreover, there are references in the passage which link it to other parts of the *Origin of Gems and Virtues*. For example, that Malpighi mentions both Steno and fossils is not unrelated. Both the Latin and English editions of Boyle's *Origin and Virtues of Gems* include a preface from the editor which quotes Steno's summation of Boyle's argument for the formation of gems from fluid via the petrifick spirit.[81] This three-page long quotation comes from the former's *Prodromus*, in which Steno

---

[78] Malpighi 1696, 90; "In lapidum quoq; concretione non rarò consimilia succedunt phænomena. Constat enim in gagate ovum ex fluida materia fieri, in qua cum adsint diversæ tincturæ, & particulæ fossilium et mineralium ex varietate gravitatis earundem, & resistentia ambientis fit ovum multiplicibus distinctum fluoribus, in quibus succedunt tandem concretione, quasi tot involucra sese contingentia ceparum instar manifestantur; quin immo & speciem simile exhibent, qualem in incubato ovo intuemur, in quo ex fermentatione circuli circa carinam velut aggeres cum interfluentibus liquoribus dilatantur, & multiplicantur."

[79] Ibid. 91.

[80] Boyle 1673, 86. This same passage is quoted in the original English earlier in this paper. See also, Boyle et al. 2000, VII: 45.

[81] Cf. Boyle et al. 2000, VII: 5–7.

describes the formation of fossils.[82] Curiously, though he published before Boyle's work on gems, Steno fairly accurately explains Boyle's thoughts on the matter in the introduction *Prodromus*.[83] Steno's dissertation was later translated into English by Oldenburg, and the two were published together in a rare 1673 edition of Boyle's *Essays of the Strange Subtilty, Great Efficacy, Determinate Nature of Effluviums*.[84]

Further still, Malpighi mentions in the passage coagulation and heterogeneity. As discussed previously in this paper, both coagulation and heterogeneous fluids are key themes for Boyle's work on minerals. Near the end of this discussion, Malpighi states that the generation of eggs requires an abundance of heterogeneous fluid, explaining the effects of the heterogeneity of matter in terms of fluids, volatile particles and motion. Thus, Malpighi's plastic power is part of a mechanical account of the generation and consistent with the observations made in his embryological treatises published several years earlier.

13.7 Malpighi, Sbaraglia, and Mechanical Explanation

The idea that fecundation occurs as the result of a mixture is by no means a new one. Indeed, even on the Galenic view the embryo is a result of a mixing of seminal fluids which produces a formative faculty, or plastic and molding power. That power organizes the matter's constitution.[85] What distinguishes Malpighi's plastic powers from those of his medical predecessors is the manner and process by which he conceives that formative agent to work. For Malpighi, as for Boyle, the plastic agent is a distinctively physical one. Like Boyle, moreover, Malpighi is interested in explaining the effects of a plastic power in terms of matter and motion (in this case, the particle accretion of fluids). In that sense, Malpighi's plastic powers are antithetical to the Galenic formative faculty. In the case of Malpighi and predecessors, it is important to remember that however similar their theories may initially seem to the present-

[82] Boyle's work was published in 1671, whereas Steno's *Prodromus* had already been published in 1668.

[83] Just how Steno was made aware of Boyle's views remains unclear. Toshihiro Yamanda has recently argued that Steno became aware of and was consequently influenced by Boyle's thesis of mineral formation through Steno's mentor, Ole Borch. Borch met Boyle in 1663 while touring Europe and thus likely transmitted Boyle's ideas later to Steno. If that it is the case, it speaks even more to the magnitude of Boyle's influence, as Steno's work on the matter is generally considered to be foundational in geomorphology (Yamanda 2009).

[84] Interestingly, Steno also wrote a dissertation on the development of the chick-egg, *De vitelli in intestina pulli transit*. See Steno 1950. That Steno wrote a dissertation on the formation of minerals so shortly after his excursus on the chick suggests that the interest in the generation of both animals and minerals was not an anomaly in the seventeenth century. Steno's work on the chick makes no reference to a plastic power or seminal principle. That he refrains from such reference might well be because his dissertation on the chick was written in 1664, approximately eight years prior to Boyle's publication of the *Origin of Gems and Virtues*.

[85] Roger 1997, 51.

day reader, that radical difference would not have been lost on their own contemporaries.

In a *Riposta* to Giovanni Sbaraglia, an Italian physician and one of Malpighi's more prolific adversaries, Malpighi maintains that the mode of operation of the soul is mechanical, arguing that, "the soul is forced to act in conformity with the machine on which it is acting; therefore, a clock or a mill could be moved by a pendulum, an animal, a man, or even an angel, but it will always be moved in the same way."[86] Thus, like Boyle, Malpighi emphasizes the mode of operation for the agents responsible for acting on matter rather than the nature of those agents themselves. Similarly, that the agent acts "in conformity with the machine" and physically upon matter is sufficient for that account to be considered mechanical for both Boyle and for Malpighi.

Sbaraglia replied with a rather harsh criticism to Malpighi's argument. His response is an important one for understanding the episode, too, because it offers the present-day reader a unique perspective from a contemporary of both Boyle and Malpighi who, in many ways, defends a distinctly Galenic conception of the body. His main complaint of Malpighi's view is the "*modus mechanicus*"—or mechanical mode—of explanation that it provides. In other words, the operation of the soul is not bound by the *faculties*, but by *local motion*.[87] From this Galenic perspective, the kind of explanation posited by Boyle and by Malpighi— one of bracketing off the agent and describing its effects in terms of matter and motion—is considered to be a distinctly mechanical one. When a seventeenth-century Galenist vehemently opposed to the mechanical philosophy criticizes the mechanical mode of explanation posited by those who *likewise* consider their explanations to be mechanical, it would be anachronistic to dismiss the accounts of either Boyle or Malpighi as simply unsuccessful or misguided.

13.8 Conclusion

Reading both Boyle and Malpighi together brings three levels of insight to the modern reader. First, it helps to clarify the concepts that each had in mind when evoking agents such as a plastic power in their accounts of generation. Analyzing Boyle's influence upon Malpighi's thoughts about generation shows that Malpighi provided a thoroughly mechanical account of generation. This account is explained in terms of the particle accretion in heterogeneous fluids. In a similar vein, Malpighi's work allows one to see Boyle's writings through the lens of a distinguished and especially relevant contemporary. This lens clarifies just how Boyle redefined a term traditionally associated with Neo-Platonism, faculties, and the soul into the context of the mechanical philosophy.

---

[86] See Adelmann 1966, vol. II: 865. Cf. Bertoloni Meli 2011, 232.

[87] See Sbaraglia 1704, 252. Cf. Bertoloni Meli 2011, 324.

Second, because of Boyle and Malpighi's influence and collaboration with others, we are informed about several other members of the intellectual community. For example, Nathaniel Highmore experimented with Boyle on the chick and dedicated his own embryological treatise to Boyle.[88] Steno, mentioned earlier, not only wrote on the generation of minerals in a treatise published with Boyle's, but also was so distinguished at anatomy that Henry Oldenburg describes him as, "a Dane living at Florence who is second to none as an anatomist…"[89] Bellini, the fellow to whom Malpighi described the crystals mailed to Boyle, expresses to Malpighi through letters a considerable amount of interest in both the generation of minerals and the works of Robert Boyle. Yet, Malpighi too, was a well-known physician in Italy who was especially dedicated to mechanical explanations. These are only a few examples that require further investigation and research, but it is important to remember that none of the figures we study were writing in isolation, but in frequent communication in a network with each other.

Finally, and most importantly, reading Boyle and Malpighi in the context of each other points to a more complex and multi-faceted understanding of both generation and explanation than previously assumed. It brings into focus questions about the mechanical philosophy and mechanical explanations as understood *by their contemporaries*. Like Boyle, Malpighi was also well known for his mitigated skepticism and emphasis on observation. Further, like Boyle, he was committed to mechanical explanations. In stipulating that the plastic power must act physically upon matter, Boyle allows for the possibility of a mechanical plastic power. Malpighi seized upon that possibility. Triangulating between Sbaraglia, Boyle, and Malpighi allows us to disentangle the complex notions of mechanical explanations regarding experimental investigations in the seventeenth century.

References

Adelmann, Howard B. 1966. *Marcello Malpighi and the evolution of embryology*. Ithaca: Cornell University Press.

Anstey, P. 2002a. Boyle on seminal principles. *Studies in History and Philosophy of Biological and Biomedical Sciences* 33(1): 597–630.

Anstey, P. 2002b. Robert Boyle and Locke's "Morbus" entry: a reply to J.C. Walmsley. *Early Science and Medicine* 7(4): 358–377.

Anstey, P. 2002c. Robert Boyle and the heuristic value of mechanism. *Studies in History and Philosophy of Science Part A* 33: 161–173.

Aristotle, and Arthur Platt. 2008. *On the generation of animals*. Whitefish: Kessinger Publications.

Bertoloni Meli, Domenico. 2011. *Mechanism, experiment, disease: Marcello Malpighi and Seventeenth-century anatomy*. Baltimore: Johns Hopkins Univ. Press.

Boyle, Robert. 1665. *Origo Formarum Et Qualitatum*. Trans. Henry Oldenburg. Lichfield, England: Oxonia Press.

[88] Highmore 1651. For a more comprehensive discussion on this topic, see Ekholm 2011.

[89] Oldenburg et al. 1969, 454–457.

Boyle, Robert. 1673. *Exercitatio De Origine Et Viribus Gemmarum*. Comp. Moses Pitt. London: Little Britain.
Boyle, Robert. 2014. *The Boyle Papers*. Comp. The Royal Society. London: The Boyle Project. http://www.bbk.ac.uk/boyle/boyle_papers/bp17_docs/bp17_115v-116r.htm.
Boyle, Robert, Michael Cyril William Hunter, and Edward Bradford Davis. 2000. *The works of Robert Boyle*. London: Pickering & Chatto.
Chalmers, A. 2002. Experiment versus mechanical philosophy in the work of Robert Boyle: a reply to Anstey and Pyle. *Studies in History and Philosophy of Science Part A* 33(1): 187–193.
Chisolm, Hugh. 1911. *The Encyclopedia Britannica: A dictionary of arts, sciences, literature and general information*, vol. 23, 11th ed, 433. New York: Encylopedia Britannica.
Clericuzio, Antonio. 1990. A redefinition of Boyle's chemistry and Corpuscular philosophy. *Annals of Science* 47(6): 561–589.
Clericuzio, Antonio. 2000. *Elements, principles, and corpuscles: a study of atomism and chemistry in the seventeenth century*. Dordrecht: Kluwer Academic Press.
Des Chene, Dennis. 2005. Mechanisms of life: Borelli, Perrault, Régis. *Studies in History and Philosophy of Biological and Biomedical Sciences* 36: 245–260.
Ekholm, Karin Jori. 2011. *Generation and its problems: Harvey, Highmore and their contemporaries*. Ph.D., History and Philosophy of Science, Indiana University.
Galen, and Arthur John Brock. 1916. *On the natural faculties*. London: W. Heinemann.
Hall, Marie Boas. 1958. *Robert Boyle and seventeenth-century chemistry*. Cambridge: Cambridge University Press.
Harvey, William. 1965. *The works of William Harvey*. New York/London: Johnson Reprint Corporation.
Hirai, Hiro. 2005. *Le concept de semence*. Turnhout: Brepols Publishers.
Hirai, Hiro. 2007. The invisible hand of God in seeds: Jacob Schegk's theory of plastic faculty. *Early Science and Medicine* 12(4): 377–404.
Hunter, William B. 1950. The seventeenth century doctrine of plastic nature. *Harvard Theological Review* 43(3): 197–213.
Malpighi, Marcello. 1696. *Opera Posthuma*. London: Royal Society.
Malpighi, Marcello, and Howard B. Adelmann. 1975. *The correspondence of Marcello Malpighi*. Ithaca: Cornell University Press.
Malpighi, Marcello. 1684. Praeclarissimo Et Eruditissimo Viro D. Jacobo Sponio Medicinae Doctori, Et Lugdunensi Anatomico Accuratissimo. Marcellus Malpighius S. P. *Philosophical Transactions* 14.155–166: 630–46.
Newman, William Royall. 2006. *Atoms and alchemy: Chymistry and the experimental origins of the scientific revolution*. Chicago: University of Chicago Press.
Newman, William R., and Lawrence M. Principe. 1998. Alchemy vs. chemistry: The etymological origins of a historiographical mistake. *Early Science and Medicine* 3: 32–65.
Newman, William R., and Lawrence M. Principe. 2002. *Alchemy tried in the fire: Starkey, Boyle, and the fate of Helmontian Chymistry*. Chicago: University of Chicago Press.
Oldenburg, Henry, A. Rupert Hall, Marie Boas Hall, and Eberhard Reichmann. 1969. *The correspondence of Henry Oldenburg*. Madison: University of Wisconsin Press.
Pagel, Walter. 1982. *Joan Baptista Van Helmont: Reformer of science and medicine*. Cambridge: Cambridge University Press.
Preus, Anthony. 2002. Plotinus and biology. In *Neoplatonism and nature: Studies in Plotinus' Enneads*, Studies in Neoplatonism, vol. 8, ed. Michael F. Wagner. Albany: State University of New York Press.
Principe, Lawrence, and Robert Boyle. 1998. *The aspiring adept: Robert Boyle and his alchemical quest: Including Boyle's "lost" dialogue on the transmutation of metals*. Princeton: Princeton University Press.
Pyle, A. 2002. Boyle on science and the mechanical philosophy: a reply to Chalmers. *Studies in History and Philosophy of Science Part A* 33(1): 171–186.

Roe, Shirley A. 1981. *Matter, life, and generation: Eighteenth-century embryology and the Haller-Wolff debate*. Cambridge: Cambridge University Press.
Roger, Jacques. 1997. *The life sciences in eighteenth-century French thought*, ed. Kieth R. Benson. Trans. Robert Ellrich. Stanford: Stanford University Press.
Sbaraglia, Giovanni Girolamo. 1704. *Oculorum et Mentis Vigilæ*. Bologna: Pietro-Maria Monti.
Steno, Nicolaus. 1671. *The Prodromus to a dissertation concerning solids naturally contained within solids*. Trans. Henry Oldenburg. Comp. Moles Pitt. London: J. Winter Press.
Steno, Nicolaus. 1950. On the passage of yolk into the intestines of the chick (De Vitelli in Intestina Pulli Transitu). Trans. Margaret T. May. J *Hist Med Allied Sci* Spring V: 119–143, With Introduction and Commentary.
Walmsley, Jonathan. 2000. Morbus– Locke's early essay on disease. *Early Science and Medicine* 5(4): 367–393.
Walmsley, Jonathan. 2002. "Morbus," Locke and Boyle – A response to Peter Anstey. *Early Science and Medicine* 7(4): 378–397.
Wilson, Catherine. 1995. *The invisible world*. Princeton: Princeton University Press.
Yamanda, Toshihiro. 2009. Hooke-Steno relations reconsidered: Reassessing the roles of Ole Borch and Robert Boyle. In *The revolution in geology from the renaissance to the enlightenment*, ed. Gary D. Rosenberg. Boulder: Geological Society of America.

Part IV
Medicalizing Philosophy?

Chapter 14
Early Modern Medical Eudaimonism

Justin E.H. Smith

Abstract In this paper, I argue that G. W. Leibniz was a representative *par excellence* of a largely forgotten tradition, which I am calling 'medical eudaimonism', and which saw medicine as entirely integral to the project of philosophy to the extent that (i) it was the key to health and longevity, and thus to the realization of the good life; and (ii) it was conceived as including rules of diet, hygiene, and bodily comportment, and to this extent was seen as nothing less than the corporeal flip-side, so to speak, of ethics. I will argue, finally, that, as much recent scholarship attests, from an initial interest in the classical philosophical problem of the mind-body connection in early modern philosophy, one is invariably compelled to take up early modern theories of the passions (as the field in which this connection most aggressively demands recognition), and from here, in turn, scholars are willy-nilly compelled by the concerns of early modern philosophers themselves to take an interest in the history of medicine, and to acknowledge its central importance to the history of philosophy.

Keywords Leibniz • Descartes • Medicine • Eudaimonism

14.1 Introduction: Early Modern Philosophy, Medicine, and the Good Life

One might, if half-jokingly, propose an alternative history of philosophy, focusing upon the preoccupations of the canonical thinkers throughout history with the state of their own bowels, from the Pythagorean interdiction on beans to Nietzsche's claim that "you will understand the origin of the German spirit—from distressed intestines."[1] In such a history, the place of René Descartes (to choose just one example) would remain no less central than it is in the canonical version. Consider the remedy

[1] Nietzsche 1967, 236.

J.E.H. Smith (✉)
Département Histoire et Philosophie des Sciences, Université Paris Diderot - Paris 7, Paris, France
e-mail: jehsmith@gmail.com

P. Distelzweig et al. (eds.), *Early Modern Medicine and Natural Philosophy*, History, Philosophy and Theory of the Life Sciences 14,
DOI 10.1007/978-94-017-7353-9_14

that Descartes recorded, and evidently used in July, 1628, in order to relieve a *trouble du transit intestinal.*

> A particularly difficult evacuation of the lower intestine after a number of meals was brought about in this way: Equal parts bull's gall, unsalted butter, black hellebore, extract of diacolocynthide, diagridion, and saffron, reduced into a single mass and heated over a flame until they have attained the consistency of honey, inserted into Italian terracotta vessels and applied to the navel. And this cataplasm is then fastened so that it does not fall; and two cataplasms of clay, filled with these potions, are applied the one after the other, one per day. The first days nothing was felt by the patient, other than agitations and murmurs; the third day, the desired evacuation arrived with great pain, but the normal excretion did not follow the very hard excrements until the abdomen of a freshly slaughtered calf, covered in aged oil sifted after cooking and heated up, was applied to the patient's stomach, and until the anus was probed by fingers covered with bile and butter.[2]

Now Charles Adam and Paul Tannery included this letter in their exhaustive edition of Descartes' writings; John Cottingham et al., by contrast, felt that they could safely leave it out of their English edition,[3] and many might agree that this is a warranted omission, that it is hardly fair to pick out this little *Zettel* as telling us much of anything about Descartes the thinker. After all, Wittgenstein surely made grocery lists, and we rightly give those no attention. But perhaps the importance of Descartes's cure for understanding the preoccupations of early modern philosophy as its practitioners themselves understood it will become clearer if we consider how this bit of the Cartesian corpus was passed down to us.

The only reason we have it is because another philosopher, G. W. Leibniz, transcribed it nearly 50 years later, when Descartes' literary executor, Claude Clerselier, gave the German philosopher one day, February 24, 1676, to look through Descartes' manuscripts and to transcribe whatever interested him. Ponder that fact for a moment: Leibniz had a single day to copy out whatever he could find of interest in the *Nachlass* of the greatest philosopher of his era, and this remedy against constipation made the cut. Clearly he was operating with a different understanding of the scope of the discipline than we are. It is this scope, and what it meant for the formulation of philosophical questions and projects in the early modern period, that I would like to investigate in the present essay.

Consider also, in this connection, Descartes's letter to William Cavendish of October, 1645: "From the beginning, the principal aim of my studies has been the

---

[2] René Descartes, *Oeuvres*, ed. Charles Adam and Paul Tannery, Paris, Léopold Cerf (Descartes 1983), XI, 644 (hereafter 'AT'). It is worth quoting this recipe in full here, if only because it has not been published in translation elsewhere. "Alvi egrestio difficillima post menses aliquot sic provocata. Fellis taurini recentis, butyri insulsi, hellebori nigri, extracti diacolocynthidis, diagridii & croci partes aequales, in unam massam redactae, & igni ad mellis consistentiam decoctae, italicae nucis testae inditae, umbilico impositae sunt. Ligataque fuit mox ne caderet, & binae testae, diebus singulis, potionibus intus assumtis, sic repleta impositae sunt. Primis diebus, nihil praeter fluctuationes & murmura a patiente sentiebantur; tertia die, cum immensis doloribus supervenit egerendi desiderium; at induratis excrementis non successit excretio, donec vituli abdomen recens, cum oleo antiquo ad ignem cribratum & calens, ventriculo induceretur, digitis que felle & butyro inunctis anus sollicitaretur."

[3] Descartes 1985–91.

conservation of health."[4] While some recent scholarship has begun to take these claims seriously, for the most part for the past several centuries scholars have chosen to selectively edit out Descartes' own characterization of his philosophical project. They have been reluctant to see that Descartes' philosophy was, to use Vincent Aucante's apt phrase, fundamentally a 'medical philosophy'.[5]

To characterize Descartes's philosophy in this way is not to say that it was itself medicine, or that it took up medicine as the focus of a detached philosophical inquiry, in the way we might today engage in the 'philosophy of medicine'. Rather, Descartes' was a medical philosophy to the extent that he saw medicine as entirely integral to the project of philosophy, insofar as it was, first of all, the key to health and longevity, and thus to the realization of the good life; and, second of all, it was conceived as including rules of diet, hygiene, and bodily comportment, and to this extent was nothing less than the corporeal flip-side, so to speak, of ethics. Medical philosophy takes care of the body and care of the soul to be two aspects of a broader eudaimonistic project. Here, medicine is not conceived as somehow subordinate to ethics, but rather as one side of a coin that includes the ethical cultivation of the self on its reverse.

This practical motivation at the foundation of his philosophy is in no way at odds with the deeper theoretical concerns of a philosopher such as Descartes, with which we are more familiar; in fact, by turning our attention to the medical eudaimonism of early modern philosophers – that is, their belief that studying the means toward health and longevity is a requisite of the good life– we are able to gain new light on some of the deepest theoretical questions with which they were concerned.

Now, the idea that early modern philosophers were concerned above all with the good life is certainly not original with me. Matthew Jones has developed it at length in his excellent book of 2006, *The Good Life in the Scientific Revolution: Descartes, Pascal, Leibniz and the Cultivation of Virtue.*[6] But Jones and others have not focused on the central role of medicine in this eudaimonistic project, taking medicine anachronistically as a branch of science. Moreover, little attention has been paid to the way in which the medical eudaimonism connects with the more familiar theoretical aims of our canonical thinkers. Now in other venues I've approached the question of this connection rather more trepidatiously, but I believe a forum explicitly dedicated to the connection between medicine and philosophy in the early modern period is an ideal place to try out a more radical claim than most historians of philosophy have been willing to accept: that at least some of the most familiar philosophical doctrines of our canonical modern philosophers may be interpreted as expressions of medical preoccupations that at first glance have little to do with philosophy as we have come to define and to bound it. We can, in other words, do the equivalent of tracing the philosophy back to the grocery list. When we do this, moreover, we gain a much richer and more adequate picture of the overarching philosophical projects of canonical figures.

[4] AT IV, 329. "La conservation de la santé a esté de tout temps le principal but de mes études."

[5] Aucante 2006.

[6] Jones 2006.

In this short essay, I would like, first of all, to illustrate how, in the case of one prominent early modern philosopher, Leibniz, the recipes and remedies, the dietetic and pharmaceutical endeavors, are directly intertwined with some of the most fundamental concerns about both the metaphysics of corporeal substance, as well as the belief that proper maintenance of the human corporeal substance constitutes a sort of corporeal flip-side, so to speak, of morality, and thus is of central concern to a philosopher. Next, I will move on to perhaps a less obvious problem that seems to have run as a sort of undercurrent through much early modern engagement with the problems of health and illness broadly construed: namely, *why* is it that a philosopher, or anyone for that matter, should have to learn how a human being must take care of his or her body? Why should this not be self-evident? That it is not self-evident seems to have been an important factor in early modern reflection on philosophical anthropology, and on the problem of the place of humanity in, or in relation to, nature and the totality of other natural beings. I will conclude with a suggestion, which will have to be fully developed elsewhere, that it is Leibniz's philosophical system that is best equipped to deal with the question as to why what is best for human health should not be self-evident, and that in consequence Leibniz's philosophy offers an elegant harmonization of elements of both rationalism and empiricism.

14.2 Leibniz Embodied

In this section I would like to focus on a few vivid examples from a philosopher, Leibniz, of whom we might say that, if the thesis of the centrality of medicine to early modern philosophy holds for anyone, it holds for him.[7] Consider Leibniz's frequent appropriation of the Hippocratic slogan, *sympnoia panta*, 'all things conspire'.[8] Here we see an extension to the cosmos of a doctrine that had originally been meant to describe the functioning of the human body, and that in Leibniz comes to serve as a sort of shorthand for his mature theory of causation, the doctrine of preestablished harmony. From the point of view of an ancient Hippocratic, of course, there would have been nothing surprising about this extension; it would not have been seen as a motion from one ontological domain to another, the body to the cosmos, but rather would have been an expression of the conviction that explanation of vital, organic processes just is the paradigm for understanding natural phenomena in general. Indeed it is worth noting in passing that the Hippocratic *De diaeta* runs through a rich compendium of pre-Socratic cosmological doctrines in the course of prescribing the best dietary regimes for maintaining health.[9]

[7] Portions of this section were previously developed, though in the course of making very different arguments than the one of the present essay, in the following publications: Smith 2011; 2012b, 203–224.

[8] Leibniz 1996, 55.

[9] See Hippocrates 1745.

Consider another example: the Leibnizian notion of 'appetite', which is ordinarily coupled with 'perception' to describe the pair of fundamental activities of the individual immaterial monad: there is nothing in the world, Leibniz says, but simple substances, and in them, perception and appetite, perception being the representation of the order of coexistence within the simple; appetite being the tendency to go from one representation to the next. Now the most common reading of Leibniz would dismiss the apparently unsophisticated reading according to which appetite here has anything to do with being hungry. But certain texts, such as the *De scribendis novis medicinae elementis* of the early 1680s, suggest that Leibniz's mature metaphysical elaboration of just how perception, and the complementary notion of appetite, work in the nonbodily monad seems to have first been worked out in the course of explaining the perception and appetite of animal bodies in straightforwardly animal-economical terms.

> The primary function of a human being is perception, but the secondary function (which exists for the sake of the first) is to procure perceptions. The advancement of human perfection consists in the same measure in the advancement of these functions. And anything that is helpful to perception or to the faculty of procuring perceptions is agreeable; anything that impedes them is disagreeable. The organs of sense exist for the sake of perception; and the organs of motion exist for the sake of procuring perceptions, which is to say for action. Both should be conserved in operation, or in the constant capacity for operation, which is brought about now by the removal of impediments, now by the increase of facilitating conditions. And the greatest increase is nutrition itself, seeing that the same individual particles cannot be conserved, but continually vanish.[10]

Here Leibniz says that perception is the primary function of the human body; and the procuring [*procuratio*] of perception is the secondary function. Here, in sharp contrast with the later writings, perception is dependent on the organs of sense; and the ordering of perceptions is dependent on the motion of the body. The greater facilitation of perceptions is the greatest end for an animal or human and is itself experienced as an agreeable perception. In this connection, the organs of generation are included among the organs of motion, as generation is the procurement of "a most agreeable perception through motion."[11]

The organs of nutrition, which humans and animals have in common with plants, must already be functioning in order for motion, and then sense, to arise. While Leibniz distinguishes these from the organs of motion, he also describes both the organs of motion and of nutrition as facilitating conditions for the procuring of perceptions (and thus as requisites for the achievement of the greatest human end), describing nutrition as the greatest facilitator of perceptions insofar as the particles of the body are continually in flux, and without nutrition neither the organs of motion nor of sense could continue to cohere for very long. The mature analysis of the faculty of appetite or *procuratio* seems unmistakably intertwined with Leibniz's early physiological interests, and only subsequently transferred out of the domain of physiology, all the while retaining its original structure. It could thus be argued that

[10] Leibniz in Smith 2011, Appendix 4, 299–300.

[11] Leibniz in Smith 2011, Appendix 4, 300.

this is a case in which Leibniz works out a general account of what's going on fairly early, but then eventually changes his idea as to what sort of entity this is to be an account *of*.

There is of course a great deal of controversy as to what Leibniz's mature metaphysical commitments were. In the example of appetite, I have been speaking as though Leibniz moves from a body-realism to a thoroughgoing monadic idealism in his mature period. Monadic idealism does become a new option for Leibniz, from roughly the late 1690s on, but there is another hallmark of his mature philosophy, long suppressed, but recently brought out in its glorious complexity by André Robinet,[12] Dan Garber,[13] and many other scholars: the doctrine of corporeal substance, according to which every true being consists in a dominating monad or entelechy and an organic body, which itself consists in an infinity of hierarchically related monads, which, finally, together give rise to the phenomenal properties of mass or body.

For Leibniz the corporeal substance is immortal, which is to say that every living being always remains embodied, from the creation to the annihilation of the universe. In the early 1670s, Leibniz had associated this doctrine with the Kabbalistic idea that there is a special bone in the body, called in Hebrew the *luz*, that perpetually serves as the anchor or fixed residence of the soul. He also elaborates the doctrine in alchemical terms by appeal, variously, to the notion of the *flos substantiae* or the *caput mortuum*, the 'flower of substance' or the 'death's head', understood as that little trace of a substance ('substance' here in the chemical, not the metaphysical, sense) that must always remain when the rest of the substance has been transformed into ash. Leibniz's early attempts to peg the soul to a particular bit of matter will quickly give way to a conception of the corporeal substance as embodied only in the same way a fountain may be said to be a 'body' of water: the particular matter that constitutes the body is always changing, and yet it can be said to be the same body over time in virtue of the fact that it preserves the same form. So a corporeal substance is an immortal, perpetually embodied, being that preserves its form through the constant exchange of matter with its environment.

Where, now, might Leibniz's mature model of corporeal substance have emerged from? Recent works from authors such as Ohad Nachtomy[14] have noted the fractal structure of the infinitely many nested monads from which the organic body results, and have attempted to root the mature theory in parallel concerns Leibniz had in his mathematics. Scholars have long cautioned that one should not skip too quickly between the theory of body and the theory of mathematical entities such as lines or the series of real numbers. The great difference is that these latter entities constitute continua, which is to say they are actually infinitely divisible, while bodies, for Leibniz, are always already actually infinitely divided. Mathematical entities are, thus, only ideal, while bodies are real. Bodies are real, unified, yet at the same time fundamentally unstable entities; their identity over time consists not in the

[12] See Robinet 1986.

[13] See Garber 2009.

[14] Nachtomy 2011, 61–80.

conservation of parts relative to one another, but rather, so to speak, in the orderly or stable flow, through nutrition, excretion, and respiration, between the body and its environment. And in order to understand where the inspiration for this comes from, we would perhaps do better to look not at the mathematics, but at the dietetics.

Over the course of his entire career, Leibniz shows a consistent interest in developing new methods for the study of the physiology of digestion. Emetics are a longstanding interest for him, from the *Directiones ad rem medicam pertinentes* of the early 1670s[15] to the 1696 publication of his treatise on ipecacuanha,[16] which we know better today as "syrup of ipecac." Leibniz is interested in vomiting not just as a pathological phenomenon, but also as a possible source of insight into the most basic processes that sustain a corporeal substance in existence. He shares in the view that digestion, the transformation of food into flesh is a sort of continuation of the process of sexual generation, to the extent that at a metaphysical level both nutrition and generation involve the transformation of aggregate matter into corporeal-substance matter. Leibniz speaks in the *Directiones* of the opening up of the human body for examination over the century prior as a "discovery" akin to that of a new continent, or even as akin to the transformation of the model of the cosmos by the work of Copernicus and Kepler.

Decades later, in the mid-1690s, Leibniz's mature theory of corporeal substance has begun to take shape. If he cannot devote all of his time to metaphysics, though, this is because Leibniz is in the course of writing his most significant contribution to medicine, his *Relation to the Illustrious Leopoldine Society of Naturalists concerning the New American Anti-Dysentery Drug, attested with Great Successes* of 1695–6, after reading extensively about, and evidently after conducting his own experiments with, the ipecacuanha root. It was Wilhelm Piso who brought the ipecacuanha root back to Europe in 1641, and wrote about it extensively in his *Natural History of Brazil,* co-authored with Georg Markgraf, in 1648. The root was introduced in Paris in 1672, and made famous when the physician Helvétius used it successfully to treat Louis XIV's dysentery. It is clear from Leibniz's study that he has a detailed knowledge of Piso and Markgraf's work, as well as of Helvétius's use of it.

By the time Leibniz presents his work on ipecacuanha to the Leopoldine society and brings it to the attention of the German public, it has been well-known for some years in France already, and it is likely that Leibniz himself has already learned of its use while in Paris a quarter of a century earlier. He is aware of its widespread use among the Brazilian natives, and of its more recent successes in France, but is nonetheless the first to give an exhaustive account of its many uses, not only against dysentery, but also as an emetic, a diaphoretic (causing one to perspire), and an expectorant (causing one to salivate). In these last three applications, it works very much as a "poison," and thus is a medicine of the sort that Leibniz had insisted in the *De scribendis novis* must be used only cautiously. It is also the very potion for which Leibniz had expressed a desire as early as the *Directiones* of 1671: one that

[15] Leibniz in Smith 2011, Appendix 1.

[16] Leibniz 1768, vol. II, 110–119.

through its ingestion can immediately make the internal external, thereby giving an instant report of the state of the body from where it cannot be directly perceived.

Leibniz's preoccupation with the proper functioning of the body, his compilation of recipes for dishes that facilitate digestion, his work on the ipecacuanha root, are not, in the end, akin to Wittgenstein's grocery lists. They are indicative of a conception of the philosophical project on which grocery lists are not entirely irrelevant. This conception of philosophy, which I have called 'medical eudaimonism', takes it for granted that the principal purpose of philosophy is to maintain oneself in a blessed and healthy state, through proper thinking, proper conduct, and yes, proper diet. I do not in any way mean to argue that this is the proper conception of philosophy, or that philosophers should begin again to medicinize; I wish only to draw our attention to the overwhelming evidence that this is in fact how a philosopher such as Leibniz conceived philosophy, and to suggest that we are missing out on important features of some of his most well known philosophical doctrines to the extent that we fail to approach him as a medical eudaimonist.

14.3 Why Must Health Be Learned?

It should by now appear at least plausible to suggest that in the early modern period, medicine was conceptualized as not just relevant to philosophy, but indeed as in large part constitutive of philosophy; philosophy was in no small measure medical philosophy. But this gives rise to another, perhaps less evident problem: why, in particular, should the project of the advancement of human knowledge have to include learning about particular dietary and hygienic rules at all? Why should these not be obvious or self-evident, or perhaps *a priori*? Even if the widespread early modern analogy between medicine and morals, to which Leibniz himself was very sympathetic, makes some intuitive sense, we are struck by at least one powerful point of difference: early modern philosophers universally supposed that knowledge of how to behave morally could be arrived at simply by use of reason; knowledge of how to be healthy, by contrast, depended on countless contingent facts about how the world happens to be: the hidden medicinal virtues of different species of plant, for example. Nobody ever supposed that the emetic effect of the *Cortex peruvianus* was something that could be deduced from *a priori* principles or that did not require an investigation into particular, contingent features of the natural world. Even Leibniz, who was unique in arguing that in metaphysical rigor there is no meaningful distinction between truths of necessity and truths of fact, understood full well that in order for our finite minds to arrive at truths of fact concerning, e.g., the medicinal virtues of quinine, one must go out and investigate.

In this respect, early modern medical philosophy was willy-nilly empirical philosophy, notwithstanding any other commitments a given philosopher may have had about the sources of other sorts of knowledge. The fact, in turn, that particular knowledge of contingent features of the world was necessary for the cultivation of the good life was itself a trigger for reflection on fundamental questions of philosophical anthropology: *why* do we find ourselves in a world whose features we need to systematically study simply in order to stay alive? What does this predicament reveal about the project of science? What, moreover, does it reveal about the place of human beings in nature or in relation to nature?

It might appear to reveal that we are a particular sort of being, so very different from the animals, to the extent that we must learn, and make learning a systematic, collective project, simply in order to live well? Of course, there had always been the model of movements such as that of the Cynics, who believed that the particulars of how one ought to live were self-evident, and this down to particular facts about what one ought to eat. Throw off the artifices of culture, and you can immediately begin living the right way, just as the animals have all along. But their approach could easily appear as a parody of the very idea that one should live in accordance with nature; indeed their supposedly 'natural' dietary choices were known to be positively deleterious to health.

Exceptions such as these aside, in antiquity and later, nearly all schools of thought supposed that one important respect in which human beings differ from animals is that in our case it is not at all obvious what is good for us. This peculiar feature of human existence seems to have been conceptualized sometimes as a sort of analogue to the problem of free will. Consider the question of human carnivorism as it was addressed by Edward Tyson and John Wallis. In a fascinating exchange published in the *Philosophical Transactions* in 1701, Wallis, having first acknowledged that "as a young Student in Philosophy, I look'd upon the *Medicinae pars Physica*, as a piece of Natural Philosophy,"[17] goes on to cite something he recollects having read in one of Gassendi's letters: that "he thought it not (originally) Natural for Man to feed on Flesh; though by long usage (at least ever since the Flood) we have been accustomed to it, and it is now familiar to us; but rather, on Plants, Roots, Fruits, Grain, &c."[18] He cites scriptural evidence both for and against this view, but quickly adds that he prefers not to "disput[e] it as a point in Divinity."[19] Instead, he wishes, along with Gassendi, to consider it "as a Question in Natural Philosophy," namely, by looking at the structure of the human body. With respect to dentition, Wallis writes that it is as if "Nature had rather furnished our Teeth, for Cutting Herbs, Roots, &c and for bruising Grain, Nuts, and other hard Fruits, than for Tearing Flesh, as Carnivorous Animals do."[20] Even more solid evidence comes from internal anatomy. There is, Wallis writes, "in Swine, Sheep, Oxen, and, I think, in most Quadrupeds that feed on Herbs or Plants, a long Colon, with a Caecum at the upper end of it, or somewhat equivalent, which conveys the Food, by a long and large progress, from the Stomach downwards, in order to a slower passage, and longer stay in the Intestines; But in Dogs, of several kinds, and I suppose, in Foxes, Wolves, and divers other Animals which are Carnivorous, such Colon is wanting; and, instead thereof, a more short and slender gut, and quicker passage through

---

[17] Wallis 1700–1701, 769–773.

[18] Wallis 1700–1701, 770.

[19] Wallis 1700–1701, 771.

[20] Wallis 1700–1701, 771.

the Intestines."[21] Wallis believes that ordinarily nature "may be reasonably presum'd to adapt the Intestines to the different sorts of aliments that are to pass through them," and that it is thus by examining intestines that we can determine "to what Animals Flesh is proper aliment."[22] But with humans it is not so easy, because the existence of human custom, as variable as it is, can actually change the physical conformation of a human body: "Tis true, that the Caecum in man is very small, and seems to be of little or no use. But in a Foetus, it is in proportion much larger than in persons adult. And it's possible, that our Customary chage of Dyet, as we grow up, from what originally would be more natural, may occasion its shrinking into this contracted posture."[23] And Wallis adds by way of conclusion "that Man's being indu'd with Reason, doth supply the want of many things, which, to other Animals may be needful."[24] Thus, he suggests, we need not be outfitted with all the defenses that will protect us against weather and predators: we can use our reason and come up with our own. In this respect, what it is a person ought to be doing – how a person ought to be dressing, or constructing armor, or eating – cannot be read off of the body as they can for other natural beings. Moreover, an attempt to read the body in this way can only give an ambiguous report, since through choice of conduct the faculty of reason can in fact alter the conformation of the body.

Leibniz, though he does not (as far as we know) ever address the topic of human carnivorism, does agree in general with Wallis that human beings are distinct in that for them there can be no easy reading of function from anatomy, of what a human being should be doing from the conformation of the human body. For Leibniz, reason is the sole characteristic function or *officium* of human beings, but this 'office' is not associated with any particular bodily organ or system, in contradistinction to a bee's honey-making office or a spider's web-weaving office. Thus Leibniz writes to A. C. Gackenholtz in 1701 that each plant and animal is a "machine able to perform certain offices," but while a squirrel is a 'jumping machine' [*machina saltatrix*], a human being is "a machine for the perpetuation of contemplation."[25] There is, for Leibniz, no particular organ that executes this function (certainly not the brain), and so there is no way of inspecting the organic structure of a human being in order to determine whether that human is adequately realizing its purpose as a human. More broadly, it is this sort of reasoning that underlies Leibniz's argument against Locke, in the *New Essays concerning Human Understanding*, that one cannot judge from external, bodily signs whether a human being, however seriously deformed, is capable of exercising reason or not. We must assume that reason, as the characteristic office of human beings, is at least latent in every human being, and in turn the sole viable criterion for determining whether a given being is a human or not, is whether this being was born of human parents.

[21] Wallis 1700–1701, 771–72.

[22] Wallis 1700–1701, 772.

[23] Wallis 1700–1701, 772.

[24] Wallis 1700–1701, 772–73.

[25] Leibniz 1768, II 2, 171.

What is more, as a result of the use, proper or poor, of human reason, the conformation of the body may change, with the result that over the long term it may become particularly difficult to determine from an investigation of the current condition of a human body what it is, besides reasoning, that that body is naturally outfitted to do. Should we be eating meat, for example? As we've seen in Wallis, human intestines, like the human faculty of willing, distinguish us from the animals in that they can go both ways, they can do one thing or the other, while animal action, it was supposed, could only flow from the singular necessity of the animal's nature. On this picture, human beings are not only distinct from other natural beings in that they have free wills and immortal souls, but also in that they have peculiar bodies, bodies that are in some sense a rupture in the natural order just as souls not governed by deterministic natural laws might be thought to be. Human bodies are the only natural bodies of which the proper usage cannot simply be read off of their contours, but must be learned, through observation and experimentation, including autoexperimentation.

Although early modern philosophers tended to conceptualize human beings as distinct from animals to the extent that the proper care of human bodies had to be learned, this did not stop them from looking to the animals as a source of learning. An interesting illustration of this is seen in the history of the medical application of quinine. In his 1695 treatise on the Brazilian ipecacuanha root,[26] Leibniz himself mentions the 'Peruvian bark', which he traces back to a discovery in Paraguay more than twenty years earlier.[27] Leibniz would not pursue the origins of the remedy any further than this, but some decades later the French traveler Charles-Marie de la Condamine would echo a very revealing tale of how the Peruvian bark first came to be used. "According to an old account, of which I do not guarantee the accuracy, the Americans owe the discovery of this remedy to the lions, which some naturalists maintain are subject to a sort of intermittent fever. It is said that the people of the country, having remarked that these animals ate the quinquina bark, made use of it for their fevers, which are quite common in that country…" [28] Thus the natives of Ecuador and Peru learned about quinine by watching sick lions chew on the bark of the *Cinchona* tree; the Spanish learn of the remedy from the natives; and the French from the Spanish, thereby completing this bit of wisdom's movement up the scale of being.[29] In more technical terms, we may say that Condamine is positing a move-

[26] I discuss this treatise at length in Smith 2012a, 377–401. Portions of that text overlap with the discussion of the *De novo antidysenterico* in the present section, though the argument here is entirely distinct.

[27] See Leibniz 1768, 113. "Cortex Peruvianus jam ante annos quadraginta quinque propemodum celebratus pene conciderat, donec per Talbotium restitutus dignitati plausum in aula invenit. Jam viginti, & amplius anni sunt, quod mire praedicari audivi herbam Paraguay a natali Provincia Paraguaria dictam, quae (rarum) ita emetica est, ut stomacho vis non fiat, magnaeque praeterea ob salutares effectus in India Hispanica famae, nec tamen, quod sciam, hactenus in nostras officinas recepta. Itaque etiam medicamenta ipsa sua fata pro captu hominum habent, ut saepe non minus restauratori, arque propalatori, quam inventori debeamus."

[28] de la Condamine 1740, 226–243; 232–33.

[29] de la Condamine 1740, 81–82.

ment from zoopharmacognosy through ethnopharmacognosy up to medical science. At each stage, there is supposed to be a greater degree of reflected choice. The widespread early modern idea that indigenous non-European languages lacked terms for abstractions, meant that, if anything could be learned from them, it would be concerning the concrete and particular, and there is perhaps no domain of knowledge more dependent on knowledge of particulars than botanical pharmaceutics. For many European explorers, as well as Europeans who interpreted the discoveries of the explorers from the comfort of home, native botanical knowledge could easily be seen as a branch of natural history complementary to botany itself, in a way that, say, ethnomathematics would not have been. Thus Bontius remarks of the Malayans that "those who in other things are illiterate have an exact knowledge of herbs and shrubs."[30] This particular combination, of botanical sophistication and illiteracy, seemed to mean that one could learn from the natives, without any need to acknowledge that the natives themselves are learned. The Dutch physician and traveller Wilhelm Piso, for example, does not put much credence in the native Brazilians' knowledge, even as he composes a massive tome on Brazilian medicinal plants that is deeply indebted to his observation of Brazilian medicinal practice.

Now it may seem that we have taken a significant detour into the history of ethnopharmacology, but in fact we have only done so in order to illuminate a central point of this investigation, to which we may now return: medicine – including dietetics and pharmaceutics – was broadly conceptualized as a part of philosophy in the early modern period, even as it was understood that the rules concerning what to eat to preserve health and what to ingest in the case of illness are, unlike other branches of philosophy, something that we cannot learn simply by reflection. We must instead go out into the world and observe and experiment, but when we go and observe, we find that there are beings – at the limit case animals, but also non-Europeans who are, it is supposed, devoid of abstract thought – that do not need to engage in systematic study in order to know how to live, what to eat, what to take for illness, and so on, but instead are able to do these things more or less spontaneously. This means that to learn dietetics and pharmacology, to learn to excel in the art and science of medical eudaimonism, is to engage in a project of systematic empirical philosophy that, in the end, does nothing more for humanity than provide what other natural beings get, so to speak, for free.

So, again, the question: why must one learn how to live? More particularly, why must we be so ill-equipped to deal with the world as bodily creatures that philosophers, the same people who spend their time coming up with proofs for the existence of God, should have to collect recipes for the cure of constipation? We can only begin to sketch out an answer here, and shall do so on the example of Leibniz, but this will already be enough to see how the inclusion of medical eudaimonism within the project of philosophy motivated what we might describe as a metaphilosophical reflection on the scope and aims of philosophy, the ultimate causes of our limited knowledge, and the proper means of overcoming these limitations.

[30] Jacob Bontius cited in Cook 2007, 203.

One long-standing concern Leibniz has, as we already began to see in the *Directiones,* and one that he thinks can be better dealt with by experimental methods, is the project of discerning the fine line between medical practice and quackery. He writes in the *De novo antidysenterico* that "medicine is an uncertain art, which sustains the credulity of men, like the great dream of the philosophers' stone." Mutatis mutandis, Leibniz believes that medicine resembles alchemy in that it is a domain of inquiry in which genuine discoveries can be made, even if human greed, impatience, selfishness, and wishful-thinking bring it about that there are more false reports of discoveries than actual ones. He recognizes medicine's inherent tendency toward fraudulence and its inherently uncertain character—uncertain because so reliant upon empirical methods and so lacking in first principles—while at the same time recognizing that, like cosmology before it, medicine may finally be on the verge of gaining secure foundations. In fact, a central theme of the *De novo antidysenterico* is the newness of medicine as a properly scientific discipline. Leibniz writes:

> We have lost the majority of ancient remedies; sometimes their salutary character resides in their names alone. It was barely two centuries ago that medicine was reborn; it has been no more than one century since anatomy has flourished. Indeed it is less than half a century since the interior constitution of the human body was revealed through the discovery of the circulation of the blood. And thus it is surprising that there are so many … efficacious remedies, given so great an ignorance of causes.[31]

Leibniz believes that the modern period has witnessed revolutionary progress in medicine, and this for reasons having to do not just with the emergence of exact anatomical study in the Paduan school in the sixteenth century, but also, perhaps most significantly, with innovations in the methods of obtaining and compiling medical data.

One idea in the *De novo antidysenterico*, which seems to have been inspired by his encounter in Italy with the physician to the Duke of Modena, Bernardino Ramazzini, concerns the improvement of medical record keeping: "I think that above all historical medical annals must be founded, … and that those who wield influence in public affairs should apportion funds from the public treasury more liberally for [medicine] than for any other art."[32] Leibniz is making these grand proposals in the course of introducing a rather concrete examination of the medicinal effects of a certain plant, and we may assume, in view of this incongruity, that he is using the pretext of his report on the root for the scientific society as an occasion for promoting some of his enduring ideas about the proper organization and practice of medicine. These proposals, in turn, are part of a much more ambitious project – one that also includes the mapping of magnetic variation across the globe, and other seemingly unrelated goals – of enhancing the predictive power of science by compiling vast data about particular cases. Leibniz understood that the best way to see the future is not through divination or augury – arts with which medicine had had a long association – but through comprehensive knowledge of particular facts

[31] Leibniz 1768, II 2, 111.

[32] Leibniz 1768, II 2, 111.

about what has already happened. This is the nascent science of statistics, of course, and through it Leibniz believes that we will someday be able to establish general laws with predictive power based on our study of what he calls *res singulares*, particular things.[33]

Recall, now, the remark made in passing above about what Leibniz believes in metaphysical rigor: that in the end there is for him no difference between truths of necessity and truths of fact. If it takes us rather longer to see the pharmacological applications of quinine than to see that 2+2=4, this is only in view of the limitations upon our perception, which is to say our capacity to represent the rational order of nature, as finite, created substances. Science, on this understanding, is the systematic project of augmenting the clarity of perception through discovering the rational order behind apparent chaos. Animals are an expression of this order, and if they are capable of doing whatever they need to do to ensure their own well-being without systematic scientific knowledge, this is only because their well-being is a rather more straightforward matter than it is for human beings. Leibniz's empiricism, to the extent that it seeks to disclose the rational order of nature, is in the end not at all in contradiction with the more familiar picture we have of him as a rationalist. In fact, his promotion of the advancement of empirical knowledge in domains such as medicine is a direct reflection of his deepest *a priori* commitments about the metaphysical order of nature. Learning about particulars is not an abandonment of the fundamental project of finding eternal truths, since particulars are in the end expressions of the same rational order as, say, the truths of mathematics are.

What Leibniz's burgeoning conception of statistics shows, moreover, is that eventually, by enough concerted attention to particulars, our understanding of them may be so to speak elevated to the level of rational knowledge, as we will be able to speak of such things as epidemics and the declination of the compass needle at different points on the globe in terms of law-governed regularities. Leibniz himself will explicitly complain that others are wrong to distinguish between the two sorts of endeavor, between the search for eternal truths and the investigation of particular facts. Thus for example he complains to the Swedish Slavist Johan Gabriel Sparwenfeld in a letter of 1698 that "people criticize me when I attempt to take leave of the study of mathematics, and they tell me that I am wrong to abandon solid and eternal truths in order to study the changing and perishable things that are found in history and its laws."[34]

14.4 Conclusion

How it is that medicine can be a part of philosophy begins to make a good deal of sense when it is Leibniz's philosophy, properly understood, that is in question. Leibniz understands himself as the very model of a medicinizing philosopher, and

[33] Leibniz 1873, 96.

[34] Leibniz 1873, 38.

he regrets that his contemporaries fail to follow this model. In the *De novo antidysenterico*, having just spent the better part of a year working assiduously on a treatise on the treatment of dysentery, Leibniz proclaims in no uncertain terms that he considers this treatise to be one of the crowning achievements of his entire career. In the same year, Leibniz also has a few other accomplishments worth mentioning: it is in 1695 that he first mentions 'monads'; in the same year, he also introduces the system of preestablished harmony, in the *Système nouveau*; and, finally, he spells out his dynamics, or theory of force, in the *Specimen dynamicum.* A busy year, indeed. Now, according to the second of these doctrines, the preestablished harmony, one need not invoke the activity of souls in order to account for processes unfolding within the organic body of a corporeal substance. Thus Leibniz famously writes in the *Système nouveau* that

> Nature has, as it were, an empire within an empire, a double kingdom, so to speak, of reason and necessity, or of forms and of the particles of matter, for just as all things are full of souls, they are also full of organic bodies. These kingdoms are governed, each by its own law, with no confusion between them, and the cause of perception and appetite is no more to be sought in the modes of extension than is the cause of nutrition and of the other organic functions to be sought in the forms or souls.

Yet recall from our discussion of *appetitus* that Leibniz appears to have initially developed his model of the activity of immaterial monads from physiological considerations, and only later changed the fundamental ontology of the entities he thought this model described. Here, now, we see that Leibniz explicitly sees the nutrition of organic bodies as the direct analog, under the aspect of corporeality, for the perceptual activity of monads. When we couple this fact with Leibniz's contemporaneous work, as a medicinizing philosopher, on the physiology of digestion, it no longer seems so implausible to suggest that the mature model of corporeal substance owes at least as much to dietetics as it does to mathematics. To speak of 'analogy' here might sound as if we are employing a loose method of '*x* sounds like *y*' inference. But this would be to miss the significance of the apparent similarities between Leibniz's different concerns, and indeed would be to fail to take up what is arguably an important technique in the interpretation of the entire corpus of an extremely inventive and polymathic author such as Leibniz. When we see apparent similarities between Leibniz's 'deep' metaphysics on the one hand and his efforts to make progress in narrow domains of the exact sciences on the other, this is often because, in the end, he takes the natural world to be an expression, in the technical sense, of the perceptual activities of monads. It is far beyond the scope of the present paper to explain how this expression works, but it is enough, here, to note that this very familiar feature of Leibniz's broad philosophical project is enough to resolve any lingering concerns about how we are to understand 'analogy' here: by 'analogy' between, say, corporeal hunger and monadic appetite, we mean simply harmonious expression of one and the same thing.

The place of dietetics in Leibniz's philosophy, moreover, is really only a particularly vivid case of a much broader tendency in early modern philosophy, evident in Descartes, Wallis, and many others, to conceptualize the project of philosophy as consisting in large part in the search after the best practices for the care of the bodily self. And this brings us some distance towards understanding why two of the great Rationalist philosophers of the 17th century were brought together not only by a shared commitment to the *a priori* grounds of certain knowledge, but also by a shared interest in remedies against constipation. By the time Nietzsche is writing, digestion and related bodily processes will come to function as mere metaphors for the activity of philosophy. Two centuries earlier, by contrast, these processes were in large part what philosophy was, literally, about.

References

Aucante, Vincent. 2006. *La philosophie médicale de Descartes*. Paris: Presses Universitaires de France.

Bontius, Jacob. 2007.*Tropische geneeskunde*. Cited in Harold J. Cook, *Matters of exchange: Commerce, medicine, and science in the Dutch golden age*. New Haven: Yale University Press.

de la Condamine, Charles-Marie. 1740. Sur l'arbre du Quinquina. In *Mémoires de l'Académie Royale, Année M.DCCXXXVIII*, 226–243. Paris: Imprimérie Royale.

Descartes, René. 1983. *Oeuvres*, ed. Charles Adam and Paul Tannery, Paris: Léopold Cerf.

Descartes, René. 1985–1991. *The philosophical writings of descartes*. Trans. John Cottingham, Robert Stoothoff, Dugald Murdoch, 3 vols. Cambridge: Cambridge University Press.

Garber, Daniel. 2009. *Leibniz: Body, Substance, Monad*. Oxford: Oxford University Press.

Hippocrates. 1745.*De humoribus purgandis liber et De diaeta acutorum libri tres*, ed. Louis Duret, Pierre Girardet, and Justus Gottfried Günz. Leipzig.

Jones, Matthew L. 2006. *The good life in the scientific revolution: Descartes, Pascal, Leibniz and the cultivation of virtue*. Chicago: University of Chicago Press.

Leibniz, G.W. 1768. G. G. Leibnitii Relatio ad inclytam Societatem Leopoldinam Naturae curiosorum, De novo antidysenterico americano magnis successibus comprobato. In *G. G. Leibnitii Opera omnia*, vol. II, ed. Louis Dutens, 110–119. Geneva: Fratres de Tournes.

Leibniz, G.W. 1873. *Sbornik pisem i memorialov Leibnitsa otnosyashchikhsya k Rossii i Petru Velikomu*, ed. V. I. Ger'e. St. Petersburg.

Leibniz, G. W. 1996. *New essays on human understanding*, ed. and tr. Peter Remnant and Jonathan Bennett. Cambridge: Cambridge University Press.

Leibniz, G.W. 2011a. Directions pertaining to the institution of medicine, Appendix 1. In *Divine machines: Leibniz and the sciences of life*, ed. Justin E.H. Smith. Princeton: Princeton University Press.

Leibniz, G.W. 2011b. On writing the new elements of medicine, Appendix 4. In *Divine machines: Leibniz and the sciences of life*, ed. Justin E.H. Smith. Princeton: Princeton University Press.

Nachtomy, Ohad. 2011. Leibniz on artificial and natural machines: Or what it means to remain a machine to the least of its parts. In *Machines of nature and corporeal substances in Leibniz*, ed. Justin E.H. Smith and Ohad Nachtomy, 61–80. Dordrecht: Springer.

Nietzsche, Friedrich. 1967. *The genealogy of morals and Ecce Homo*. Trans. Walter Kaufmann and R. J. Hollingdale. New York: Vintage Books.

Robinet, André. 1986. *Architectonique disjonctive, automates systémiques et idéalité transcendantale dans l'œuvre de Leibniz*. Paris: Vrin.

Smith, Justin E.H. 2011. *Divine machines: Leibniz and the sciences of life*. Princeton: Princeton University Press.

Smith, Justin E.H. 2012a. Leibniz on natural history and national history. *History of Science* 1: 377–401.

Smith, Justin E.H. 2012b. 'Spirit is a Stomach': The Iatrochemical roots of Leibniz's theory of Corporeal substance. In *Matter and form in early modern science and philosophy*, ed. Gideon Manning, 203–224. Leiden: Brill.

Wallis, John 1700–1701. A letter of Dr Wallis to Dr Tyson, concerning Mens feeding on Flesh, February 3, 1700. In *Philosophical transactions of the Royal Society of London*, vol. 22, 769–773. London.

Chapter 15
Tres medici, duo athei? The Physician as Atheist and the Medicalization of the Soul

Charles T. Wolfe

> *With us there was a doctor of Physic …*
> *Well read was he in Esculapius, and … Hippocrates,*
> *Galen … Serapion, Rhazes, and Avicen, Averrhoes,*
> *Gilbert, and Constantine, Bernard and Gatisden, and John Damascene.*
> *… It's no libel/To say he read but little in the Bible.*
>
> (Chaucer 1933, Prologue, l. 411–438)

> *medicus est physicus sensualis*
>
> (Bezançon 1677, 335)

Abstract Until recently, examinations of the 'mind-body problem' in historical context paid only cursory attention to its specifically medical dimension, if at all. At best, some 'folk physiology' was entertained, usually to laugh at it (the pineal gland, animal spirits). Conversely, historians of neuroscience or of artificial intelligence (Jeannerod M, The brain machine. The development of neurophysiological thought, trans. D. Urion, Harvard University Press, Cambridge, 1985; Dupuy J-P, The mechanization of the mind: on the origins of cognitive science, trans. M.B. DeBevoise, Princeton University Press, Princeton, 2000) often present figures like La Mettrie as heroic early cases of 'naturalization', giving an experimental basis to materialism: their symmetrically inverse mistake is to take professions of medical authority too literally (although there are genuine cases where all of the above does coalesce – where 'actors' categories mysteriously transcend historiographic projections –, such as Hieronymus Gaub's reflections on the 'regimen of the mind' in the mid-eighteenth century, or, more theoretically, Guillaume Lamy's Epicurean-inflected *Anatomical Discourses on the Soul*, eighty years earlier). Contrary to the denial of the relevance of medicine in early modern philosophy, as regards issues such as the body-soul (then body-mind) relation among others, it seems patently difficult to separate medical theory, medically nourished philosophical speculation, and

C.T. Wolfe (✉)
Ghent University, Ghent, Belgium
e-mail: ctwolfe1@gmail.com

P. Distelzweig et al. (eds.), *Early Modern Medicine and Natural Philosophy*, History, Philosophy and Theory of the Life Sciences 14,
DOI 10.1007/978-94-017-7353-9_15

metaphysics. This is the case, whether in Descartes, Gaub, the 'animist' Georg-Ernest Stahl, or materialists such as Guillaume Lamy and La Mettrie: medicine, or rather 'a certain idea of medicine', is everywhere.

Here I focus on the motif of a radical medicine – a medical precursor of the Radical Enlightenment (Israel J, Radical enlightenment. Philosophy and the making of modernity, 1650–1750, Oxford University Press, Oxford, 2001; Israel J, Enlightenment contested. Oxford University Press, Oxford, 2006, Israel J, Enlightenment, radical enlightenment and the "medical revolution" of the late seventeenth and eighteenth centuries. In: Grell OP, Cunningham A (ed) Medicine and religion in enlightenment Europe. Ashgate, Aldershot, pp 5–28, 2007), symbolized negatively by the slogan, *tres medici, duo athei*, or 'where there are three doctors, there are two atheists', i.e. medicine as a basis for atheism. This theme runs through various works of medical or medico-theological propaganda: Thomas Browne's 1643 *De religio medici* begins with Browne regretting rumors of doctors being atheists as the "general scandal of my Profession"; Germain de Bezançon's 1677 *Les médecins à la censure* works hard at rebutting the saying, "Bon Physicien, mauvais chrétien." But these are examples of the *fear* of a radical medicine – a medicine that denies the existence of an immortal soul, or even defends materialism and atheism. Are there positive statements of this doctrine? Indeed, attacks on it are much more common than statements identifying with it, like medical versions of natural theology in general.

In fact, just as there were theologically motivated medical works, there were also medically motivated works of radical or heretical theology, like William Coward's *Second Thoughts on the Human Soul* (Coward W, Second thoughts on the human soul. R. Basset, London, 1702, building on Overton 1644), which engaged in polemics concerning the nature of the soul – mortal or immortal? (Thomson A, Bodies of thought: science, religion, and the soul in the early enlightenment. Oxford University Press, Oxford, 2008). Parallel to the mortalist trend, but flowing into a common genre of radical, medico-materialist texts (sometimes anonymous, such as *L'Âme Matérielle*, from the 1720s) are at least two other strands of radical medicine: a post-Cartesian focus on *medicina mentis* and the nature of the mind (Henricus Regius, Hieronymus Gaub, Antoine Le Camus), and an Epicurean medicine, in which mind and body are organismically united, with an additional hedonistic component, notably in Lamy, Mandeville and La Mettrie (Wright JP, Locke, Willis, and the seventeenth-century epicurean soul. In: Osler MJ (ed) Atoms, Pneuma, and Tranquillity: Epicurean and stoic themes in European thought. Cambridge University Press, Cambridge, pp 239–258, 1991; Wolfe CT, van Esveld M, The material soul: strategies for naturalising the soul in an early modern epicurean context. In: Kambaskovic D (ed) Conjunctions: body, soul and mind from Plato to the enlightenment. Springer, Dordrecht, pp 371–421, 2014). The focus on a medicine of the mind (Corneanu, (ms. 2013), The care of the whole man: medicine and theology in the late renaissance, 2013) is obviously connected to a 'medicalization of the soul': there was a body-soul problem *in and for* medicine, a sort of medicalized 'pneumatology'. Radical medicine is located somewhere in between the early forms of 'naturalization' or 'medicalization' of the soul and the pose of scientific neutrality that is characteristic of early nineteenth-century medicine (as in Cabanis, Bichat or Bernard): it is a short-lived episode. I seek to reconstruct this intellectual figure, in which mortalist, post-Cartesian and Epicurean strands intersect and sometimes come together. I suggest that medically influenced materialism in the Radical Enlightenment (e.g. in the later French cases, La Mettrie, Ménuret and Diderot), is different from later, more experimentally focused and more quantitatively oriented forms of medical materialism, precisely because of its radical dimension. This radical medicine often insists on vitality, as opposed to "anatomie cadavérique": it is vital and hedonistic, a medicine concerned with maintaining bodily pleasure.

Keywords Radical enlightenment • Heterodoxy • Medical materialism • Mortalism • Atheism • Soul

15.1

The interrelation between the genres, discourses or practico-conceptual clusters called 'medicine' and 'philosophy' in early modernity is more apparent now than it was even ten years ago. A number of interesting works have appeared which demonstrate in diverse ways, bristling with erudition, that we ignore this interrelation – and thus all sorts of 'contamination' of apparently neutral, austere metaphysics by messy, fluid, embodied discourse, from animal spirits, fermentation, mental pathology and fevers (notably in the work of Thomas Willis: Willis 1659/1681; Willis 1684), to venery or vertigo (e.g. La Mettrie's 1737 *Traité du vertige*) – at our peril.[1] Much as Katharine Park and Lorraine Daston noted, with a degree of insider humour, that unlike their classmates in the 1970s, when they opened works of early modern philosophy, they saw monsters everywhere (Park and Daston 1998, 9), I observe that, contrary to the standpoint of scholars of early modern philosophy who sometimes go out of their way to deny that medicine could have played a role, let alone a constitutive role, in classic metaphysical debates (Garber 1998, citing Henry 1989[2]), on issues such as the body-soul (then body-mind) relation among others, it seems patently hard to separate medical theory, medically nourished philosophical speculation, and metaphysics. This is the case, whether in Descartes, in the 'animist' Georg-Ernest Stahl, or materialists such as Guillaume Lamy and La Mettrie: medicine is everywhere, certainly in materialist and other heterodox approaches to body-soul and body-mind relations (Thomson 2008). Not only are medical doctrines cited for philosophical purposes (and 'physician-philosophers', *médecins-philosophes*, were eager to intervene in quarrels of the former sort); the

[1] This is, *grosso modo*, the topic of Wolfe and Gal eds., 2010. See also more generally Smith, ed. 2006; Thomson 2008.

[2] To be more precise, Garber brackets off the medical influence (1998, 764), and refers to Henry, for whom medicine had a limited impact compared to philosophy (1989, 93f.). In earlier work Henry allowed for a more broad influence of medicine on metaphysical debates ...

reconstruction of the genesis of what we understand as pure philosophical doctrines sometimes reveals explicitly medical elements therein. We witness a significant interplay in this period between physicians, natural philosophers, and medically 'influenced' philosophers – both physicians writing 'philosophy', like Guillaume Lamy and Abraham Gaultier, in his 1714 *Parité de la vie et de la mort* (*Réponse à un théologien*),[3] but also Bernard Mandeville, in his *Treatise of the Hypochondriack and Hysterick Diseases* (1711, revised 1730); physicians who become philosophers, like La Mettrie; and philosophers whose career reflects a continuous engagement with developments in medicine and physiology, like Diderot, for whom "there are no works I read with more pleasure than medical works" (Diderot 1975-, XVII, 510).

This is patent in the case of debates on the status of the soul: mortal or immortal? material or immaterial? Beginning in the late Renaissance, through the sixteenth and seventeenth centuries, there is a clear sense in some authors, not that medical observations e.g. at patients' deathbeds, necessarily 'prove' materialism, but rather, the slightly weaker and more flexible claim that physicians have a special kind of expertise in dealing with matters concerning body and soul. As Rob Illife put it suggestively, "In an important sense, the soul – its location and its function as the active and moral essence of the individual – should be seen as the product of … forensic and physiological knowledge" (Iliffe 1995, 434). From the medical mortalists in late seventeenth-century England (including William Coward, discussed in Thomson 2008) to Guillaume Lamy's *Anatomical Discourses* (1675, revised 1679) and *Mechanical and Physical Explanation of the Functions of the Sensitive Soul* (1677), and in the eighteenth century, La Mettrie's *Natural History of the Soul* (1745), to which we could add some of the medical entries in the *Encyclopédie* by the Montpellier physician Jean-Joseph Ménuret de Chambaud (e.g. the entry on Death, "Mort," 1765), medicine matters very much in the approach to and construction of the problem of the soul.[4]

This connection between medicine and a traditional metaphysical problem could be articulated on the basis of varying historical sources: it could stem from late Aristotelian-Averroist discourses on 'the organic soul' or the soul as 'life principle', as in Pomponazzi or, more classically, it could build on Galen's treatise *Quod animi mores*, on the soul's dependence on the body. For here, Galen had argued that "it is

[3] It is thanks to Olivier Bloch that we are familiar with the work of the physician Abraham Gaultier, a Protestant turned atheist, and author of an 'epigenetic materialist' treatise (Gaultier 1714/1993) that derives some of its ideas from Lamy and Harvey (his materialist reading of epigenesis is not unlike Diderot's later articulation of epigenesis and Spinozism (Wolfe 2014), although ironically, Gaultier denounces Spinoza as well as Descartes and Malebranche as metaphysicians, even though the subtitle of his work includes the statement that "Life and Death are … modifications of one Substance"). Gaultier's original work was almost entirely unknown, but large portions of it were excerpted in a clandestine manuscript entitled *Parité de la vie et de la mort*, which did circulate.

[4] On Lamy see Thomson 1992 and Anna Minerbi Belgrado's extensive and extremely informative introduction to her edition of Lamy's *Discours anatomiques & Explication méchanique et physique des fonctions de l'âme sensitive* (Lamy 1996), as well as the section on Lamy in Wolfe and van Esveld 2014. On Ménuret see Rey 2000a and Wolfe and Terada 2008.

preferable to say ... that the mortal part of the soul is the mixture of the body" (Galen 1997, 153, 157); even if there were a "separate substance" for the soul, it would still be dependent on ("a slave to") the mixtures of the body (155). That is, given the presence of these various humoral mixtures in any part of our body, the soul "cannot but be a slave to the body." While Galen himself did not make any overt *philosophically* reductionist claims (as distinct from a *medically* reductionist account), in an early modern context these ideas, whether directly quoted or modified, could sound quite different, as in the Gassendist François Bernier's suggestion that "it would appear that Galen was persuaded the Soul was a spirit that emerged out of the blood" (Bernier 1678, vol. V, 452) – even if some other, earlier authors had sought to integrate, not the teleological Galen of *De usu partium* but indeed the 'materialist' Galen of *Quod animi mores* with "the notion that the soul is a separate substance with agentive autonomy, although one vitally united with the body," as Corneanu puts it.[5]

With some reservations, because these contexts do not lend themselves to an overall theoretical 'systematization', one can say that there was a body-soul problem *in and for* medicine, growing out of some of the medicalization of 'pneumatology' in natural philosophy (which of course, arguably, had been 'physiological' and/or 'medical' in earlier Stoic contexts before it got turned into a 'mental' or 'mind-body' issue[6]). This is not restricted to, or explainable in terms of any particular medical tradition exclusively (the invocation of Galen's *Quod animi mores* fuels one strain of physicians claiming authority on the question of body-soul relations, but Cartesian-inspired physicians from Regius to Boerhaave and Gaub do so as well, as do 'Epicurean' physicians like Lamy). In that sense, historians of philosophy claiming that medicine is irrelevant to early modern disputes on atheism or the status of the soul, which are purely metaphysical (Henry 1989, 92–93; Garber 1998, 764), would be well advised to leaf through any of the writings of the above-mentioned physicians. It would then be harder to claim that (here with the specific case of England) physicians "failed to make a substantial contribution ... on the propagation and spread of radical ideas between 1660 and 1700" (Elmer 2007, 225); or that "The proverbial atheist, the physician, hardly appears at all in learned attacks on atheism," because "mechanist versions of atheism drew all the fire" (Henry 1989, 91). This would have been news to anyone reacting to provocative claims concerning the mortality of the soul, some of which emanated from physicians claiming a medical 'foundation' for such claims, e.g. William Coward's *Second Thoughts on the Human Soul* (Coward 1702). There are actually many examples of this – to pick one, Thomas Browne, in his 1643 *De religio medici*,

[5] For the former view, see generally Bernier 1678, vol. V, book VI, ch. iii: "What the animal soul is." Laurentius (Du Laurens) in his 1597, argues vehemently against Galen for having a 'deterministic' vision of the functioning of body-soul interaction. Bezançon's sentiment is that we should not follow Galen in everything: "that he was nourished in the darkness of paganism, and consequently was not enlightened by the heavenly rays of faith, is his own personal misfortune" (Bezançon 1677, 330). For the latter view, see Corneanu (ms., 2013).

[6] I thank Brooke Holmes for this point, in conversation.

alludes to precisely the 'wrong' sort of influence Galen could have: "I remember a Doctor in Physick, of Italy, who could not perfectly believe the immortality of the soul, because Galen seemed to make a doubt thereof" (Browne 1892, § XXI, 45).[7]

That physicians played a prominent role in challenges to body-soul dualism, or could be quoted and used, sometimes liberally, in reductive materialist arguments (most characteristically in La Mettrie: Wolfe 2009) is itself an outgrowth of a much more general and amorphous cultural 'construct': the physician as atheist (Kocher 1947; Mothu 2010). This is not to say that debates on the soul were necessarily debates on atheism, but that both reveal more of a medical coloration than the historian of philosophy generally acknowledges. Examples of the 'physician as atheist' abound, but I shall mention just a few. At least as early as John of Salisbury's 1156 *Polycratus*, the warning was heard that physicians "encroached" on the territory of religion:

> Physicians however, placing undue emphasis upon nature, in general encroach upon the rights of the author of nature by their opposition to faith. I am not accusing all of them of error although I have heard very many of them arguing about the soul, virtue and its works, growth and decay of body, the resurrection of the same, and creation in a manner contrary to the tenets of faith (John of Salisbury, cit. Kocher 1947, 232–233).

With more gentle irony, Chaucer, in the *Canterbury Tales* (published gradually during the later fourteenth century), warns regarding the "Doctour of Phisike" that "His study was but little on the Bible" (Chaucer 1933, Prologue, portrait, l. 440). In early modern Europe, accusations that physicians were atheists – that they favored Hippocrates or Galen over the revealed Word of God – were common. Alain Mothu mentions an anecdote from the thirteenth-century French king, Saint Louis, who is said to have healed a physician of the sickness of atheism, "body and soul," by extracting through the nose a "putrid humor" preventing the physician from "knowing his Maker"; and the Scottish physician Marc Duncan, who taught philosophy at the Protestant academy at Saumur in the first third of the seventeenth century, was credited with a witticism that was widely quoted, according to which the physician is *animal incombustibile propter religionem*, a modern rendering of which might be that the physician is not combustible in religion.[8]

---

[7] Of course, one should not confuse the interpenetration of medicine with metaphysics (something of an *auberge espagnole*, where many tendencies coexisted and evolved: see Edwards 2012 on how anatomical material was present even in the "scholastic philosophical mainstream" [44]; he speaks of the relation between anatomical studies and the soul as like "cross-border traffic," 46), with the more specific case of an *atheist medicine*, that is, a type of argument and/or rhetorical figure in which a certain idea of medicine was made to (a) play a role traditionally devolved to metaphysics or (b) have a deflationary impact on traditional metaphysics, notably as regards the soul (Iliffe 1995). But both of these (the general and the specific forms) run counter to assertions in Henry 1989 and Garber 1998 regarding the distance between medicine and metaphysics.

[8] Mothu 2010, 317–318, 319. I am indebted to Mothu's piece for its portrayal of several different 'atheist medicine' figures. My analysis differs from his in that I add the question of a medicine of the mind and its relation to materialism (both of which can be seen as giving rise to a 'positive' version of what the polemical figure of the doctor as atheist sketched out 'negatively').

One factor in these accusations was the figure of the physician as enemy of superstition, including demonic possession: the critique of superstition could of course slide into the critique of Christianity. The Constitution of the Society of Jesus (the Order of the Jesuits), written in 1558, forbids the study of medicine, and the study of law, as neither of these was seen as contributing to the ultimate goal of Jesuit study, the glory of God (Edwards 2012, 56). Mothu mentions the case that can serve as indicative of many others, of one Saporta, a Montpellier physician who in 1608 gave a speech "to prove that in Lazarus's resurrection, there was no miracle" (Mothu 2010, 324). This tension between religion and medicine traces back to the doctrine of original sin (Epistle to Romans, V, 12), from which death and by extension all forms of bodily corruption flow. Perhaps better put, just as there are powerful early figures such as 'Christ the healer', 'Christus medicus', conversely, the status of the body, its health and disease as possibly reflecting original sin could be the source of a destabilization. On the first side of the balance sheet, Augustine and others taught, famously, that it was first and foremost the soul that had to be healed, in order to heal all other disorders. And the greatest healer of all was, again, Christ; as the physician Germain de Bezançon put it in his interesting 1677 work *Les Médecins à la censure ou Entretiens sur la médecine*, "it is God who properly heals, not the physician" (Bezançon 1677, 30). Clerics were prominently represented on the faculty of medical schools until the turn of the sixteenth century (see also the impressive list in Bezançon 1677, 351–353 of prominent physicians who were also clerics). Indeed, faced with this situation there was a real need to defend the ethical legitimacy of calling for a doctor when sick.[9]

The figure of the physician as atheist runs through various works of seventeenth- and eighteenth-century medicine or medico-theological propaganda (that is, with a strong, ideologically suffused polemical intent): Browne's *De religio medici* begins with him regretting rumors of doctors being atheists as the "general scandal of my Profession" (Browne 1892, § I, 1) and he expected his book would improve this bad image of the medical profession[10]; a 1707 issue of the London *Weekly Comedy* claimed that "Physicians are … generally accounted Atheists" (Porter 2001, 296 n. 39), a concern that also motivates Bezançon's *Les Médecins à la censure*, a dialogue partly in defense of medicine including its 'ethical' character (part VIII is devoted

[9] Maria Conforti observes that most histories of medicine written between the sixteenth and eighteenth centuries make no mention of the Christian religion, not even to gesture towards the relation between illness and original sin (Conforti 2007, 77). One can add that histories of medicine written in the past two centuries never mention the atheist motif, perhaps also because they were usually written by physicians who did not desire to highlight the more controversial aspects of their profession.

[10] Ironically, Browne's project backfired in various ways, both in England and on the Continent, where his book was at times equated, sadly for its author, with skepticism or even atheism, and ended up on the Index. Some commentators, including even Pierre Bayle, joked that rather than 'religio medici', the religion of the physician, the book was more of a 'medicus religionis' or 'médecin de la religion', the work of a physician thinking himself above religion (Bayle 1740, IV, 646; Mothu 2010, 324, 330). On the 'religio medici' theme more generally see Cunningham and Grell eds. 1996).

to a defence of the 'religio medici', or at least to addressing the question, 'are medicine and Christianity compatible?' [338]); the influential cleric and natural philosopher Marin Mersenne devoted an entire chapter of his 1624 *L'Impiété des déistes, athées et libertins de ce temps* (I, ch. VI) to discussing and rebutting the charge that physicians were atheists, including by an appeal to Galen (Mothu 2010 observes that Mersenne earlier had denounced physicians as potential atheists, so this is something of a reversal; but in the second part of the 1624 work, Mersenne adds that not all physicians are decent individuals; some are libertines, servants of the devil who corrupt women, and so on). These concerns are also at work in varied works of 'medical natural theology', like Philippe Hecquet's 1733 *Médecine théologique*, which has the explicit subtitle: *La Médecine créée, telle qu'elle se fait voir ici, sortie des mains de Dieu* (Hecquet 1733; on Hecquet see Brockliss 1989).

In *Les Médecins à la censure*, through the voice of the character Cariste, a cleric and lawyer who is hostile to medicine (as per the 'Avertissement'), Bezançon reflects more specifically on the psychological processes at work as physicians become detached from 'fundamentals' (including the existence of design in the universe and thus a Designer-God). Physicians in their professional activity focus so much on 'sensibles' that their minds become accustomed to merely grasping "crude ideas of bodies," as they deny anything supernatural, since flesh and blood do not reveal it. Here a Hippocratic motif is impugned: these physicians explain everything on the basis of 'temperaments', body and matter, and are thus at best proto-materialists (if not materialists outright). In an intriguing turn of phrase, Bezançon says these physicians are justified in calling themselves "sensual physicians": *medicus est physicus sensualis.*[11] Physicians have always had a strong "antipathy towards religion," in his view (329). But what is it to be a 'sensual physician'? We can surmise it is something like being an empiricist – not an 'empirick' but the type of thinker who relies on the evidence of the senses. Indeed, when Bezançon glosses on the phrase 'sensual physicians' a few pages later (339), he confirms that it means physicians who fully rely on causes derived from sensation (one can also point to the presence of the empiricist slogan, '*nihil est in intellectu quod non fuerit in sensu*' in a variety of medical texts of the period, from Harvey and Sylvius to Mandeville and Ménuret, which has been discussed elsewhere: Cranefield 1970; Wolfe 2010). Indeed, Ménuret de Chambaud, who I have mentioned, uses almost the same phrase, but in a positive sense, in his entry "Mort" in the *Encyclopédie*:

> The separation of the soul from the body, a mystery which may be even more incomprehensible than its union, is a theological dogma certified by religion, and consequently is uncontestable. But it is in no way in agreement with the lights of reason, nor is it based on any medical observation; hence we will not mention it in this purely medical article, in which we will restrict ourselves to describing the changes of the body, which, as they alone fall under the senses, can be grasped by the physicians, those sensual artisans, *sensuales artifices* (Ménuret de Chambaud 1765/1966, 718b).

[11] Bezançon 1677, 8th dialogue, 334–335. The phrase is not original to Bezançon: it occurs notably in Riolan's commentary on Fernel on temperaments (*Praelectiones in libros physiologicos & de abditis rerum caussis*, Paris, 1602, 43); thanks to Sorana Corneanu for pointing this out.

A "purely medical article" should restrict itself to dealing with changes in the body that "fall under the senses" and thereby can be grasped (especially) by physicians, who are *sensuales artifices*, craftsmen dealing with the sensory world of bodies. So, reflecting on this and the above cases: the physician is a potential atheist, a proto-materialist, and (more than either of these), a kind of empiricist.

Elsewhere, I have examined aspects of this medico-philosophical doublet with respect to the idea of medical empiricism (Wolfe 2010), the 'material soul' as in part a medical construct (Wolfe and van Esveld 2014), or the project of a 'medical Epicureanism' in La Mettrie (Wolfe 2009). Here, my concern is with the specifically *radical* motif: the idea of a radical medicine – a kind of medical precursor but also actor of the Radical Enlightenment, as noted by Jonathan Israel (Israel 2001). He notes the presence of physicians in his 'Spinozist' narrative, although here (and also in Israel 2006 and 2007) he runs the risk of overstating or taking too literally the implication that, e.g., members of Spinoza's circles such as van den Enden, Meyer and, differently, Adrian Koerbagh were all physicians: what is overstated is the specifically *medical* dimension. It is not clear that any of these figures, and certainly not Spinoza himself, emphasize either the atheistic or otherwise radical implications of medicine. This radical figure of medicine was symbolized negatively by the slogan, *tres medici, duo athei* (a shortened version of *ubi tres medici, ibi duo athei*, or *Ubi sunt tres medici ibi sunt duo athei*[12]), a variant of which was "Bon Physicien, mauvais chrétien" (Busson 1948, 144). La Mettrie will quote the formula *tres Medici, duo Athei*, ironically (in *Les animaux plus que machines*, in La Mettrie 1987, I, 328). He also declared, in a kind of mock warning: "the first years of medical study are the first step leading physicians towards irreligion" (La Mettrie 1749, I, 255).

But most often these expressions (the non-ironic ones) are instances of the *fear* of a radical medicine – a medicine that denies the existence of an immortal soul, or even defends materialism and atheism. Attacks on this doctrine are vastly more common (like medical versions of natural theology in general) than statements identifying with it. Where are we to find positive statements of this doctrine?

In fact, just as there were theologically motivated medical works like Browne's *Religio medici* or better, Hecquet's *Médecine théologique*, there were also medically motivated works of radical or heretical theology, like Coward's *Second*

[12] White 1898, II, ch. vii, citing the bull of Pius V (*Bullarium Romanum*, ed. Gaude, Naples, 1882, VII, 430, 431); Nutton 2001, 32. Sometimes it was presented as a positive claim, e.g. 'radicals' like Vanini were said to claim happily that philosophers and physicians were generally atheists: in his influential attack on free-thinkers, the 1623 *Doctrine curieuse des beaux esprits de ce temps*, the Jesuit François Garasse took Vanini as a target for this reason: "the miserable Vanini, a charlatan by profession and … an atheist in religion, tried to show both by example and in his wicked doctrine, that *philosophers and physicians are ordinarily atheists*"; Garasse retorted that we know many "able physicians who are even better Catholics," "wholeheartedly so" (Garasse 1623, III, § 9, 255–256). Among other notorious examples, Rabelais himself, something of a father figure for libertines, free-thinkers and other defenders of heterodoxy, was a practicing physician, including at the Hôtel-Dieu between 1532 and 1534. For more on these accusations of atheism specifically in the English context, see Kocher 1947.

Thoughts on the Human Soul, which engaged in polemics concerning the nature of the soul – mortal or immortal? (Thomson 2008). Mortalism was a recognized heresy, the view that the universe is entirely material, and the soul is as mortal as the body, sometimes on the definition that the soul is just a mode of the body … yet the universe is nevertheless a divine creation. Mortalists generally held that the word (and idea) of an immortal soul was a Catholic invention, nowhere to be found in Scripture. The most intriguing brand of mortalism for the present discussion was that professed by some English physicians in the late seventeenth and early eighteenth centuries (from Richard Overton's *Man's Mortality*, 1644 to Coward's *Second Thoughts on the Human Soul*, 1702), who felt that their medical expertise, specifically at patients' deathbeds, gave them the authority to state that the soul died with the body. As Mandeville, himself a practicing physician, put it, "Nor is it clashing with Christianity to affirm … that Man is wholly mortal. …The Resurrection of the same person … must necessarily include the Restitution of Consciousness" (Mandeville 1711/1730, 51). The additional medical emphasis of Coward and others was to claim an empirical foundation for the mortality of the soul, which dies with the body (or 'sleeps' depending on the theological variants), until resurrection on the day of the Last Judgment. Sometimes it was additionally argued that the soul has been mortal ever since the Fall.

Parallel to the mortalist trend, but flowing into a common genre of radical, medico-materialist texts (sometimes anonymous, such as *L'âme matérielle*, from the 1720s) are at least two other strands of radical medicine: one we might term a post-Cartesian focus on *medicina mentis* and the nature of the mind (Henricus Regius, Hieronymus Gaub, Antoine Le Camus, not to be assimilated directly to earlier, more 'therapeutic' and less empirical versions of a *medicina mentis* like Tschirnhaus'), and the other an Epicurean medicine, in which mind and body are united, but the vision of organism is also one which emphasises a hedonistic motivational process, notably in Lamy, Mandeville and La Mettrie (Wright 1991; Wolfe and van Esveld 2014). In both cases, medicine has a rhetorical dimension tailored to lend support to various kinds of heterodoxy or radicalism, whether these are presented as overtly atheist or not, and whether the author is happy with the implications of his own view or not (witness the case of Gaub's integrated mind-body medicine and how easily La Mettrie can appropriate it).

I shall discuss each in turn before turning to the way these discourses of a 'medical atheism' end up in the service of a materialization of the soul.

15.2

The post-Cartesian strand and the Epicurean strand appear relatively distinct if we consider some representative seventeenth- and early eighteenth-century figures, but they were also sometimes combined, as in the case of Cyrano de Bergerac's physics and cosmology, or La Mettrie's idea of the man-machine (Wolfe 2004), or, more germane to the present essay, the theme of a 'materialization of the soul'. But first I shall quickly examine the Cartesian and the Epicurean elements (or 'players') in turn.[13]

On the Cartesian side (increasingly broadly construed as a mechanist project for an integrated mind-body medicine), we first find Henricus Regius or Hendrik de Roy, a physician and Professor of Theoretical Medicine at the University of Utrecht, who was often called the 'first apostle of Cartesianism' (e.g. in the *Nouvelles de la république des lettres* in October 1686), asserting that the soul could be a mode of the body, with the body being understood as a machine, and that the human mind, inasmuch as it exists in a body, is organic.[14] Regius caused a celebrated scandal and brought much ideological discredit to his mentor Descartes, who had to deny any paternity of such ideas. But there could also be more concrete causes for worry: not just that physicians are proto-empiricists, or tend to view the soul-body relation in very 'sensory' terms, but also, that Descartes' consideration of particular mechanisms such as the heartbeat could have deflationary consequences on the nature and existence of the soul. Thus the Leuven theologian Libertus Fromondus, in his controversy with Descartes and Plempius over the heartbeat, felt, as Lucian Petrescu puts it, that if some of the operations normally attributed to the soul are presented as actually taking place as a result of the functioning of a mechanism, then we are in danger of explaining all operations of the soul, including its purely intellectual ones, through this mechanism. Reflecting on the causes of the heartbeat, Fromondus writes that Descartes' (mechanistic) approach "opens the way to the atheists, so that similar causes [motion provoked by heat] are assigned to the rational soul" (Fromondus' articles in reply to Descartes' *Discourse* and *Essays*, AT I 403, discussed in Petrescu 2013, 419–420). However, the danger of atheism can come from anywhere: physics and cosmology, bedside medical practice, or armchair metaphysics. Of primary interest here is the question of the soul or mind (the reconfiguration of *anima* as *mens*, which is occurring in this period, is part of the issue).[15]

This question occupies center stage in the lecture given at Leiden in 1747 by Hieronymus Gaub, *De regimine mentis quod medicorum est* (translated in Rather 1965). Gaub had been Herman Boerhaave's student, and took over his Chair in Leiden. Here, Gaub suggests a clinical perspective on the problem of mind-body interaction (for he is speaking of *mens* rather than *anima*; Wright 2000, 249), in which the metaphysical distinction between mind and body is irrelevant. "Although

[13]As was observed by several reviewers of this chapter, this story could also be told as a tale of post-Galenic medicine of the mind. I emphasize 'post-Cartesian' here because of the emergence of works specifically entitled *Medicina mentis, La médecine de l'esprit*, etc., in early modern Europe, which, however critically or eclectically, actively engaged with a Cartesian picture – of medicine, of an anthropology of the passions, of a mechanistic project in which health was also a paramount value.

[14]Regius 1646, 248, 246. See Alexandrescu 2013.

[15]There is of course earlier discussion (often based on Cicero) of the relation between *animus* and *anima*, where the former becomes interchangeable with *mens* and the latter with the notion of the organic soul (I thank an anonymous reviewer for this observation.) I simply mean that in texts of the period I am discussing, there is an increasing insistence that 'soul' just means 'mind', as Charles Bonnet often puts it.

the healing aspect of medicine properly looks toward the human body only, rather than the whole man, it does refer to a *body closely united to a mind* and, by virtue of *their union*, almost continually acting on its companion as well as being itself affected in turn" (Gaub 1747, in Rather 1965, 70, emphasis mine). Gaub refers to the authority of Descartes, "the most ingenious philosopher of his age," who "yielded to physicians" regarding the priority of medicine in these matters (74),[16] and states that due to the variability of temperaments, itself explainable in humoral (and hence medical) terms, the philosopher "cannot dispense with the aid of the physician" where the mind is concerned (86).

Interestingly, La Mettrie seems to have attended Gaub's lecture, some months prior to finishing *L'Homme-Machine* (Gaub mentions his presence), and spoke very favourably of it, carrying these ideas to what may seem (to us) their obvious materialist conclusion. Gaub did not appreciate La Mettrie's materialist appropriation of his ideas, and in 1763 included a short essay against him in his new edition of *De regimine mentis* (Rather 1965, 115–117), calling him "a little Frenchman" who produced a "repulsive offspring … his mechanical man" (Gaub 1763, in Rather 1965, 115). More significantly, Gaub also denounced La Mettrie in a 1761 letter to Charles Bonnet (in Caraman 1859, 172), referring to the bad materialist usage of the "physics of the soul" undertaken by physicians. He worries that Bonnet's essay on the soul will have the same negative effect as La Mettrie's work, but concludes that this is not the case: the study of the "mechanism of the soul" (*ibid.*) need not entail materialism, since in Bonnet's case (and Gaub approves) it supports the claim that thought is not a mere "effect" of this mechanism. Gaub even begs Bonnet to publish an additional treatise on this topic, "so as to demonstrate that the mechanism of the operations of the soul is so far from favoring materialism, being instead the most convincing proof of the opposite system" (in Caraman 1859, 173).

It must thus have been disturbing to Gaub that La Mettrie spoke so favourably about the ideas he heard at the 1747 lecture, not least since his enthusiasm makes sense: Gaub had defended the view that for the physician, the metaphysical distinction between mind and body is irrelevant. Faced with the consequences, Gaub has to demand in 1761 that someone of Bonnet's stature and scientific competence write a treatise to show that the naturalistic study of the "physics of the soul" (presumably some combination of psycho-physiology, psychology and of course the 'medicine of the mind') does not entail materialism. To be clear, there is an interesting duality between Gaub's position in his own eyes, and for us. From his standpoint, he is carrying on a project of the scientific study of the living body which is both distinct from and in fact serves to refute materialism (not unlike Boylean physico-theology, but in a vaguer sense). From our standpoint, Gaub's assertions about bodies "closely united" to minds as an object (a) of medical study and (b) of philosophical reflection

[16] Gaub has in mind the passage from Part VI of Descartes' *Discourse on Method* where Descartes notes the interpenetration of mind and the organs of the body, so that medicine is the best way to render people wiser than they have hitherto been (AT VI, 62). In La Mettrie, this becomes: "medicine alone can change mind and behavior [*les esprits et les mœurs*] along with the body" (La Mettrie 1987, I, 67), and "the best philosophy is that of the doctors" (La Mettrie 1987, II, 36).

enhanced by such medical study sound very close to materialism indeed, especially to the medically nourished form of materialism defended by someone like La Mettrie.

The soul also undergoes a gradual process of naturalization in a medical context with figures of Epicurean medicine as the seventeenth-century physician Guillaume Lamy (Del Lucchese 2010). Here, anatomical arguments about body and soul are used in support of an anti-finalism, an Epicurean appeal to chance, and a more or less overt atheism. Guillaume Lamy (1644–1683) was a self-proclaimed Epicurean, a philosopher and physician based in Paris, who published his major works between the late 1660s and the late 1670s. His first work, the 1669 *De Principiis rerum*, was an explicit piece of early modern Epicurean atomism, favoring Gassendi over Descartes (who was also viewed as a covert supporter of atomism), to show that Epicurus was right in the first place (although, in a gesture we will find often in works of this period, e.g. those of Cyrano de Bergerac as mentioned above, he also seeks to present these theories as complementary or compatible). He discusses atoms and the nature of matter at some length, hesitating as to which theory he finds most convincing, but we shall chiefly focus on his medical-materialist approach to the soul. In *De Principiis* (I, v) and in his later works, the *Discours anatomiques* (1675, 2nd revised edition 1679) and the *Explication méchanique et physique des fonctions de l'âme sensitive* (1677), he claims that the soul and animal spirits are actually identical.[17]

Lamy identifies the functions of the soul with (a) the nervous centres that receive impulses from external stimuli, and which ensure consciousness,[18] and (b) the animal spirits which carry the "agitation" produced by the objects to the brain, which is the "source" or "reservoir" of the soul (Lamy 1996, 152 153), and then return to the heart, where they give rise to the passions, and to the muscles (which Lamy, following Galen, views as the instruments of voluntary motion); he says that the soul "flows" from the brain like rivers flowing through the "canals" of the nerves (153, 160, 142). Lamy verbally still maintains a difference between the sensitive soul and the rational soul but ultimately locates all of these distinctions within a physiological frame (104 f.). This idea that the soul is, in the end, a medical matter is prominently taken up by La Mettrie in his 1745 *Histoire naturelle de l'âme* (the revised version, which he entitled *Traité de l'âme*, appeared in 1750: see especially ch. VIII), but was also featured earlier in the century, in the famous anonymous, clandestine manuscript, *Treatise of the Three Impostors* (chapters XIX and XX) and later, in the article "Âme" of the *Encyclopédie*, in an atheist framework (Thomson 1992; Thomson 2008).

In the fascinating anonymous manuscript *L'Âme Matérielle* (*The Material Soul*),[19] which can be dated to approximately 1725–1730 based on some of its citations, we witness a first, programmatic attempt at the *naturalisation of mental phe-*

[17] *Discours anatomiques*, in Lamy 1996, 102, 105.

[18] *Explication*, in Lamy 1996, 142–143, 160–161.

[19] Its modern editor, Alain Niderst, considers various possible candidates for authorship, which have evolved since his first edition of the work in the 1970s.

nomena, that is, at locating mental phenomena within an integrated corporeal and cognitive scheme (the distinction between these two levels being both anachronistic and irrelevant here), without restricting the analysis to a (Galenic?) medical context. The text speaks explicitly of the "materiality of the soul" (Anon. 2003, 222). Thanks to Alain Niderst's research, we know that this text is an ingenious patchwork of Spinoza via Bayle (particularly his article "Buridan" in the *Dictionnaire*, the *Pensées diverses sur la comète*, but also the *Réponse aux questions d'un provincial)*, Malebranche's psychophysiology, the doctrine of the (material) soul as a "fiery soul" from Gassendi as mediated through Bernier, Epicurean physiology (particularly borrowed from Guillaume Lamy), travel narratives, and various materialist prodromes from Lucretius to Vanini and Hobbes, typically using the analyses and summaries given in anti-materialist works. This means that the physiological portions of the work are based on older notions such as the "innate fire" in the soul, and animal spirits (which by the 1720s is no longer exactly state-of-the-art neurophysiology), to which the author adds the idea of cerebral traces as a basis for memory and association.

L'Âme Matérielle uses the argument (explicit in Galen's *Quod animi mores*, section 6, but considerably expanded on by La Mettrie, especially in *L'Homme-Machine*) that states of disease are evidence for the interaction of soul and body – and further, that they establish that both are composed of one and the same substance (Anon. 2003, 56), such that "the mind is subject to the law of all corporeal beings" (*ibid.*). We are also told in the last sentence of the treatise (236) that it is the "matter of which the brain is composed" that thinks. But even if the soul is material, the ontological status of this materiality is not generic: "The human soul is material, and is made up of the most subtle parts of the blood" (228). To use a distinction suggested by Wright (2000), but in a different sense (he was more concerned to chart the workings of diverse kinds of 'functional dualisms' understood in a post-Cartesian sense as dualisms of vital function versus cognitive function), we can say that the soul as locus of mental activity is here being conceptualised both as *substance* – as a material substance subject to physical and biological laws – and as *function* – belonging to medicine in general and *medicina mentis* in particular – that which Gaub feared while also being an actor in its unfolding. [20] As Aram Vartanian put it, "if one conceives of the soul as the effect or function as certain structures of organised matter, it is inevitable that the more legitimate articulation of such a concept occurs through the progress of knowledge regarding anatomical structures themselves, and their modes of operation" (Vartanian 1982, 159–160).

From Regius and Gaub (on the one hand) to Lamy and *L'Âme Matérielle* (on the other hand), we witness one of the greatest fears described in the earlier part of this essay, come true: the *materialization of the soul* through its inscription, either in an actual medical set of concerns (most patent in Gaub), or in a complex combination of medical and metaphysical reflections, sometimes taking the form of a matter theory (Lamy), sometimes of a reductionist materialist program for reconfiguring

[20] On the naturalisation of the soul as 'substance' or as 'function', see Vartanian 1982 and Wright 2000.

mental properties as physical or corporeal properties (*L'Âme Matérielle*). One might ask, how medical is this, then?

15.3

I shall mention two features in which this radical medicine retains at the very least a strong rhetorical usage of the medical motif: the idea of a medicine of the mind (*medicina mentis*), which existed in different varieties, some more materialist than others, and that of a medicalization of morals, which in our early modern context is very much a late-Epicurean radical project – with a surprising invocation of Machiavelli (the figure of the physician as Machiavellian)[21] – although Renaissance versions existed, appealing to Galen, precisely. But this is, of course, rhetoric, not clinical or experimental medicine. When, at opposite ends of the ideological spectrum, a Thomas Browne or a La Mettrie invoke medical authority in their metaphysical pronouncements, it is just that: an argument by authority – but, as I hope will be clear by the close of the essay, the radical medicine rhetoric may have had a certain productivity.

To be clear, I am not arguing either (a) that all forms of medicine of the mind are materialist or crypto-materialist, or (b) that the project of a medicine of the mind ends up leading to a materialist philosophy of mind. Rather, (c) I argue that of the various forms of medicine of the mind, its more reductionist or reduction-friendly versions end up, through eclectically composed paths of articulation and development (elements of Galen, Cartesianism, Epicureanism, etc.) articulating a medicalized and materialized account of mind, which mutatis mutandis, matches rather well the figure of atheist medicine that earlier commentators feared.

Medicine of the mind can be broadly distinguished into two strands: a non-reductionist (indeed, holist) project of a 'medicine of the whole man', e.g. in Erasmus: "Now the physician is concerned not only with the care of the body, the lower element in man, but with the treatment of the entire man, and just as the theologian takes the soul as his starting point, the physician begins with the body" (Erasmus 1989, 39–41; see Corneanu ms. 2013). But this non-reductionist version could also take the form of treatises on the passions (like Juan Luis Vives) or of a *medicina mentis* understood as the logic or pedagogy of the understanding (of a mixed Cartesian-Spinozist sort), as in Tschirnhaus' *Medicina mentis et corporis*

---

[21] La Mettrie has a satirical medical work entitled *The Politics of Machiavelli's Physician* [La Mettrie 1746], which presents itself as the translation of a Chinese original, and in other related writings such as La Mettrie 1749–1750 he exploits at length the figure of the corrupt or deceitful physician as a 'Machiavellian'; as early as 1588, the surgeon John Read associates atheist physicians with Machiavellians: in his "Complaint of the abuse of the noble Arte of Chirurgerie," he deplores that some in his profession are "papists, nulli fidians, atheists temporizers, and some machiavells" (cit. Kocher 1947, 230).

(1695).[22] Contrast this non-reductionist version with two later variants: Gaub's project for a medical 'regimen of the mind' which would deal with "a body closely united to a mind and, by virtue of their union, almost continually acting on its companion as well as being itself affected in turn" (Gaub 1747, in Rather 1965, 70) and with the equally explicitly titled 1753 *Médecine de l'esprit* by the Paris physician Antoine Le Camus (who credits Gaub in the final pages, while also criticizing him for remaining too abstract – "general axioms without practical consequences": Le Camus 1753, II, 335). In Gaub and Le Camus, it is a full mind-body integration to be studied by the physician (with implications for materialist philosophy, as La Mettrie saw in Gaub and Diderot did in Le Camus).

Le Camus notes that most people would not deny medicine's expertise when it comes to the body, but they would be reluctant to grant it authority over the mind, and he wants to remedy this situation; his program is indeed that "to remedy the vices of the mind is nothing other than to remedy the vices of the body" (Le Camus 1753, I, 7). Le Camus's program for medicine holds that it is the science which has equal knowledge of mind and body, and hence can treat their "abstract combinations," and their "relations" (*commerce*). While terminologically he still refers to these as two substances, in practice he gives, e.g. an integrated account of "virtues" and "passions" as being as much part of body as they are of the soul (Le Camus 1753, I, 111 f.; II, 239). Indeed, "God only excites ideas in our souls relative to the dispositions in our bodies" (*ibid.*, I, ch. III, § 2, 49).

Here, medicine of the mind is a reductionist project (with a good deal of flexibility, depending on which states, processes, and organ systems are serving as the basis for the 'reducing theory' – a humorally reductionist medicine of the mind is different from one in which the reducing theory/entity is the passions, or an iatromechanist-flavoured reduction of mental states to states of the body-machine). It is neither purely medical nor purely philosophical: on the latter side, consider La Mettrie's statements – which themselves invoke the authority of the physician (both his own and that of more celebrated figures like Boerhaave, Haller and Gaub):"s/he who wishes to know the properties of the soul must first search for those which manifest themselves clearly in the body" (*Traité de l'âme*, ch. I, in La Mettrie 1987, I, 125: this is not an assertion that *there is no such thing as the soul*, but rather the advice to start with the body), or "The soul is just a pointless term of which we have no idea and which a good mind should only use to refer to that part of us which thinks" (*L'Homme-Machine*, in La Mettrie 1987, I, 98). In this case, soul is being construed as a functional definition: it is neither eliminated in favour of a hypothetical 'basic physics' or the properties of matter in general, nor asserted as substantivally unique in its own right.

---

[22] To be clear (and for more discussion see Corneanu, ms. 2013), treatises of the passions could be entirely non-medical, like Cicero's *Tusculan Disputations*, III and IV (Cicero 1927); or strongly medically oriented, like Juan Huarte's *Examen de ingenios para las sciencias* (Huarte 1575/1989 – again a case of the influence of Galen's text on body and soul) but also late Renaissance humoral anthropologies such as Timothy Bright's 1586 *Treatise of Melancholie* (see Henry 1989 and Wright 2000), or – somewhere in between, like Descartes' *Passions de l'âme* (1649). I thank Sorana Corneanu for her help and many fruitful discussions on these topics.

The more explicitly medical statements of a (reductionist) medicine of the mind, however, often extend beyond their province, whether accidentally or deliberately, as in the dismissive comment in Tarin's article "Physiologie" in the *Encyclopédie*: "If the body is healed, one need not worry about the soul" (XII, 1765, 538a). And we have already seen Ménuret, in the article "Mort," apparently granting the authority of theology ("The separation of the soul from the body ... is a theological dogma certified by religion, and consequently is uncontestable") but then immediately asserting the authority of "the lights of reason" and "medical observation," which entail, in his view, that the soul need not be further mentioned "in this purely medical article, in which we will restrict ourselves to describing the changes of the body, which, as they alone fall under the senses, can be grasped by the physicians" (Ménuret de Chambaud 1765/1966, 718b).

In either case, the soul is being reconfigured as a part of medico-natural discourse, and nothing more: it is "that part of us which thinks," for any additional, e.g. theological or ontological determinations have been naturalized.[23] But none of these medical reconfigurations of soul (a.k.a. 'medicalization') are reducing it to size, shape and motion, or the functioning of pulleys, funnels and sieves; they are integrating the soul *into an embodied framework* (again, my point is not that this is not a reduction, but that it is not a reduction to 'the physical', or the 'physico-mechanical'). This implies conversely that the type of medicine at issue is, if not 'ensouled', at least somehow enhanced. The physician as atheist may be a dangerous character – including to morals – but this can only be true if this danger lies elsewhere than sheer physicalism or mechanism (yet another reason why Henry and Elmer's belief that the danger of atheism had nothing, or hardly anything, to do with medicine, and was restricted to mechanism, is very strange indeed).

Diderot, in his late manuscript on 'physiology' and its conceptual ramifications, the *Éléments de physiologie* (unpublished in his lifetime; finished by the 1770s), imagines what he calls a "physical medicine": since "every sensation and every affection is corporeal, it follows that there is a physical medicine which is equally applicable to the body and the soul" (Diderot 1975-, vol. XVII, 512). It is not clear what exactly this "physical medicine" might be (Rey 2000b suggests it would be more organismic, less reductive), but it hints (as does Le Camus' book which lacks any philosophical sophistication) at the emergence of a scientific psychology – at a naturalisation of mental phenomena and beyond, which is both quantitative and experimental (looking at the role of poisons and hallucinogens, at the organic dimensions of mental illness, etc.), and at the same time squarely focused on an embodied solution. Why such a medicine might be expanded, and thereby also more of a danger, starts to become clear in other proclamations, by Diderot and La Mettrie notably (extending a tendency already present in Lamy and *L'Âme Matérielle*). Diderot asserts the primacy of medicine over both medicine and morals: "it is quite

[23] On the idea of a 'naturalisation of the mind' in the early radical Enlightenment which is not quite an elimination of mental or animate properties in favour of the basic properties of matter, but rather a bracketing-off of ontological considerations in order to treat mental faculties through their empirical manifestations, see Hatfield 1995 (e.g. 188).

difficult to be a good metaphysician and a good moralist, without being an anatomist, a naturalist, a physiologist and a physician" (*Réfutation d'Helvétius*, in Diderot 1975-, XXIV, 555); referring to Locke, Diderot also wrote that "only he who has practiced medicine for a long time is entitled to write [works of] metaphysics" (Diderot 1765, 625b). Diderot's friend and executor Naigeon repeated this:

> [Diderot] was rightfully convinced, and often repeated that only he who has practiced medicine for a long time is entitled to speak and write on metaphysics, because he alone has seen the phenomena: the body (*machine*; 'machine' being a common French term for the body at the time, CW) peaceful or furious, weak or strong, healthy or broken, raving or regulated, in turn moronic, enlightened, stupid, noisy, silent, lethargic, active, living and dead (Naigeon 1821: 217).

If the good moralist has to be a physician, we are no longer dealing just with a medicine of the mind or earlier types of medico-atheistic radicalism; we are dealing with a *medicalization of morals*.

The radical dimension of the figure I have been discussing here – not the 'physician of the soul' but the *physicus sensualis*, the *animal incombustibile propter religionem* – makes him (since it was inevitably a man) a different figure from the medical-materialist who ushers in the pose of scientific neutrality that is characteristic of early nineteenth-century medicine (as in Cabanis, or Claude Bernard, for whom "Physiology today is becoming an exact science, it must rid itself of the philosophical and theological ideas which for a long time were mingled with it"; Bernard 1878, I, 44. Cf. also Bernard 1937).

Medicine can *replace* morals; it can serve as a reducing theory which will provide the constraints for a revised moral theory, as in La Mettrie or Diderot (a viewpoint already expressed in Gaultier's essentially unknown 1714 work, in the last chapter). La Mettrie creates a conceptual equivalence between *médecin* and *moraliste*: "It would doubtless be desirable for there to be only excellent Doctors to serve as Judges, for only they could distinguish the innocent criminal from the guilty one (*L'Homme-Machine*, in La Mettrie 1987, I, 91). That is, he deplores the fact that judgments of life and death are typically made without any knowledge of the physiological level of determination of action. The physician is the "only philosopher to whom her country should be grateful" (*ibid.*, 62), also because, if we recall that the traditional task of philosophy is to meditate on life and death, the physician engages in this much more directly, "delivering certificates (*brevets*) of life and death" (La Mettrie 1747, 100). La Mettrie's medicalized morals does away with the dilemma of happiness and virtue by invoking 'deep structure', as opposed to a surface ethics, which relies on the dualism of body and soul. It is the deep structure of the man-machine, which is organic and follows the norm of health: "Of all kinds of happiness, I prefer that which develops with our organs, and seems to be more or less present, like strength, in all animate bodies" (*Discours sur le Bonheur*, in La Mettrie 1987, II, 247).

The medicalization of morals here takes on the form of radical hedonism, with potentially immoralist consequences (for more discussion, see Wolfe 2009). But even here, the normative force of medical authority still makes itself felt. Even some twentieth-century scholars (e.g. Desné 1963; Rétat 1975 and more cautiously,

Wellman 1992), faced with La Mettrie's potential immoralism and how closely it resembles that of his admirer, the Marquis de Sade, retort that La Mettrie was a physician, concerned with the overall physical and moral well-being of the organism, not someone preaching the destruction of a society's moral code. Nevertheless, there is something surprising in thus turning a thinker's professional status into a philosophical argument, and moreover, into a source of moral credibility.

15.4

Focusing on the figure of the physician as atheist allows one to highlight some aspects of the interpenetration (or "cross-border traffic," in Michael Edwards' expression) of medicine and radical doctrines in early modern Europe, which are not otherwise immediately apparent. Notably, it shows that medicine played a role in debates on the soul, both in an earlier 'heterodox' context, such as the mortalists (Thomson 2008) and in later variations on either 'the material soul' (Wolfe and van Esveld 2014), the medicine of the mind (in its post-Cartesian and Epicurean forms), or the medicalization of morals (an explicitly Epicurean project in La Mettrie: Wolfe 2009); thereby, it was mistaken of some scholars to claim that medicine played no role in metaphysical debates. Further, focusing on this figure brings to light a context in which the embodied, 'sensual physician' is neither subservient to natural theology, nor a way station onto positive, experimental medicine of the nineteenth century.

Does this idea of the atheist physician, which clearly was both 'ideology-laden', 'theory-laden', and connected to a diverse set of references in the ongoing negotiation of medicine-philosophy boundaries, have specifically philosophical consequences? Sometimes it is clear that the critics of the danger of atheism have in mind the problem of 'naturalism' in general. Recall John of Salisbury warning, in the twelfth century, that physicians placed "undue emphasis upon nature." The editor of a later edition of Browne's *Religio Medici*, Thomas Keck, observed more precisely that "The vulgar lay not the imputation of Atheism only upon Physicians, but upon Philosophers in general, who for that they give themselves to understand the operations of Nature, they calumniate them, as though they rested in the second causes without any respect to the first" (cit. in Mothu 2010, 346n.). Resting on second causes is also what Bezançon described as the 'empiricist' danger of medicine, given the predilection of physicians to rely on "sensory causes" (Bezançon 1677, 339).

Yet the problem of naturalism, as Keck indicates, is not just that physicians have a dangerous tendency to place Hippocrates and Galen above Christ, or that their 'sensory practice' makes them *de facto* empiricists; the claim is also that they are 'naturalists' in the sense that they grant too much to the power of Nature rather than to God (Mersenne).[24] In his attempt at a medical natural theology, Hecquet tries to

[24] Mothu 2010 cites a variety of texts which criticize physicians for relying on 'secondary' rather than 'primary' causes.

rebut this kind of charge, asserting that even if Nature *is* appealed to by doctors and philosophers, it is not Nature in the sense of "what is merely material in objects," but rather a finalistic Nature understood as the work of the Creator, whose presence is evident "even in the least of man's organs" (Hecquet 1733, 279). Similarly Hecquet observes that the practice of the physician is not just a "practice" of design ("the contemplation of the wonders of the Creator in the design [*ordonnance*] of the parts of the human body is likely to increase the Faith of a Physician, to exercise and excite it") but also a meditation on death and hence on the 'ends' of man ("what is a better source of piety than the continual consideration of death, and the uninterrupted presence of the ultimate end of man?") (Hecquet 1733, Preface, xlix-xli). Bezançon had argued earlier that medicine is, of all the sciences, that which most strongly reminds us of our frailty and finitude, and thereby brings us closer to God:

> Far from Medicine leading us to atheism and libertinage, I argue on the contrary that of the natural sciences, there is none that elevates man to the knowledge of God more than Medicine. Nothing detaches us more from the creature, and carries us further to God, than the perfect knowledge of our own weakness and nothingness; nothing commits us more to consider another life, than the consideration of our death (Bezançon 1677, 342–343).

But as I indicated above, my concern is not strictly with the 'danger' of the physician as atheist, or with arguments intended to rebut such charges (such as Hecquet's and Bezançon's above). It is also with the positive side of the issue, namely, what we can see as an original intellectual outcome of the physician as atheist (not in a causal or linear sense, but in the sense that was feared in the polemics was indeed in part a medicalized, naturalized conception of soul, itself facilitating or 'boosting' the expression of a materialist picture of the mind). The medicalization of the soul, in its post-Cartesian, Epicurean and otherwise materialist forms (from Regius and Lamy to Gaub and La Mettrie) was, of course, not a self-consciously unified theoretical project. But, if earlier forms of *medicina animi* and *medicina mentis* were primarily intended as pedagogical emendations of the mind (whether of a Cartesian-Spinozist sort, like Tschirnhaus's *Medicina mentis*, or other), or aimed at 'the medicine of the whole man' rather than a reductionist project, this more materialistic medicalization of the soul (which could also appeal to Galen's *Quod animi mores*) sought to *integrate the soul into an embodied framework, the latter being chiefly specified by medicine*. And this marks a difference between the 'negative' claim dating back at least to the Middle Ages (*tres medici, duo athei*) and the 'positive' claim emerging in the later seventeenth century: the latter seeks to articulate a new kind of knowledge, rather than just being purely destructive. Namely, it seeks to articulate a specifically medical body-soul problem – a different scenario, or different treatment at least of what we have come to call the 'mind-body' problem, whether it appeals to a concept of 'material soul', or more naturalistically, insists that the physician has the expertise to address such questions which were traditionally the bailiwick of metaphysics. The medicalization of the soul is, to use a different vocabulary, the articulation of an *embodied* concept of soul, which opens onto a materialist medicine devoted to the organic fulfilment of our 'machine': "Our organs are capable of experiencing feelings or undergoing modifications which delight us and makes us love life" (La Mettrie, *Discours sur le Bonheur*, in La Mettrie 1987, II, 238). Worse even than Bezançon's fear of the physician as *physicus sensualis*, this brand of radical medicine, without being especially 'vitalist' in any technical sense, regularly insists on vitality, as opposed to "anatomie cadavérique": it is corporeal and hedonistic, a medicine for pleasure, in which the physician seeks to promote the 'organic happiness' of the individual (recall La Mettrie's articulation of medicine and morals, and his insistence that "Of all kinds of happiness, I prefer that which develops with our organs"; *Discours sur le Bonheur,* in La Mettrie 1987, II, 247).[25]

References

Alexandrescu, Vlad. 2013. Regius and Gassendi on the human soul. *Intellectual History Review* 23(4): 433–452.

Anon. 2003. L'Âme Matérielle (approx. 1725–1730). Edited by A. Niderst, 3d edition. Paris: Champion.

Bayle, Pierre. 1740. *Dictionnaire historique et critique*. Amsterdam: Libraires associés.

Bernard, Claude. 1878. *Leçons sur les phénomènes de la vie communs aux animaux et aux végétaux*, vol. I. Paris: Germer Baillière.

Bernard, Claude. 1937. *Pensées, notes détachées*, ed. L. Delhoume. Paris: Baillière.

Bernier, François. 1678. *Abrégé de la philosophie de Mr. Gassendi*, 8 vols. Lyon: Anisson et Possuel.

Bright, Timothie. 1586. *A treatise of melancholie: containing the causes thereof, and reasons of the strange effects it worketh in our minds and bodies....* London: Vautrollier.

Brockliss, Lawrence. 1989. The medico-religious universe of an early eighteenth-century Parisian doctor: the case of Philippe Hecquet. In *The medical revolution of the seventeenth century*, ed. R. French and A. Wear, 191–221. Cambridge: Cambridge University Press.

Browne, Thomas. 1892. Religio Medici (1643). In *Writings of Sir Thomas Browne*, ed. D. Lloyd Roberts. London: David Stott.

Busson, H. 1948. *La Religion des Classiques*. Paris: PUF.

Caraman, A., Comte de. 1859. Charles Bonnet, philosophe et naturaliste. Sa vie et ses œuvres. Paris: A. Vaton.

Chaucer, Geoffrey. 1933. The Canterbury tales. In *The complete works of Chaucer*, ed. F.N. Robinson. Boston: Houghton Mifflin.

Cicero. 1927. *Tusculan disputations*. Trans. J.E. King. Cambridge: Harvard University Press.

Conforti, Maria. 2007. Medicine, history and religion in Naples in the seventeenth and eighteenth centuries. In *Medicine and religion in enlightenment Europe*, ed. O.P. Grell and A. Cunningham, 63–78. Aldershot: Ashgate.

Corneanu, Sorana. 2013 (ms.). *The care of the whole man: Medicine and theology in the Late Renaissance*.

Coward, William [Estibius Psychalethes]. 1702. *Second thoughts on the human soul*. London: R. Basset.

Cranefield, P.F. 1970. On the origin of the phrase *nihil est in intellectu*. *Journal of the History of Medicine and Allied Sciences* 25: 77–80.

Cunningham, Andrew, and Ole Peter Grell (eds.). 1996. *Religio Medici: Medicine and religion in seventeenth-century England*. Aldershot: Scholar Press.

---

[25] I wish to thank Sorana Corneanu for her valuable input – both critical and constructive. I benefited additionally from comments from the anonymous reviewers, and the editors of the volume.

de Bezançon, Germain. 1677. *Les médecins à la censure, ou, Entretiens sur la médecine*. Paris: Louis Gontier.

Del Lucchese, Filippo. 2010. Winged men and the cast of dice: Anti–Finalism and radical materialism in Guillaume Lamy. *Dialogue* 49(4): 527–546.

Desné, Roland. 1963. L'humanisme de La Mettrie. *La Pensée* 109: 93–110.

Diderot, Denis. 1765. Locke. In *Encyclopédie des arts et des métiers*, vol. IX, ed. D. Diderot and J. le Rond D'Alembert, 625b–627a. Paris: Briasson. (Reprint, Stuttgart/Bad Cannstatt: Frommann, 1966).

Diderot, Denis. 1975-. *Œuvres complètes*, ed. H. Dieckmann, J. Proust and J. Varloot. Paris: Hermann.

Du Laurens, André. 1597. *Discours de la conservation de la vue*. Paris: J. Mettayer.

Dupuy, Jean-Pierre. 2000. *The mechanization of the mind: On the origins of cognitive science*. Trans. M.B. DeBevoise. Princeton: Princeton University Press.

Edwards, Michael. 2012. Body, soul, and anatomy in late Aristotelian psychology. In *Matter and form in early modern science and philosophy*, ed. G. Manning, 33–75. Leiden: Brill.

Elmer, Peter. 2007. Medicine, witchcraft and the politics of healing. In *Medicine and religion in enlightenment Europe*, ed. O.P. Grell and A. Cunningham, 223–241. Aldershot: Ashgate.

Erasmus. 1989. *Oration in praise of the art of medicine* (1518), translated and annotated by Brian McGregor, in *The collected works of Erasmus*, vol. 29, ed. E. Fantham and E. Rummel with J. Ijsewijn, 31–50. Toronto: University of Toronto Press.

Galen. 1997. *Selected works*, ed. & trans. P.N. Singer. Oxford: Oxford University Press.

Garasse, François. 1623. *La Doctrine curieuse des beaux esprits de ce temps ou prétendus tels*. Paris: Sébastien Chappelet.

Garber, Dan. 1998. Soul and mind: Life and thought in the seventeenth century. In *The Cambridge history of seventeenth-century philosophy*, vol. 1, ed. D. Garber and M. Ayers, 759–795. Cambridge: Cambridge University Press.

Gaultier, Abraham. 1993. *Parité de la vie et de la mort. La Réponse du médecin Gaultier* (1714), ed. O. Bloch. Paris/Oxford: Universitas/Voltaire Foundation.

Hatfield, Gary. 1995. Remaking the science of mind: Psychology as natural science. In *Inventing human science. Eighteenth-century domains*, ed. C. Fox, R. Porter, and R. Wokler, 184–231. Berkeley: University of California Press.

Hecquet, Philippe. 1733. *La Médecine théologique. La Médecine créée, telle qu'elle se fait voir ici, sortie des mains de Dieu*, 2 vols. Paris: G. Cavelier.

Henry, John. 1989. The matter of souls: Medical theory and theology in seventeenth-century England. In *The medical revolution of the seventeenth century*, ed. R.K. French and A. Wear, 87–113. Cambridge: Cambridge University Press.

Huarte de San Juan, J. 1989. *Examen de ingenios para las sciencias* (1575), ed. G. Serés. Madrid: Cátedra.

Iliffe, Rob. 1995. 'That puzleing problem': Isaac Newton and the political physiology of self. *Medical History* 39: 433–458.

Israel, Jonathan. 2001. *Radical enlightenment. Philosophy and the making of modernity, 1650–1750*. Oxford: Oxford University Press.

Israel, Jonathan. 2006. *Enlightenment contested*. Oxford: Oxford University Press.

Israel, Jonathan. 2007. Enlightenment, radical enlightenment and the "medical revolution" of the late seventeenth and eighteenth centuries. In *Medicine and religion in enlightenment Europe*, ed. O.P. Grell and A. Cunningham, 5–28. Aldershot: Ashgate.

Jeannerod, Marc. 1985. *The brain machine. The development of neurophysiological thought*. Trans. D. Urion. Cambridge: Harvard University Press.

Kocher, Paul H. 1947. The physician as atheist in Elisabethan England. *The Huntington Library Quarterly* 10(3): 229–249.

La Mettrie, Julien Offray de. 1746. *Politique du médecin de Machiavel ou le chemin de la fortune ouvert aux médecins. Ouvrage réduit en forme de conseils par le Dr Fum Ho Ham, traduit sur l'original chinois par un nouveau maître es arts ... contient les portraits des plus célèbres médecins de Pékin*. Amsterdam: n.d.

La Mettrie, Julien Offray de. 1747. *La faculté vengée*. Paris: Quillau.

La Mettrie, Julien Offray de. 1749–1750. *L'Ouvrage de Pénélope ou Machiavel en médecine*, "par Aletheius Demetrius." Cramer et Philibert: Berlin, 3 vols.

La Mettrie, Julien Offray de. 1987. *Œuvres philosophiques*, ed. F. Markovits, 2 vols. Paris: Fayard-Corpus.

Lamy, Guillaume. 1996. *Discours anatomiques. Explication méchanique et physique des fonctions de l'âme sensitive*, ed. Anna Minerbi Belgrado. Paris/Oxford: Universitas/Voltaire Foundation.

Le Camus, Antoine. 1753. *Médecine de l'esprit, où l'on traite des dispositions et des causes physiques qui sont des conséquences de l'union de l'âme avec le corps, influant sur les opérations de l'esprit; et des moyens de maîtriser ses opérations dans un bon état ou de les corriger quand elles sont viciées*. Paris: Ganeau.

Mandeville, Bernard. 1730. *A treatise of the hypochondriack and hysterick diseases, in three dialogues* (1st edition 1711), 2nd corrected edition. London: Tonson.

Ménuret de Chambaud, Jean-Joseph. 1765/1966. Mort (Médecine). In *Encyclopédie ou Dictionnaire raisonné des arts et des métiers*, ed. D. Diderot and J. Le Rond D'Alembert, X, 718–727. Paris: Briasson, David, Le Breton & Durand. (Reprint, Stuttgart/Bad Cannstatt: Frommann)

Mersenne, Marin. 1624. *L'Impiété des déistes, athées et libertins de ce temps, combatue et renversée de point en point par raisons tirées de la philosophie et de la théologie*. Paris: P. Bilaine.

Mothu, Alain. 2010. *Animal incombustibile*. La rumeur du médecin athée. *La Lettre clandestine* 8: 317–358.

Naigeon, J.-A. 1821. *Mémoires historiques et philosophiques sur la vie et les ouvrages de D. Diderot*. Paris: Brière.

Nutton, Vivian. 2001. God, Galen and the Depaganization of Ancient Medicine. In *Medicine and religion in the middle ages*, ed. Peter Biller and Joseph Ziegler, 15–32. Woodbridge: York Medieval Press.

Overton, Richard. 1968. *Man's mortality* (1644). Amsterdam; reprint, ed. H. Fisch. Liverpool: Liverpool University Press.

Park, Katharine, and Lorraine Daston. 1998. *Wonders and the order of nature*. New York: Zone Books/MIT.

Petrescu, Lucian. 2013. Descartes on the heartbeat: The Leuven affair. *Perspectives on Science* 21(4): 397–428.

Porter, Roy. 2001. *Bodies politic: Disease, death and doctors in Britain, 1650–1900*. Ithaca: Cornell University Press.

Rather, L.J. 1965. *Mind and body in eighteenth-century medicine. A study based on J. Gaub's De regimine mentis*. Berkeley: University of California Press.

Regius, Henricus. 1646. *Fundamenta physices*. Amsterdam: Elzevier.

Rétat, Pierre. 1975. Le "cœur" de La Mettrie. In *Mélanges de littérature française offerts à René Pintard*, ed. N. Hepp, R. Mauzi, and C. Pichois, 533–545. Strasbourg: Centre de Philologie et de Littératures Romanes.

Rey, Roselyne. 2000. *Naissance et développement du vitalisme*. Oxford: Voltaire Foundation.

Rey, Roselyne. 2000b. Diderot and the medicine of the mind. In *The renewal of materialism*, ed. C.T. Wolfe, *Graduate Faculty Philosophy Journal* 22(1): 149–159

Riolan, Jean. 1602. *Praelectiones in libros physiologicos & de abditis rerum caussis*. Paris: A. Perier.

Smith, Justin E.H. (ed.). 2006. *The problem of animal generation in early modern philosophy*. Cambridge: Cambridge University Press.

Thomson, Ann. 1992. Guillaume Lamy et l'âme matérielle. *Dix-huitième siècle* 24: 63–71.

Thomson, Ann. 2008. *Bodies of thought: Science, religion, and the soul in the early enlightenment*. Oxford: Oxford University Press.

Vartanian, Aram. 1982. Quelques réflexions sur le concept d'âme dans la littérature clandestine. In *Le matérialisme du XVIIIe siècle et la littérature clandestine*, ed. O. Bloch, 149–165. Paris: Vrin.

Wellman, Kathleen. 1992. *La Mettrie: Medicine, philosophy and enlightenment*. Durham: Duke University Press.

White, Andrew Dickson. 1898. *A history of the warfare of science with theology in Christendom*, 2 vols. New York: Appleton and Co.

Willis, Thomas. 1681. *A medical-philosophical discourse of fermentation; or, Of the intestine motion of particles in every body* (1659) and *Of Feavers*. Trans. S. Pordage. London: Dring, Harper, Leigh, and Martin.

Willis, Thomas. 1684. *An essay of the pathology of the Brain and Nervous stock: In which convulsive diseases are treated of* (Translation of *De Morbis Convulsivis*), In *Dr. Willis's practice of Physick*. Trans. S. Pordage. London: T. Dring, C. Harper & J. Leigh.

Wolfe, Charles T. 2004. 'Epicuro-Cartesianism': La Mettrie's materialist transformation of early modern philosophy. In *La Mettrie. Ansichten und* Einsichten, H. Hecht, Hrsg., 75–95. Berlin: Berlin Wissenschafts-Verlag.

Wolfe, Charles T. 2009. A happiness fit for organic bodies: La Mettrie's medical Epicureanism. In *Epicurus in the enlightenment*, ed. N. Leddy and A. Lifschitz, 69–83. Oxford: Voltaire Foundation.

Wolfe, Charles T. 2010. Empiricist heresies in early modern medical thought. In *The Body as object and instrument of knowledge. Embodied empiricism in early modern science*, ed. C.T. Wolfe and O. Gal, 333–344. Dordrecht: Springer.

Wolfe, Charles T. 2014. Epigenesis as Spinozism in Diderot's biological project. In *The life sciences in early modern philosophy*, ed. O. Nachtomy and J.E.H. Smith, 181–201. Oxford: Oxford University Press.

Wolfe, Charles T., and Motoichi Terada. 2008. The animal economy as object and program in Montpellier vitalism. *Science in Context* 21(4): 537–579.

Wolfe, Charles T., and Ofer Gal (eds.). 2010. *The body as object and instrument of knowledge. Embodied empiricism in early modern science*. Dordrecht: Springer.

Wolfe, Charles T., and Michaela van Esveld. 2014. The material soul: Strategies for naturalising the soul in an early modern Epicurean context. In *Conjunctions: Body, soul and mind from Plato to the enlightenment*, ed. D. Kambaskovic, 371–421. Dordrecht: Springer.

Wright, John P. 1991. Locke, Willis, and the seventeenth-century Epicurean soul. In *Atoms, Pneuma, and Tranquillity: Epicurean and Stoic themes in European thought*, ed. M.J. Osler, 239–258. Cambridge: Cambridge University Press.

Wright, John P. 2000. Substance vs. Function dualism in eighteenth-century medicine. In *Psyche and Soma. Physicians and Metaphysicians on the Mind–Body problem from antiquity to the enlightenment*, ed. J.P. Wright and P. Potter, 237–254. Oxford: Clarendon.

Index

P. Distelzweig et al. (eds.), *Early Modern Medicine and Natural Philosophy*, History, Philosophy and Theory of the Life Sciences 14,
DOI 10.1007/978-94-017-7353-9

MIX
Papier aus verantwortungsvollen Quellen
Paper from responsible sources
FSC® C105338

If you have any concerns about our products,
you can contact us on
ProductSafety@springernature.com

In case Publisher is established outside the EU,
the EU authorized representative is:
Springer Nature Customer Service Center GmbH
Europaplatz 3, 69115 Heidelberg, Germany

Printed by Libri Plureos GmbH
in Hamburg, Germany